Modelling, Analysis and Simulation of Thermo-Elasto-Plasticity with Phase Transitions in Steel

von Sören Boettcher

Dissertation

zur Erlangung des Grades eines Doktors der Naturwissenschaften
– Dr. rer. nat. –

Vorgelegt im Fachbereich 3 (Mathematik & Informatik)
der Universität Bremen

im Februar 2012

Datum des Promotionskolloquiums: 28. September 2012

Gutachter: Prof. Dr. Michael Böhm (Universität Bremen)
Prof. Dr. Alfred Schmidt (Universität Bremen)

Bibliografische Information der Deutschen Nationalbibliothek

Die Deutsche Nationalbibliothek verzeichnet diese Publikation in der Deutschen Nationalbibliografie; detaillierte bibliografische Daten sind im Internet über http://dnb.d-nb.de abrufbar.

ISBN 978-3-8325-3296-3

Logos Verlag Berlin GmbH
Comeniushof, Gubener Str. 47,
10243 Berlin
Tel.: +49 (0)30 42 85 10 90
Fax: +49 (0)30 42 85 10 92
INTERNET: http://www.logos-verlag.de

Abstract

Steel is an important material in various applications and has a complex material behaviour which is, in addition to elastic and plastic effects, basically characterised by phase transitions. An important feature in this context is the transformation-induced plasticity, which arises during phase transitions and leads to a permanent deformation at deviatoric stress (below the yield point). Moreover, the material properties of steel often depend on the carbon content, which makes the whole situation even more complex, in particular, if an inhomogeneous carbon distribution is considered in the context of special heat treatment processes like the so-called carburisation.
Modelling the main phenomena of the material behaviour of steel leads to a system of coupled nonlinear partial and ordinary differential equations for the time- and space-dependent temperature, displacement and phase fractions, coupled with a variational inequality which takes classical plasticity into account.
Coupled models for the material behaviour of steel, which describe the phase transitions in addition to the temperature and the deformation, have been insufficiently investigated in a strict mathematical and numerical context so far. A mathematical investigation of this problem is not only of independent mathematical interest, but also of great importance for the numerical simulation of the material behaviour of steel workpieces in certain applications, such as quenching or case hardening.
This thesis deals with mathematical modelling, analysis and numerical simulation of processes involved in the quenching of steel components. The work presented integrates the complex behaviour of steel materials (in particular phase transitions and transformation-induced plasticity) in general models of (linear) thermo-elasto-plasticity and deals with the analysis of the corresponding mathematical problem.
The main objective of this thesis is to formulate a model for the coupled problem of linear thermo-elasto-plasticity including phase transitions and the transformation-induced plasticity in steel and to prove existence and uniqueness of a (global-in-time) solution of the corresponding initial-boundary value problem under appropriate conditions for mixed boundary conditions. Three different settings are considered: In the first one, a Steklov regularisation of the fully coupled problem is investigated. In the second one, a visco-elastic regularisation of the fully coupled problem is studied and in the third setting, a quasi-static model for the displacement is considered.
A simple numerical implementation that works with all settings and the fully coupled problem is suggested and validated. The qualitative behaviour of the solution of the fully coupled model is illustrated for the Jominy-End-Quench-Test.
To sum up, the results provide a theoretical basis for further mathematical investigation or the efficient implementation of numerical algorithms suitable for real-world applications.

Zusammenfassung

Stahl besitzt ein komplexes Materialverhalten, das neben elastischen und plastischen Effekten vor allem durch Phasenumwandlungen gekennzeichnet ist. Ein wichtiges Phänomen ist dabei die Umwandlungsplastizität, die auftritt, wenn Phasenumwandlungen unter deviatorischen Spannungen (unterhalb der Fließgrenze) stattfinden und zu bleibenden Verformungen führen, z.B. häufig im Rahmen von Abkühlungsprozessen. Überdies hängen die Eigenschaften von Stahl oft auch vom Kohlenstoffgehalt ab, wodurch sich das Geschehen noch komplexer gestaltet, wenn eine inhomogene Kohlenstoffverteilung im Bauteil vorliegt, wie etwa nach dem sogenannten Aufkohlen.

Eine Modellierung der wesentlichen Phänomene des Materialverhaltens von Stahl führt auf gekoppelte Systeme von nichtlinearen partiellen und gewöhnlichen Differentialgleichungen, sowie zur Kopplung mit einer Variationsungleichung bei Berücksichtigung der klassischen Plastizität.

Insbesondere gekoppelte Modelle zum Materialverhalten von Stahl, die neben der Temperatur und den Verschiebungen auch die Phasenumwandlungen sowie die Umwandlungsplastizität beschreiben, sind bislang im engeren mathematischen und numerischen Kontext nur wenig untersucht worden. Eine mathematische Untersuchung dieser Modelle ist nicht nur von eigenständigem mathematischem Interesse, sondern ebenso von enormer Wichtigkeit für die Simulation und Numerik dieser Modelle, um das Verhalten von Bauteilen in bestimmten Situationen, z.B. beim Abschrecken oder Einsatzhärten, vorherzuberechnen.

Diese Arbeit beschäftigt sich mit der mathematischen Modellierung, der Analysis und der numerischen Simulation von Abschreckprozessen bei Stahlbauteilen. Die Aufgabe besteht darin, das komplexe physikalische Materialverhalten von Stahl, insbesondere die Phasenumwandlungen und die Umwandlungsplastizität, in ein allgemeines Modell der Thermo-Elasto-Plastizität einzubinden.

Wesentliches Ziel dieser Arbeit ist es, die entsprechende Anfangs-Randwert-Aufgabe für das gekoppelte Problem der „Linearen Thermo-Elasto-Plastizität mit Phasenumwandlungen und Umwandlungsplastizität", welches ein prototypisches Beispiel für einen Wärmebehandlungsprozess von Stahlbauteilen beschreibt, zu formulieren und unter geeigneten Bedingungen die Existenz und Eindeutigkeit einer (zeitlich globalen) Lösung dieser Anfangs-Randwert-Aufgabe mit gemischten Randbedingungen nachzuweisen. Es werden drei unterschiedliche Fälle betrachtet: Zunächst wird eine Steklov-Regularisierung des vollständig gekoppelten Systems untersucht. Im zweiten Szenario wird eine viskoelastische Regularisierung des vollständig gekoppelten Systems analysiert und zuletzt wird ein quasistatisches Modell für die Verschiebungen betrachtet.

Eine einfache numerische Implementierung des gekoppelten Systems wird erarbeitet und

validiert. Das qualitative Verhalten der Lösung der gekoppelten Aufgabe wird für den Stirnabschreckversuch nach Jominy dargestellt.
Zusammenfassend kann man sagen, dass die erzielten Resultate ein grundlegendes theoretisches Fundament für weitere mathematische Untersuchungen oder die effiziente numerische Umsetzung für reale Anwendungen bilden.

Acknowledgements

This thesis could not have been composed without the support of several people. First of all, I wish to thank my supervisor Prof. Dr. Michael Böhm for his guidance, supervision and motivation in writing this thesis. I am also grateful to Prof. Dr. Alfred Schmidt for his interest in my work and for examining my dissertation thesis. Besides, I would like to thank PD Dr. Michael Wolff for his advice help and many helpful suggestions. Furthermore, I would like to thank my family and all my colleagues and friends who have contributed to the realisation of this thesis. Special thanks to Sabrina Fröhlking who encouraged and supported me while writing this thesis. In particular, I thank Sören Dobberschütz for various beneficial hints and endless fruitful discussions – both mathematically and personally. I also thank Simone Bökenheide, Nils H. Kröger, Hari Shankar Mahato, Dr. Sebastian A. Meier for proofreading and Dr. Jonathan Montalvo Urquizo for helpful comments.
This thesis was written during my membership in the PhD program 'Scientific Computing in Engineering' (SCiE) at the Centre for Industrial Mathematics (ZeTeM) located at the University of Bremen. I thank all people from ZeTeM for mathematical and technical support, especially the AG Böhm for a very pleasant working environment and a nice time. My special thanks go to the members and the organizers of the SCiE program for the great working conditions and to the state of Bremen for the financial support.
Finally, I would like to acknowledge that this work has partially been supported by the German Research Foundation (DFG) via the Collaborative Research Centre (SFB) 570 'Distortion Engineering' at the University of Bremen, Germany.

Bremen, February 2012 Sören Boettcher

Every human activity,
good or bad,
except mathematics,
must come to an end.

Paul Erdős[1]

[1]Mathematician, ⋆March 26, 1913 in Budapest (Hungary), †September 20, 1996 in Warsaw (Poland)

Contents

List of Figures

List of Tables

Chapter 1

Introduction

Today, steel is one of the most common materials in industry. The world steel production, with more than 1.3 billion tons produced annually, takes the second place behind cement. Modern steel is generally identified by various grades defined by several standards organisations. Steel and iron materials are used in various areas, very often with high quality requirements according geometry, structure, surface finish, hardness and other properties of the components (cf. e.g. [Bleck et al. 2003] for questions of production and processing of modern steels). Some examples worth mentioning are major components in buildings, infrastructure, tools, ships, automobiles, machines, appliances and much more.[1]

The special characteristics of iron-carbon alloys (cf. Section 1.2 or [Schröter et al. 1995] for instance) result in a great variability in the properties and complexity in the material behaviour. Moreover, the production process of steel components usually takes place in several stages. All this creates a major challenge for the prediction of the material behaviour, i.e. stress, strain and phase composition. The modelling and the simulation of essential effects can help to gain a deeper understanding of the material behaviour. An important goal is to minimise the distortion at the end of the production process (i.e. the unwanted or undesirable deviation from the norm geometry, cf. the next section and e.g. [Hoffmann et al. 2002]).

Section 1.1 explains the term 'Distortion Engineering' and the motivation for this work, given by the SFB 570. An example of use is given in Section 1.2, discussing the material behaviour of steel in the context of heat treatment processes. In Section 1.3 the scope and the outline of this thesis is presented. Moreover, a few hints to the reader are given. Section 1.4 collects the notation used within this thesis.

[1] Extensive information are available at the Stahl-Informations-Zentrum or at the Stahl-Zentrum in Düsseldorf.

1.1 Distortion Engineering

The Collaborative Research Centre (SFB) 570 'Distortion Engineering – Distortion Control in the Production Process'[2] at the University of Bremen works on the intrinsic reasons for the distortion of steel components during the production process. The subject covers a broad field and it has become almost a brand name for scientific approach to the understanding and control of distortion phenomena. Here is a precise definition, which has been added to the International Federation for Heat Treatment and Surface Engineering (IFHTSE) Multilingual Glossary.

1.1.1 Definition (Distortion Engineering)
The scientific and technological investigation of the origins and causes of geometrical inadequacy during the manufacture of engineering components, and the measures and procedures that can be taken to compensate for, or minimise them.

Controlling distortion in the manufacturing process, especially in the hardening process of steel, remains a complex problem even today (cf. Figure 1.1 for the significant distortion potential of steel bars), and the costs of compensating distortion failures of wrought component parts are considerably large. Therefore it is a major interest to understand the causes of distortion in every production step in detail.
So far hardening was considered to be responsible for distortion, but it is state-of-the-art today, that this consideration is not correct, because in the hardening process of heating up and quenching only the distortion potential is released which has been accumulated in various previous processing stages in the component part along a series of fabrication steps building the manufacturing process chain.
The main concept of the long-time research project SFB 570, established by the German Research Foundation (DFG) in 2001, is a system-orientated view on distortion failures of component parts. By applying the holistic view to the production process of a component part, the optimisation of a single fabrication step are less important than the optimisation of the whole process. The integrated analysis, evaluation and optimisation considering all types of influence factors is superior to single solutions and partial optimisation approaches. For a comprehensive explanation we refer to [Thoben et al. 2002].

The SFB 570 has also a great significance for the economy, because millions of tons of steel are processed in Germany every year and many thousands of tons have to be melted down again as result of distortion.

[2]The term 'Distortion Engineering' was established by Prof. Dr.-Ing. Peter Mayr and the project team at the Institute for Materials Science (IWT) in Bremen.

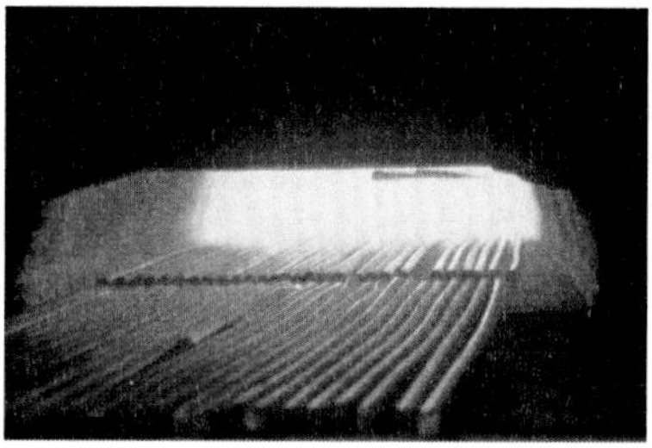

(a) Entering the furnace.

(b) After hardening.

Figure 1.1: Distortion behaviour of bars (length: 6m), Source: Krupp Edelstahl Profile (Siegen, Germany).

1.2 Application: Quenching of Steel Workpieces

The SFB 570 focuses on the determination of significant factors and interactions in the overall production process of three selected model components, namely bearing ring, shaft and gear wheel, cf. Figure 1.2, to identify relevant distortion mechanisms. In [Suhr 2010; Kern 2011] heat treatment processes, involved in the quenching of steel, are modelled and simulated for conical rings in order to analyse and describe the problem of distortion within the framework of the SFB 570.

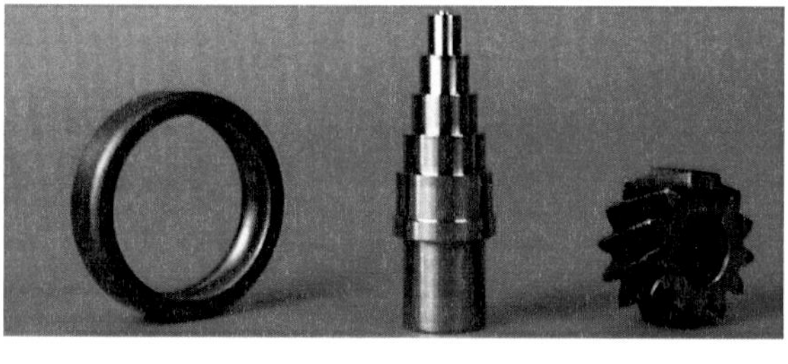

Figure 1.2: Three selected component parts differing in the form of appearance of distortion: ring, shaft, gear wheel, from [Thoben et al. 2002].

In this thesis we consider the cooling of a steel workpiece with occur phase transformations as well as thermal and mechanical effects.

Steel (in the solid state) is a polycrystalline material that consists mostly of iron and carbon (cf. Subsection 1.2.1). The polymorphism of the iron lattice and the different forms of alloys effect a different construction of the structure of steel materials and give, depending on mechanical properties, opportunities for systematic heat treatment.
Special features in the material behaviour of steel are the possible solid-solid phase transformations, which are strongly influenced by the history of the temperature and the carbon content (cf. Subsection 2.1.2 and [Horstmann 1992; Dahl 1993; Kohtz 1994; Seidel 1999; Bleck 2001; Bleck et al. 2003; Wegst and Wegst 2004; Berns and Theisen 2006]).
At the macroscopic level, steel is generally present as a mixture of its phases, which differ in their micro-structure and have different material parameters. These phases are assumed to be continuously distributed in the model, so that the steel appears as a co-existing mixture of its phases (or components), while diffusion processes (of the phases) are neglected. Moreover, we do not consider carbon diffusion.
Phase transitions, which occur during the quenching of steel materials, are connected with volume changes and heat effects. This results in a time- and temperature-dependent stress distribution and deformation, which are crucial for the result of heat treatment.
Another important phenomenon is the transformation-induced plasticity (TRIP) caused by phase transformations. TRIP already occurs under deviatoric stresses and already leads to permanent deformation at relatively low stress (cf. Sections 2.4 and 2.5), even if the yield stress of the softer phase is not reached. This effect cannot be explained by classical plasticity at the macroscopic level of modelling.

The described problem is strongly coupled. Several effects are involved and interact with each other during the thermal processing (cf. the sketch in Figure 1.3):

- Temperature changes cause phase transitions.
- The phase transformations influence the temperature because the thermal parameters (specific heat capacity and thermal conductivity) depend on the so-called latent heat and the phase fractions.
- Inhomogeneous temperature distributions within the sample cause stresses due to thermal expansion and the temperature dependence of the mechanical parameters (Young's modulus and Poisson's ratio).
- Phase changes influence directly the mechanical behaviour (strain and stress state), because of transformation-induced stress and the phase dependence of the mechanical parameters (in particular the density of the material depends on both temperature and phase fractions) as well as the transformation-induced and classical plasticity.

- Furthermore, a back coupling of the stress on the temperature and on the phase transformation by the dissipation process can be observed, i.e. mechanical energy is transformed into thermal energy.

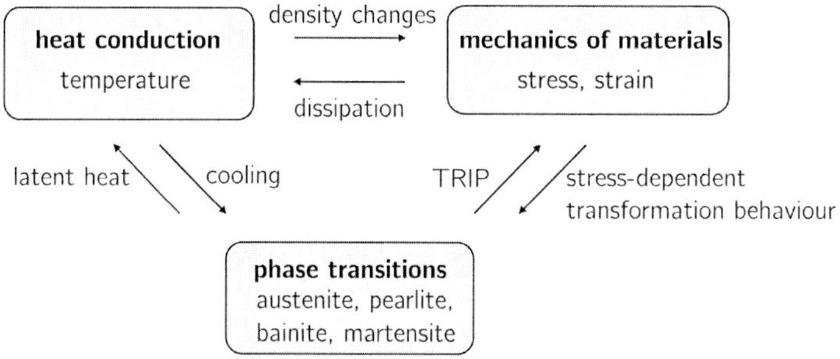

Figure 1.3: Interactions in the process of quenching of steel workpieces, from [Inoue et al. 1981].

In summary, depending on the quenching process, the interactions between temperature, stress, deformation and structural developments play a decisive role. The numerical simulation of the complex interaction of the material behaviour offers the possibility to predict the stress-strain distribution. Thus, degraded material and a cost-intensive processing in order to correct the heat-treatment-related changes in dimension and shape are avoided. However, the simulation of the resulting material and component properties requires not only a deep understanding, but also a sufficiently accurate modelling of the relevant processes involved and the profound knowledge of the required material and process parameters.

1.2.1 Material Behaviour of Steel

Steel is an alloy that consists mostly of iron and has a carbon content between 0.2% and 2.1% by weight. Carbon is the most common alloying material for iron, but various other alloying elements are used, such as nickel, manganese, chromium, vanadium and tungsten. Carbon and other elements act as a hardening agent, preventing dislocations in the iron atom crystal lattice from sliding past one another. Varying the amount of alloying elements and the form of their presence in the steel controls properties such as hardness, ductility, and tensile strength of the resulting steel. Steel with increased carbon content can be made harder and stronger than iron, but such steel is also less ductile than iron.

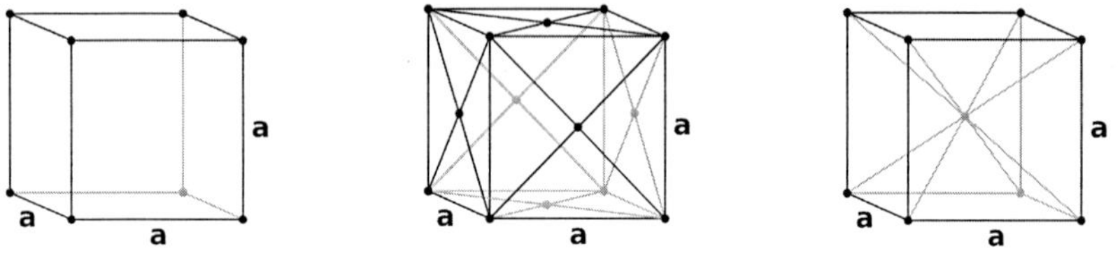

(a) Simple cubic.

(b) Face-centered cubic, γ-iron, lattice constant $\mathbf{a} = 0.364\,\mathrm{nm}$ at 900° C.

(c) Body-centered cubic, α-iron, lattice constant $\mathbf{a} = 0.286\,\mathrm{nm}$ at room temperature.

Figure 1.4: Crystal structure (of iron), from [Bleck 2001; Horstmann 1992].

The density of steel varies based on the alloying constituents, but usually ranges between $7750\frac{\mathrm{kg}}{\mathrm{m}^3}$ and $8050\frac{\mathrm{kg}}{\mathrm{m}^3}$. The physical properties of various steel types and of any given steel alloy at varying temperatures depend primarily on the amount and distribution of carbon present in the iron.
Before heat treatment, most steels are a mixture of three substances: ferrite, pearlite, and cementite. Ferrite is iron containing small amounts of carbon (no more than 0.02% at 723°C and only 0.005% at 0°C) and other elements in solution and has the properties of being soft and ductile. It has the most stable form of iron, which is the body-centered cubic (BCC) structure. Cementite, a compound of iron with the chemical formula of Fe_3C containing about 7% carbon, is extremely brittle and hard. Pearlite is an intimate mixture of ferrite and cementite having a specific composition and characteristic structure, and physical characteristics intermediate between its two constituents. Toughness and hardness of a steel that is not heat treated depend on the proportions of these three ingredients. As the carbon content of a steel increases, the amount of ferrite present decreases and the amount of pearlite increases until it is entirely composed of pearlite, when the steel has a carbon content of 0.8%. Steel with still more carbon is a mixture of pearlite and cementite. Raising the temperature of steel changes ferrite and pearlite to an allotropic form of iron-carbon alloy known as austenite (or γ-iron), which has the property of dissolving all the free carbon present in the metal. It has a face-centered cubic (FCC) structure. If the steel is cooled slowly the austenite reverts to ferrite, pearlite and bainite (consists of cementite and ferrite). If cooling is sudden, the austenite changes to martensite, because the atoms 'freeze' in place when the cell structure changes from

FCC to BCC. Martensite is an extremely hard allotropic modification that resembles ferrite, but contains carbon in solid solution. It is significantly stronger than other steel phases, see Figure 1.5 for a light-microscopical image of martensite. Depending on the carbon content the martensitic phase takes different forms. Below approximately 0.2% carbon it takes an α-ferrite BCC crystal form, but higher carbon contents take a body-centered tetragonal (BCT) structure. There is no thermal activation energy for the transformation from austenite to martensite.
Martensite has a lower density than austenite, so that transformation between them results in a change of volume. In this case, expansion occurs. Internal stresses from this expansion generally take the form of compression on the crystals of martensite and tension on the remaining ferrite, with a fair amount of shear on both constituents. If quenching is done improperly, the internal stresses can cause a part to shatter as it cools. At the very least, they cause internal work hardening and other microscopic imperfections.

For more information on steel we refer to [Bleck 2001; Berns and Theisen 2006; Horstmann 1992].

1.2.2 Heat Treatment of Steel

The condition in which workpieces and tools made of steel are processed fulfills rarely the needs arising from the intended purpose. Therefore, it is necessary to change the condition of the steel material, e.g. by heat treatment, in order to modify the material properties w.r.t. the different conditions needed in the specific application. There are

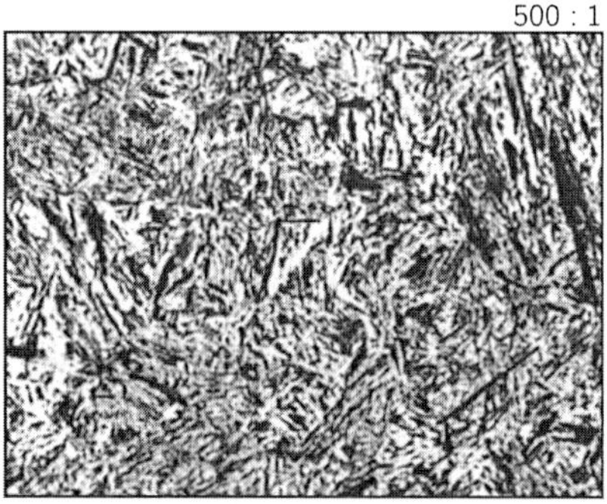

Figure 1.5: Light-microscopical image of martensite, from [Liedtke 2005b].

many types of heat treatment processes available for steel. The purpose of heat treating carbon steel is to change the mechanical properties of steel, usually ductility, hardness, yield strength, or impact resistance. The most common are annealing, quenching and tempering. Annealing is the process of heating the steel to a sufficiently high temperature to soften it. This process occurs through three phases: recovery, recrystallisation, and grain growth. The temperature required to anneal steel depends on the type of annealing and the constituents of the alloy.

The basic process of hardening steel by heat treatment consists of heating the metal to a temperature at which austenite is formed, usually about 760°C to 870°C (steel has a higher solid solubility for carbon in the austenite phase) and then rapidly cooling (quenching) it in water, oil, salt bath or with gas. The rate at which the steel is cooled through the eutectoid reaction affects the rate at which carbon diffuses out of austenite. Generally speaking, cooling swiftly will give a finer pearlite (until the martensite critical temperature is reached) and cooling slowly will give a coarser pearlite. Cooling a hypo-eutectoid (less than 0.8% carbon) steel results in a pearlitic structure with α-ferrite at the grain boundaries. If it is hypereutectoid (more than 0.8% carbon) steel then the structure is full pearlite with small grains of cementite scattered throughout.

Such hardening treatments, which form martensite, set up large internal strains in the metal, and these are relieved by tempering, or annealing, which consists of reheating the steel to a lower temperature.

Tempering results in a decrease in hardness and strength plus an increase in ductility and toughness. Quenching and tempering first involves heating the steel to the austenite phase, then quenching it in water, oil and salt bath or with gas. This rapid cooling results in a hard and brittle martensitic structure. The steel is then tempered, which is just a specialized type of annealing. In this application the annealing (tempering) process transforms some of the martensite into cementite or spheroidite to reduce internal stresses and defects, which ultimately results in a more ductile and fracture-resistant metal.

Many variations of the basic process are practiced. Metallurgists have discovered that the change from austenite to martensite occurs during the latter part of the cooling period and that this change is accompanied by a change in volume that may crack the metal, if the cooling is too swift. Three comparatively new processes have been developed to avoid cracking. In time-quenching the steel is withdrawn from the quenching bath when it has reached the temperature at which the martensite begins to form, and is then cooled slowly in air. In martempering the steel is withdrawn from the quench at the same point, and is then placed in a constant-temperature bath until it attains a uniform temperature throughout its cross section. The steel is then allowed to cool in air through the temperature range of martensite formation, which for most steels is the range from about 288°C

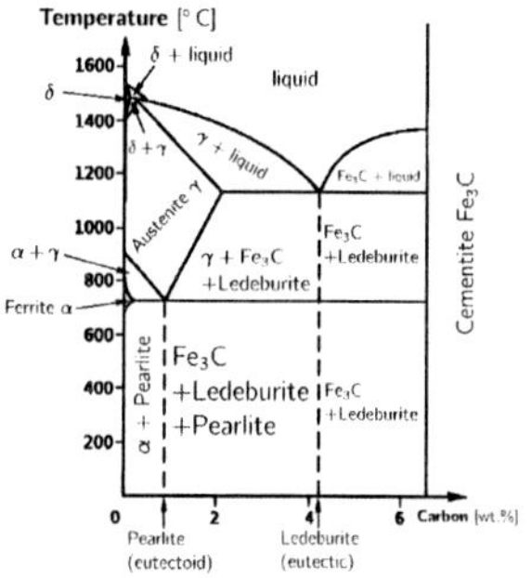

(a) Iron-carbon diagram,
α – alpha-ferrite, γ – austenite,
δ – delta-ferrite, Fe_3C – cementite.

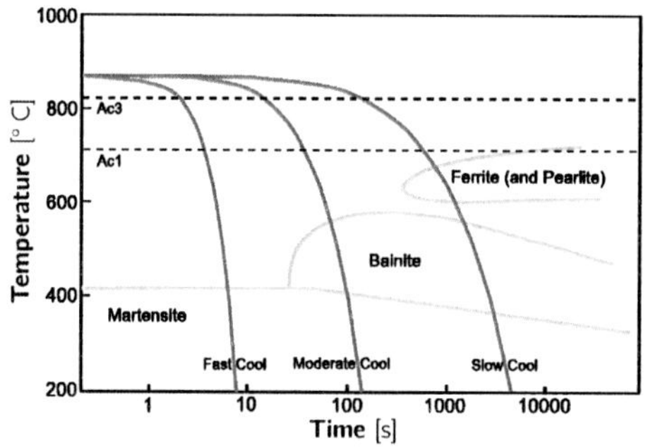

(b) Continuous Cooling Transformation (CCT) diagram, $Ac1$ is the lower transition point of the α-γ-transformation (start of austenite) and $Ac3$ is the upper transition point of the α-γ-transformation (end of austenite).

Figure 1.6: Phase diagrams, from [Horstmann 1992].

to room temperature. In austempering the steel is quenched in a bath of metal or salt, maintained at the constant temperature at which the desired structural change occurs and is held in this bath until the change is complete before being subjected to the final cooling.

Some other methods of heat treating steel are used to harden it, cf. Remark 8.1.1 for hardenability of steel. In case hardening or surface hardening, a finished workpiece of steel, often a low carbon steel, is given an extremely hard surface by heating it with carbon, called carburising, or nitrogen compounds, called cyaniding or nitriding, while leaving the interior soft and therefore tougher and more fracture-resistant. These compounds react with the steel, either raising the carbon content or forming nitrides in its surface layer and make the steel products more resistant to abrasion and wear, while leaving the interior soft and therefore tougher and more fracture-resistant.
In carburising, the workpiece is heated in charcoal or coke, or in carbonaceous gases such as methane or carbon monoxide, from which it absorbs carbon, in a furnace at a temperature of 800°C to 900°C for periods varying from several hours to several days. After that the steel is suddenly immersed in cold water. Cyaniding consists of hardening in a bath of molten cyanide salt to form both carbides and nitrides. In nitriding, steels of special composition are hardened by heating them in ammonia gas to form alloy nitrides. Case hardening is important in the manufacture of gears, axles and other machine parts subject to much mechanical wear.

(a) Gleeble™ Machine.

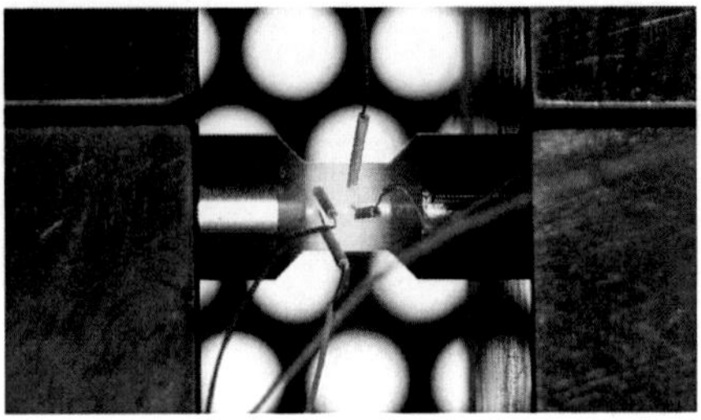

(b) Gleeble™ Sample.

Figure 1.7: Gleeble™ Testing Device, Source: Official Material Testing Laboratory in Bremen (MPA Bremen).

For the standard range of application of the materials these properties are determined and available in the literature. During the manufacturing or during a heat treatment, temperatures and micro-structures are set, which lie clearly outside of the technical area of application for a material. Here the material data is often not available and in many cases can not be determined with conventional testing methods. For this reason, it is possible to get the material data necessary for the calculation with the help of a Gleeble™ testing machine (see Figure 1.7), whereby these material data are determined at the same heat, which is used in the other experiments.

For a comprehensive explanation on heat treatment processes we refer to [Liedtke 2005a,b, 2009], although most information of this section is descended from [Smith and Hashemi 2006].

1.3 Scope and Outline of the Thesis

This thesis deals with mathematical modelling, analysis and numerical simulation of processes involved in the quenching of steel workpieces. Therefore, it is positioned in the overlap of two scientific disciplines: applied mathematics and materials science. The considered physical processes are heat conduction, phase transformations, thermo-elasticity, classical plasticity and TRIP. Although all three parts modelling, analysis and simulation are discussed, the scientific contribution of this work is in the field of mathematical analysis.

1.3.1 Scope of the Thesis

As it has been mentioned before, steel has a complex material behaviour. Stress- and strain-dependent phase transformations, TRIP and its interactions with (classical) plasticity are important phenomena in the material behaviour of both theoretical and practical interest, as they may cause distortion of steel work-pieces.
The modelling of the relevant interactions between temperature, mechanical behaviour and phase transitions leads even in the case of simplified assumptions to an initial boundary value problem for a system of coupled nonlinear partial and ordinary differential equations and -inequalities for the time- and space-dependent temperature, displacement and phase fractions, which is not only of an independent mathematical interest, but also of great importance for the numerical simulation of the material behaviour of steel workpieces in certain highly relevant situations, such as quenching or case hardening.
Coupled models for the material behaviour of steel, which describe phase transitions in addition to the temperature and the deformation, have been insufficiently investigated in a strict mathematical and numerical context so far. Therefore the idea came up to integrate the complex behaviour of steel materials (especially the phase transformations and TRIP) in general models of thermo-elasticity.
The model of thermo-elasto-plasticity in the present work takes phase transitions and TRIP into account. It is a prototypical example for heat treatment processes of steel components (cf. [Wolff et al. 2008c, 2011b] for instance). The model extensions, especially dealing with transformation-induced and classical plasticity, are a major challenge for the modelling itself as well as for the functional analytic and numerical investigation of the mathematical problem which occurs.

The main objective of this thesis is the formulation and the analysis of the mathematical problem of TRIP and their interaction with the classical plasticity. To the best of the author's knowledge, this new aspect has not been considered in the literature so far.

1.3.2 Outline of the Thesis

This thesis consists of nine chapters. After the introduction in the first chapter, the modelling, analysis and simulation is contained in Chapters 2 – 8, followed by a discussion and outlook (Chapter 9) and concluded with an appendix.
In Chapter 1, some well-known properties of the material behaviour of steel and its heat treatment processes are introduced and explained.
The aim of Chapter 2 is to formulate the model of linear thermo-elasto-plasticity with phase transitions and TRIP describing the material behaviour of steel in the context of macroscopic continuum mechanics and to discuss the capabilities. Due to the

possible interaction (coupling) of transformation-induced and classical plasticity, the usual approach in plasticity without phase transformations has to be modified substantially. In Chapter 3, we summarise and discuss the resulting complex system of coupled nonlinear partial and ordinary differential equations for the time- and space-dependent temperature, displacement and phase fractions, coupled with a variational inequality which takes the classical plasticity into account. Great difficulties arise by dealing with the fully coupled problem. Therefore we discuss some slightly different (simplified) problems, but nevertheless, it remains a great challenge to prove existence and uniqueness of a solution of the corresponding initial-boundary value problem for the coupled problem of linear thermo-elasto-plasticity including phase transitions and TRIP under appropriate conditions.
The mathematical analysis of this initial-boundary value problem or rather of the three modified problems is given in Chapters 4 – 7.
After briefly qualifying the research context, Chapter 4 gives an overview of the function spaces, the (mathematical) assumptions, the main results and the related literature.
Existence and uniqueness results are given for the following three different settings:

- The Steklov regularisation of the fully coupled problem is investigated in Chapter 5.
- In Chapter 6, a visco-elastic regularisation of the fully coupled problem is studied.
- Finally, a quasi-static model for the displacement is considered in Chapter 7.

Numerical simulations of a simplified model of the fully coupled problem are presented in Chapter 8 for the Jominy-End-Quench-Test.
In Appendix A, we cite a few results for Lebesgue spaces, Sobolev spaces and Bochner spaces and collect some inequalities that are important for the proofs in Chapters 5 – 7.
In Appendix B, we show some basic results for the difference quotient and the Steklov average and cite an existence and uniqueness result for systems of elliptic equations. In Appendix C, we show the basic techniques needed for proving existence, uniqueness and regularity for (parabolic) differential inclusions. Moreover, we explain the connection to (parabolic) variational inequalities and give some references.

1.3.3 A few Hints to the Reader

The three main parts modelling, analysis and simulation can be read independently from each other. One section of each part provides references for related work on the particular topic. The chapters or sections denoted by an asterisk '*' placed at their end provide additional information or present related topics, which are not continued in detail within this work.

The fully coupled model of thermo-elasto-plasticity with phase transitions and TRIP is given in Section 3.1. The three investigated modifications are discussed extensively in Section 3.3. The three chapters investigating the modified problems (which are the regularisation via Steklov averaging, the visco-elastic regularisation and the quasi-static setting) can also be read independently from each other. The main results are collected in Section 4.4.
Moreover, we remark that we (principally) consider the quenching process of steel workpieces in this thesis.
Finally, we remark that we use British English in this work.

1.4 Notation

We fix some notation and mention some conventions in this section: Vectors and tensors are expressed in bold-face. In estimates, we frequently use a generic constant $c > 0$. The value of this constant depends on the context and may vary in different steps of the particular estimate.

1.4.1 Nomenclature

$\mathbb{N}$	–	set of natural numbers including zero
$\mathbb{R}$ (resp. $\mathbb{R}^+$, $\mathbb{R}_0^+$)	–	set of (resp. positive, non-negative) real numbers
$\mathbb{R}^{n\times n}$ (resp. $\mathbb{R}^{n\times n}_{\mathrm{sym}}$)	–	set of (resp. symmetric) $n \times n$ matrices
$\mathbf{u}^T$ (resp. $\mathbf{A}^T$)	–	transpose of a vector $\mathbf{u} \in \mathbb{R}^n$ (resp. matrix $\mathbf{A} \in \mathbb{R}^{n\times n}$)
$\mathbf{I}$	–	identity matrix (or unity tensor)
$\mathrm{tr}(\mathbf{A})$	–	trace of a matrix $\mathbf{A} \in \mathbb{R}^{n\times n}$
$\mathbf{A}^*$	:=	$\mathbf{A} - \frac{1}{n}\mathrm{tr}(\mathbf{A})\,\mathbf{I}$, deviator of a matrix $\mathbf{A} \in \mathbb{R}^{n\times n}$
$\mathrm{int}(K)$	–	interior of a set K
$\mathrm{meas}(K)$	–	Lebesgue measure of a set K
$\mathrm{h}(A, B)$	–	Hausdorff distance between the two sets A and B, cf. Appendix C.2.2 for details
χ_K	–	indicator function of a set K

1.4.2 Differential Expressions

$\frac{\partial}{\partial t}\left(\text{resp. } \frac{\mathrm{d}}{\mathrm{d}t}\right)$	–	partial (resp. total) derivative w.r.t. t
∇	$:=$	$\left(\frac{\partial}{\partial x_1}, \dots, \frac{\partial}{\partial x_n}\right)^T$, Nabla operator
Δ	$:=$	$\sum_{i=1}^{n} \frac{\partial^2}{\partial x_i^2}$, Laplace operator
$\mathrm{grad}(\mathbf{u})$	$:=$	$\nabla \mathbf{u}$, gradient of a function $\mathbf{u} : \mathbb{R}^n \to \mathbb{R}^n$
$\mathrm{div}(\mathbf{u})$	$:=$	$\sum_{i=1}^{n} \frac{\partial u_i}{\partial x_i}$, divergence of vector $\mathbf{u} \in \mathbb{R}^n$
$\mathrm{Div}(\mathbf{A})$	$:=$	$\left(\sum_{j=1}^{n} \frac{\partial}{\partial x_j} A_{ij}\right)_{i=1,\dots,n}$, divergence of a matrix $\mathbf{A} \in \mathbb{R}^{n \times n}$
$\partial \Phi$	–	subdifferential of the convex functional $\Phi : X \to \mathbb{R}$,

In this case, X denotes a real Banach space. The symbol $'$ is sometimes used instead of $\frac{\partial}{\partial t}$. The symbols ∇, Δ, div and Div refer to derivatives w.r.t. $\mathbf{x}$.

1.4.3 Function Spaces

We use specific notation for space-time domains: If $S :=]0, T[$ is a finite time interval and $\Omega \subset \mathbb{R}^n$ is a domain, then we denote $\Omega_T := S \times \Omega$. For a measurable part of the boundary $\Gamma \subset \partial\Omega$, we denote $\Gamma_T := S \times \Gamma$.
In the following, we mention the function spaces needed in this thesis. An introduction to these function spaces, especially to spaces on space-time domains is given in Appendix A. Table 4.3 provides an overview of the used function spaces on space-time domains (and its abbreviations).

In the sequel, let X, Y be (real) Banach spaces.

$C^k(\Omega)$	set of functions with continuous derivatives up to order $k \in \mathbb{N}$
$C_0^k(\Omega)$	subspace of $C^k(\Omega)$ of functions with compact support
$L^p(\Omega)$	standard Lebesgue space over Ω, $p \in [1, \infty]$
$W^{k,p}(\Omega)$	standard Sobolev space over Ω, $k \in \mathbb{N}$, $p \in [1, \infty]$
$W_0^{k,p}(\Omega)$	subspace of $W^{k,p}(\Omega)$ of functions with zero boundary values

$\|\cdot\|_\infty$	maximum norm on $\mathbb{R}^n$
$\|\cdot\|_X$	norm on X
$(\cdot,\cdot)_X$	scalar product in X
X^*	dual space of X
$\langle\cdot,\cdot\rangle_{X^*X}$	dual pairing in $X^* \times X$
$C^k(\bar{S};X)$	set of X-valued functions with continuous derivatives up to order $k \in \mathbb{N}$
$L^p(S;X)$	standard Bochner-Lebesgue space, $p \in [1,\infty]$
$W^{k,p}(S;X)$	standard Bochner-Sobolev space, $k \in \mathbb{N}$, $p \in [1,\infty]$
$W^{1,p}(S;X,Y)$	Bochner-Sobolev space, $p \in]1,\infty[$
$\mathcal{C}^{k,p}$	class of domains whose boundary is locally representable as a graph of a $C^{k,p}$-function, cf. [Adams and Fournier 2003; Sohr 2001; Wloka 1987] for details

The scalar products of vectors $\mathbf{u},\mathbf{v} \in \mathbb{R}^n$ and tensors $\mathbf{A},\mathbf{B} \in \mathbb{R}^{n\times n}$ are defined (in Cartesian coordinates) as

$$\mathbf{u}\cdot\mathbf{v} := \sum_{i=1}^{n} u_i v_i \qquad \text{and} \qquad \mathbf{A}:\mathbf{B} := \sum_{i,j=1}^{n} A_{ij}B_{ij}, \qquad n \in \mathbb{N}.$$

1.4.4 Material Parameters

We give a list of symbols including all used material parameters and model variables. Following units derived from SI base units are used: $\mathrm{Pa} = \frac{\mathrm{N}}{\mathrm{m}^2}$, $\mathrm{W} = \frac{\mathrm{J}}{\mathrm{s}}$, $\mathrm{J} = \mathrm{N\,m}$ and $\mathrm{N} = \frac{\mathrm{m\,kg}}{\mathrm{s}^2}$. Moreover, it holds [°C] = [K] − 273.15.

Parameter or Variable	Unit	Description	Page
ρ_0	$\frac{\mathrm{kg}}{\mathrm{m}^3}$	bulk density in the reference configuration	19
$\boldsymbol{\sigma}$	Pa	(Cauchy) stress tensor	19
$\boldsymbol{\sigma}_{vM}$	Pa	von Mises stress	21
$\boldsymbol{\sigma}_m$	Pa	mean (principal) stress	21

continued on next page

continued from previous page

Parameter or Variable	Unit	Description	Page
$\mathbf{f}$	$\frac{\mathrm{N}}{\mathrm{m}^3}$	volume density of external forces	19
$\mathbf{u}$	m	displacement vector	19
$\boldsymbol{\varepsilon}$		(linearised Green) strain tensor	19
$\boldsymbol{\varepsilon}_{te}$		thermo-elastic strain	21
$\boldsymbol{\varepsilon}_{trip}$		strain due to TRIP	21
$\mathbf{X}_{trip}$	Pa	backstress due to TRIP	27
$\boldsymbol{\varepsilon}_{cp}$		strain due to (classical) plasticity	21
$\boldsymbol{s}_{cp}$		accumulated plastic strain	27
$\mathbf{X}_{cp}$	Pa	backstress due to (classical) plasticity	25
μ	Pa	shear modulus	22
K	Pa	compression modulus	22
α	$\frac{1}{\mathrm{K}}$	heat dilatation coefficient	22
λ	Pa	Lamé's first coefficient	23
ν		Poission's ratio	23
E	Pa	Young's modulus	23
θ	K	(absolute) temperature	19
θ_0	K	initial temperature	22
θ_Γ	K	ambient temperature	28
c_d	$\frac{\mathrm{J}}{\mathrm{kg\,K}}$	specific heat	30
λ_θ	$\frac{\mathrm{W}}{\mathrm{m\,K}}$	heat conductivity	28
δ	$\frac{\mathrm{W}}{\mathrm{m}^2\,\mathrm{K}}$	heat exchange coefficient	28
r	$\frac{\mathrm{W}}{\mathrm{m}^3}$	volume density of heat supply	19
e	$\frac{\mathrm{J}}{\mathrm{kg}}$	mass density of internal energy	19
Ψ	$\frac{\mathrm{J}}{\mathrm{kg}}$	mass density of free (Helmholtz) energy	19
η	$\frac{\mathrm{J}}{\mathrm{kg\,K}}$	mass density of entropy	19
$\mathbf{p}$		vector of phase fractions	20
Φ_i		saturation function of the ith phase	36
L_i	$\frac{\mathrm{J}}{\mathrm{kg}}$	latent heat of the ith phase	31
κ_i	$\frac{1}{\mathrm{Pa}}$	Greenwood-Johnson parameter of the ith phase	36
F	Pa	yield function	25
R_0	Pa	yield stress	25
R	Pa	increment of the yield stress	25
Λ	$\frac{1}{\mathrm{Pa}}$	plastic multiplier	26

Table 1.1: List of symbols.

Chapter 2

Modelling of Steel Material Behaviour

In this chapter we deal with a complex (macroscopic) model of the material behaviour of steel including specific phenomena like stress-dependent phase transitions, TRIP and the possible interaction with classical plasticity developed in [Wolff et al. 2008c, 2011b] for small deformations. Our aim is to formulate a mathematical model, which is suitable for further investigations and numerical simulations.
Section 2.1 provides the governing equations. In Sections 2.2 and 2.3, we derive the dissipation inequality and the heat-conduction equation for a particular case. In Sections 2.4 and 2.5, phase transitions and, as a special feature of the material behaviour of steel, TRIP are discussed in detail. Sections 2.6, 2.7 and 2.8 address some additional aspects like the relation for the free energy, thermodynamic consistency and non-dimensionalisation. In Section 2.9, the model equations will be summarised.

2.1 General Scheme of Mathematical Modelling

We summarise the proposed small-deformation model of complex material behaviour of steel, developed in [Wolff et al. 2008c, 2011b]. The fully coupled (macroscopic) model is composed of

- the deformation equations for the displacement vector,
- strain and stress tensor,
- yield criterion and plastic flow rule,
- the heat equation,
- the evolution of phase transitions and
- the evolution of TRIP.

2.1.1 Balance Equations and Clausius-Duhem-Inequality

A model for the material behaviour of steel in the framework of continuum mechanics includes the balance equations for mass, momentum, angular momentum, energy and possibly also an inequality for the entropy (and possibly for the carbon concentration and other quantities if they play a role). The mass balance leads to the time-constant density in the reference configuration, the balance of angular momentum to the symmetry of the Cauchy stress tensor. Since only 'small' deformations are considered, which correspond to a geometric linearisation, the reference and current configuration differ only infinitesimal, and the stress tensors coincide.

2.1.1 Remark (Small Deformations)
The small deformation theory deals with infinitesimal deformations of a continuum body. For an infinitesimal deformation the displacements $\mathbf{u}$ *and the displacement gradients* $\nabla\mathbf{u}$ *are small compared to unity, allowing the geometric linearisation of the Lagrangian finite strain tensor and the Eulerian finite strain tensor, i.e. the non-linear or second-order terms of the finite strain tensor can be neglected. The assumption of small deformations is feasible in heat treatment processes (e.g. quenching) of steel components, if no additional (large) mechanical forces, such as rolling or forging, act from outside.*
An introduction to large deformation mechanics can be found in [Bertram 2005; Lubliner 2006; Hallberg et al. 2007].

As a result of the above 'restrictive considerations in this situation', the two balance equations for momentum and energy in local form read as: A material body will be identified with its reference configuration $\Omega \subset \mathbb{R}^3$, where Ω is a bounded domain with Lipschitz boundary. In the framework of small deformations we have the well-known balance equations for linear momentum and energy:

$$\rho_0 \frac{\partial^2 \mathbf{u}}{\partial t^2} - \mathrm{Div}(\boldsymbol{\sigma}) = \mathbf{f}, \tag{2.1}$$

$$\rho_0 \frac{\partial e}{\partial t} + \mathrm{div}(\mathbf{q}) = \boldsymbol{\sigma} : \frac{\partial \boldsymbol{\varepsilon}}{\partial t} + r. \tag{2.2}$$

Moreover, the second law of thermodynamics is applied in the form of the Clausius-Duhem inequality:

$$-\rho_0 \frac{\partial \Psi}{\partial t} - \rho_0 \eta \frac{\partial \theta}{\partial t} + \boldsymbol{\sigma} : \frac{\partial \boldsymbol{\varepsilon}}{\partial t} - \frac{1}{\theta}\mathbf{q} \cdot \nabla\theta \geq 0. \tag{2.3}$$

The relations (2.1) – (2.3) have to be fulfilled in the space-time domain Ω_T. The notations are standard: ρ_0 – bulk density in the reference configuration, i.e. for $t = 0$,

$\mathbf{u}$ – displacement vector (cf. Figure 2.1 for an explanation), $\boldsymbol{\varepsilon}$ – linearised Green strain tensor, θ – absolute temperature, $\boldsymbol{\sigma}$ – Cauchy stress tensor, $\mathbf{f}$ – volume density of external forces, e – mass density of internal energy, $\mathbf{q}$ – heat flux, r – volume density of heat supply, Ψ – mass density of free (or Helmholtz) energy, η – mass density of entropy.

Note the relations:

(2.4) $$\boldsymbol{\varepsilon} = \boldsymbol{\varepsilon}(\mathbf{u}) := \frac{1}{2}\left(\nabla \mathbf{u} + \nabla \mathbf{u}^T\right), \qquad \mathbf{u} = (u_1, u_2, u_3)^T \qquad \text{and}$$

(2.5) $$\Psi := e - \theta\eta.$$

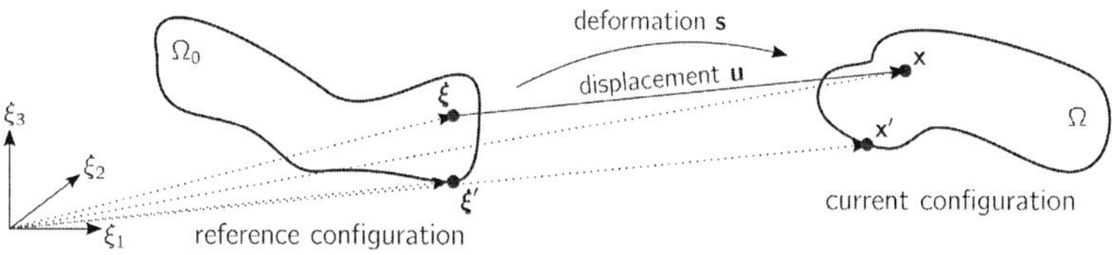

Figure 2.1: Deformation and displacement, from [Eck et al. 2008].

2.1.2 Phase Transitions

Phase transitions during heat treatment are an important phenomenon in the material behaviour of steel. Together with the thermal and mechanical behaviour of steel workpieces they cause the change of volume/mass and shape (distortion). Furthermore, stresses based on density changes in the differentiated microstructure of the steel occur. Moreover, the phase transformation also influences the temperature profile by the so-called latent heat. The latent heat is the heat released or absorbed by a thermodynamic system during a change of state that occurs without a change in temperature. For example, if a steel sample is cooled down until a phase transition takes place, energy is produced, which heats the steel sample, because atoms are detached from the lattice structure of the previous phase. This occurring energy is called latent heat. Conversely, the steel sample has no interior heat source during heating, but a heat sink. The heat equation has to be completed with an additional term (cf. Section 2.3).
In order to gain a macroscopic model for the material behaviour of steel, we have to make some additional assumptions:

- At the macroscopic level, steel is a solid body and a coexisting mixture of its phases, which are considered as constituents.
- A macroscopic diffusion of these (individual) phases – such as in general mixtures – is excluded. The phases stay at their place of origin, i.e. it is a given constant macroscopic (homogeneous) carbon concentration (distribution) assumed and possible segregation effects are neglected. Moreover, there are no outer sources and sinks for the phase transitions. As a consequence, the transformation equations of the phases do not include diffusion terms (cf. Equation (2.6)).

The evolution of phase fractions is given by the ODEs (cf. e.g. [Hömberg 1995; Hüßler 2007; Visintin 1987; Wolff et al. 2007a, 2006a])

$$\text{(2.6)} \qquad \frac{\partial p_i}{\partial t} = \gamma_i(\mathbf{p}, \theta, \boldsymbol{\xi}) \qquad (i = 1, \ldots, m) \qquad (m \geq 2)$$

In general, the phase evolution may depend on internal variables labelled by $\boldsymbol{\xi}$ (cf. Section 2.4). Here, $\mathbf{p}$ denotes $(p_1, \ldots, p_m)$. Moreover, the phase (mass) fractions p_i of the ith phase ($i = 1, \ldots, m$) have to fulfill the subsequent balance and non-negativity relations

$$\text{(2.7)} \qquad \sum_{i=1}^{m} p_i = 1, \qquad p_i \geq 0 \qquad \text{for} \qquad i = 1, \ldots, m$$

which implies

$$\text{(2.8)} \qquad \sum_{i=1}^{m} \gamma_i = 0.$$

In materials science one traditionally works with volume fractions when dealing with steel. In the mixture theory, however, mass fractions are usually preferred.
We do not distinguish between mass and volume fraction of a steel phase, because their difference is small (the absolute error is ranging from 0.01 to 0.02, cf. discussion in [Wolff et al. 2003a] and experimental results in [Wolff and Suhr 2003; Wolff et al. 2007d]) due to small density differences of the steel phases (at the same temperature). However, it is possible, that even these small differences in density may have a significant influence on the distortion behaviour of steel components.
The number of phases depends on the sort of steel and on the circumstances (cf. [Wolff et al. 2007d,a,c, 2006a]).

2.1.2 Remark (Dependence on carbon content and stress)
The transformation behaviour depends on the carbon content of the parent phases, i.e. the right-hand side of Equation (2.6) *additionally depends on the carbon concentration, cf. e.g. [Hüßler 2007; Wolff et al. 2006a].*

The phase evolution (kinetics of phase transition) is affected by the present stress. Some modelling ansatz for the stress-dependent transformation behaviour assume the dependency of the phase transitions on the mean principal stress $\boldsymbol{\sigma}_m$ and/or the von Mises stress $\boldsymbol{\sigma}_{vM}$, where

(2.9)

$$\boldsymbol{\sigma}_{vM} := \sqrt{\frac{3}{2}\boldsymbol{\sigma}^* : \boldsymbol{\sigma}^*} \text{ (von Mises stress)}, \qquad \boldsymbol{\sigma}_m := \frac{1}{3}\operatorname{tr}(\boldsymbol{\sigma}) \text{ (mean (principal) value)}.$$

This approach leads to the dependence on the displacement $\mathbf{u}$ because of the representation of the stress tensors $\boldsymbol{\sigma}_m$ and $\boldsymbol{\sigma}_{vM}$. Thus, the influence of stress (or possibly effective stress) on phase transformation (cf. [Wolff et al. 2003b]) can be described by its (principle) invariants: $\operatorname{tr}(\boldsymbol{\sigma})$, $\frac{1}{2}((\operatorname{tr}(\boldsymbol{\sigma}))^2 - \operatorname{tr}(\boldsymbol{\sigma} : \boldsymbol{\sigma}))$, $\det(\boldsymbol{\sigma})$.[1] *Moreover, the explicit dependence on* $(\mathbf{x}, t)$ *is considered in [Hüßler 2007].*

More detailed explanation about phase transformation will be given in Section 2.4.

2.1.3 Decomposition of the Strain Tensor

As usual in the theory of small deformations (using geometric linearisation, cf. e.g. [Eck et al. 2008]), the linearised Green strain tensor $\boldsymbol{\varepsilon}$ is decomposed into the following three components

$$\boldsymbol{\varepsilon} = \boldsymbol{\varepsilon}_{te} + \boldsymbol{\varepsilon}_{trip} + \boldsymbol{\varepsilon}_{cp}, \tag{2.10}$$

where $\boldsymbol{\varepsilon}_{te}$ – thermoelastic strain (including isotropic density variations due to temperature changes and phase transformations), $\boldsymbol{\varepsilon}_{trip}$ – (non-isotropic) strain due to TRIP and $\boldsymbol{\varepsilon}_{cp}$ – strain due to (classical) plasticity. Sometimes, $\boldsymbol{\varepsilon}_{te}$ is also split up into a pure elastic part, a pure thermal part and a part only due to phase changes.
Viscosity effects are not taken into account. In principal, they could be considered analogously with some modifications (cf. [Haupt 1977] for continuum mechanical modelling of visco-elasticity and [Bonetti and Bonfanti 2003, 2005, 2008; Duvaut and Lions 1976; Gawinecki 2003] for mathematical analysis of thermo-visco-elastic systems).
As usual, we assume that the inelastic strains are volume-preserving, i.e.

$$\operatorname{tr}(\boldsymbol{\varepsilon}_{trip}) = 0 \implies \boldsymbol{\varepsilon}_{trip} = \boldsymbol{\varepsilon}^*_{trip}, \qquad \operatorname{tr}(\boldsymbol{\varepsilon}_{cp}) = 0 \implies \boldsymbol{\varepsilon}_{cp} = \boldsymbol{\varepsilon}^*_{cp}. \tag{2.11}$$

Therefore, classical plasticity and TRIP act by the stress tensor on the deformation.

[1] It holds $\frac{1}{2}((\operatorname{tr}(\boldsymbol{\sigma}))^2 - \operatorname{tr}(\boldsymbol{\sigma} : \boldsymbol{\sigma})) = \operatorname{tr}(\operatorname{adj}(\boldsymbol{\sigma}))$, where $\operatorname{adj}(\boldsymbol{\sigma}) = \det(\boldsymbol{\sigma})\boldsymbol{\sigma}^{-1}$. We remark that some authors distinguish between invariants and principle invariants (cf. e.g. [Zeidler 1988]).

2.1.3 Remark (Two-Mechanism Model)
Because of the two mechanisms for inelasticity (interaction between classical plasticity and TRIP), the considered model is an concrete example of a so-called two-mechanism model (cf. [Wolff and Taleb 2008; Wolff et al. 2011c, 2010a; Saï 2011] for multi-mechanism models). The spin-off project 'Mehr-Mechanismen-Modelle: Theorie und ihre Anwendung auf einige Phänomene im Materialverhalten von Stahl' (DFG BO 1144/4-1) follows up this multi-mechanism approach.

2.1.4 Thermo-Elasticity Relation

The material law (2.12) is a generalisation of the so-called Duhamel-Neumann's law (or generalised Hooke's law) of the classical (linear) thermo-elasticity for isotropic bodies, cf. e.g. [Wilmanski 1998]. (Isotropy means that the response to a force is independent of the direction of the force in elastic behaviour.) The last term in (2.12) takes the density changes as a result of phase transitions into account. In order to separate this part from the thermal expansion, the phase densities appear at the initial temperature. Alternatively, it would also be possible, to sum up the last two terms in Equation (2.12), cf. Remark 2.1.4.
The stress tensor $\boldsymbol{\sigma}$ and the thermo-elastic part $\boldsymbol{\varepsilon}_{te}$ of the strain tensor are connected by the law of thermo-elasticity taking density changes due to phase transformations into account:

$$\boldsymbol{\sigma} = 2\mu\boldsymbol{\varepsilon}^*_{te} + K\,\mathrm{tr}(\boldsymbol{\varepsilon}_{te})\,\mathbf{I} - 3K_\alpha(\theta-\theta_0)\,\mathbf{I} - K\sum_{i=1}^{m}\left(\frac{\rho_0}{\rho_i(\theta_0)}-1\right)p_i\,\mathbf{I} \tag{2.12}$$

where μ – shear modulus, K – compression (bulk) modulus, $K_\alpha := K\alpha$ – modulus taking compression and linear heat-dilatation of the bulk material into account, $\rho_i(\theta_0)$ – density of the ith phase phase at initial temperature θ_0, e.g. $t = 0$. The deviator $\boldsymbol{\varepsilon}^*_{te} := \boldsymbol{\varepsilon}_{te} - \frac{1}{3}\mathrm{tr}(\boldsymbol{\varepsilon}_{te})\,\mathbf{I}$ can be written as $\boldsymbol{\varepsilon}^*_{te} = \boldsymbol{\varepsilon}^* - \boldsymbol{\varepsilon}_{trip} - \boldsymbol{\varepsilon}_{cp}$, using Equation (2.10).

2.1.4 Remark (Density Changes)
The last two terms in Equation (2.12) may be written as a sum

$$\frac{1}{3}\frac{\rho_0-\rho(\theta)}{\rho_0}\,\mathbf{I} = \alpha(\theta-\theta_0)\,\mathbf{I} + \frac{1}{3}\frac{\rho_0-\rho(\theta_0)}{\rho_0}\,\mathbf{I}$$

where $\rho(\theta)$ – current density related to the current temperature using a mixture rule for density and the balance equation (2.7), cf. [Wolff et al. 2003a]. This is useful for processing of dilatometer data, cf. [Hömberg et al. 2009] for instance.

2.1.5 Remark (Lamé Parameters)
The stress in Equation (2.12) can be alternatively defined by using Young's modulus and Poisson's ratio or the Lamè constants instead of using bulk and shear modulus. We have

$$\lambda := K - \frac{2\mu}{3}, \qquad \nu := \frac{3K - 2\mu}{2(3K+\mu)}, \qquad E := \frac{9K\mu}{3K+\mu} \tag{2.13}$$

where λ – Lamé's first coefficient, ν – Poisson's ration and E – Young's modulus.

2.1.5 Plasticity

In contrast to elastic material behaviour, which describes a functional correlation between stress and strain, i.e. the deformation under internal and external forces returns (on the same path) to its original shape if the forces are removed, plasticity describes the deformation of a material undergoing non-reversible changes of shape in response to applied forces. A graphical representation of the relationship between stress and strain is given in Figure 2.2. For example, a steel workpiece is deformed into a new shape, then

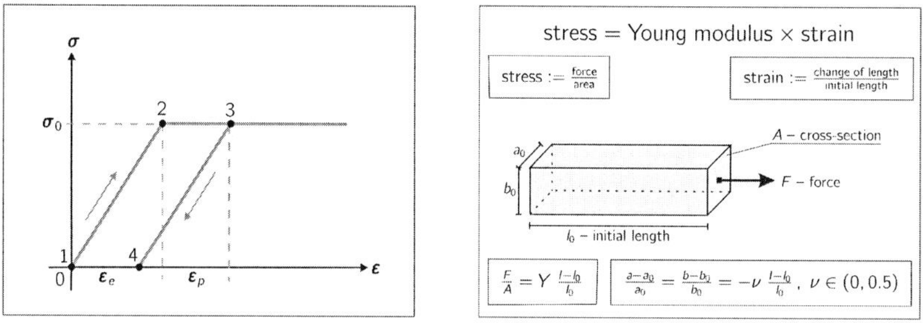

(a) Stress-strain diagram. (b) Tensile test.

Figure 2.2: Tensile testing and stress-strain diagram showing typical yield behaviour for non-ferrous alloys. Stress is shown as a function of strain, from [Duvaut and Lions 1976].

plasticity displays as a permanent strain which is left after unloading within the material itself. In engineering, the transition from elastic behaviour to plastic behaviour is called yield (increase of load beyond the yield stress, which is a material parameter).
Plastic deformation occurs in many metal-forming processes (rolling, pressing, forging) and in geologic processes (rock folding and rock flow within the earth under extremely

high pressures and at elevated temperatures). The physical mechanisms that cause plastic deformation can vary widely. At the crystal scale, plasticity in metals is usually a consequence of dislocations. In most crystalline materials such defects are relatively rare. But there are also materials where defects are numerous and are part of the very crystal structure, in such cases plastic crystallinity can result.

Consider a metal rod (e.g. steel) that is subjected to a traction force $\boldsymbol{\sigma}$ and that in consequence undergoes a relative elongation $\boldsymbol{\varepsilon}$. In a system of two orthogonal axes ($\boldsymbol{\varepsilon}$ – abscissa, $\boldsymbol{\sigma}$ – ordinate) we trace the graph of the relation $(\boldsymbol{\varepsilon}, \boldsymbol{\sigma})$, cf. Figure 2.2(a). When $\boldsymbol{\varepsilon}$ increases, starting from zero, $\boldsymbol{\sigma}$ increases and the point $(\boldsymbol{\varepsilon}, \boldsymbol{\sigma})$ describes a straight line segment from the origin 1 (let $\frac{1}{E}$ be the slope of this segment). If we continue to increase $\boldsymbol{\varepsilon}$, the curve described by the point $(\boldsymbol{\varepsilon}, \boldsymbol{\sigma})$ starting from the point 2 becomes a parallel to the line $\overline{1\boldsymbol{\varepsilon}}$ (perfectly plastic material behaviour). Thus, when $\boldsymbol{\varepsilon}$ increases from 0 to $+\infty$, the graph for the relation $(\boldsymbol{\varepsilon}, \boldsymbol{\sigma})$ is composed of a straight line segment $\overline{12}$ and a parallel to the line $\overline{23}$. In the region $\overline{12}$ the behaviour of the material is linear and reversible, i.e. elastic. At the point 2 the behaviour of the material is no longer reversible, i.e. the plastic region is starting from this point.

The constitutive law for such a perfectly plastic metal rod can be written as, cf. [Duvaut and Lions 1976]:

$$\boldsymbol{\varepsilon}' = E\,\boldsymbol{\sigma}' + \zeta$$

where

$$\begin{array}{llll} \zeta = 0, & \text{when} \quad \boldsymbol{\sigma} < R & \text{or when} & \boldsymbol{\sigma} = R,\ \boldsymbol{\sigma}' < 0 \\ \zeta \geq 0, & \text{when} \quad \boldsymbol{\sigma} = R & \text{and} & \boldsymbol{\sigma}' = 0 \end{array}$$

which is equivalent to

$$\zeta(\tau - \boldsymbol{\sigma}) \leq 0 \qquad \forall\, \tau \leq R \qquad \zeta\boldsymbol{\sigma}' = 0 \tag{2.14}$$

(if the function $t \mapsto \boldsymbol{\sigma}(t)$ is differentiable in t, the equation $(2.14)_2$ can be derived from the inequality $(2.14)_1$).

One can generalise the obtained constitutive law to three-dimensional bodies. If the stress exceeds a critical value, as was mentioned above, the material will undergo plastic, or irreversible, deformation. This critical stress can be tensile or compressive. We assume that the elastic behaviour is characterised by

$$F(\boldsymbol{\sigma}) < 0$$

and the plastic region is described by

$$F(\boldsymbol{\sigma}) = 0.$$

The plastic deformation is isochoric or volume preserving. Hydrostatic pressure (i.e. isotropic stress) leads to no plastic deformation and the material behaviour in plasticity is scleronomic (i.e. not explicitly depending on time). The von Mises criteria is commonly used to determine whether a material has yielded.

2.1.6 Remark (Perfect Plasticity (Prandtl-Reuss Law))
Perfect plasticity is a property of materials to undergo irreversible deformation without any increase in stresses or loads. Plastic materials with hardening necessitate increasingly higher stresses to result in further plastic deformation. Generally plastic deformation is also dependent on the deformation speed, i.e. usually higher stresses have to be applied to increase the rate of deformation and such materials are said to deform visco-plastically. Prandtl-Reuss plasticity is a relatively simple material behaviour. The associative plasticity model with von Mises criterion, i.e. plastic flow rule and hardening law are determined by derivatives of a yield function F, i.e. $\boldsymbol{\varepsilon}'_{cp} = \Lambda \frac{\partial F}{\partial \boldsymbol{\sigma}^*}$, *cf. Subsection 3.2.3.*

The von Mises yield surfaces in principal stress coordinates circumscribes a cylinder around the hydrostatic axis. The von Mises ansatz is based on effective stress under uniaxial loading, subtracting hydrostatic stresses, and claims that all effective stresses greater than that which causes material failure in uniaxial loading will result in plastic deformation, i.e. the material behaviour is constrained by

$$F(\boldsymbol{\sigma}, \mathbf{X}_{cp}, R_0, R) := \sqrt{\frac{3}{2}\left(\boldsymbol{\sigma}^* - \mathbf{X}^*_{cp}\right) : \left(\boldsymbol{\sigma}^* - \mathbf{X}^*_{cp}\right)} - (R_0 + R) \leq 0, \tag{2.15}$$

which is a restriction on the deviator of the effective stress $\boldsymbol{\sigma}^* - \mathbf{X}^*_{cp}$. The notations are: F – yield function, $\boldsymbol{\sigma}^*$ – stress deviator, $\mathbf{X}_{cp}$ – back-stress associated with plasticity, R_0 – initial radius of the yield sphere in the stress space (or initial yield stress), R – its possible increment due to isotropic hardening (cf. Subsection 2.6.3).
Thus, the sum $R_0 + R$ is the radius of the yield sphere (resp. the stress under which the material fails in uniaxial loading). For simplicity, we restrict ourselves to the von Mises criterion (cf. [Lemaitre and Chaboche 1990; Lemaitre 2001] for generalisations).
A visual representation of the yield surface may be constructed using the above equation, which takes the shape of an ellipse. Inside the surface, materials undergo elastic deformation. Reaching the surface means the material undergoes plastic deformations. It is physically impossible for a material to go beyond its yield surface.

2.1.7 Remark (Tresca Yield Criterion)
This criterion is based on the notion that when a material fails, it does so in shear, which is a relatively good assumption when considering metals. Given the principal stress state, we can use Mohr's circle to solve for the maximum shear stresses the material will experience and conclude that the material will fail if:

$$F(\boldsymbol{\sigma}, R_0, R) := \left(\boldsymbol{\sigma}_{\max} - \boldsymbol{\sigma}_{\min}\right) - \sqrt{\frac{3}{2}}(R_0 + R) \leq 0, \tag{2.16}$$

where $\boldsymbol{\sigma}_{\max} := \max\{\boldsymbol{\sigma}_1, \boldsymbol{\sigma}_2, \boldsymbol{\sigma}_3\}$, $\boldsymbol{\sigma}_{\min} := \min\{\boldsymbol{\sigma}_1, \boldsymbol{\sigma}_2, \boldsymbol{\sigma}_3\}$ *and* $\boldsymbol{\sigma}_1, \boldsymbol{\sigma}_2, \boldsymbol{\sigma}_3$ *– eigenvalues of the stress tensor.*

The evolution of the plastic strain $\boldsymbol{\varepsilon}_{cp}$ is governed by the flow rule:

$$\boldsymbol{\varepsilon}'_{cp} = \Lambda \left(\boldsymbol{\sigma}^* - \mathbf{X}^*_{cp}\right), \tag{2.17}$$

where the plastic multiplier Λ has to fulfill

$$\Lambda = 0, \quad \text{if} \quad F(\boldsymbol{\sigma}, \mathbf{X}_{cp}, R_0, R) < 0 \quad \text{and} \tag{2.18}$$

$$\Lambda \geq 0, \quad \text{if} \quad F(\boldsymbol{\sigma}, \mathbf{X}_{cp}, R_0, R) = 0 \quad \text{(yield condition)}. \tag{2.19}$$

Thus, plastic deformation is only possible, if the yield condition (2.19) is fulfilled.

2.1.8 Remark (Alternative Formulation of the Flow Rule)
Sometimes, especially in numerics, an alternative ansatz for the flow rule is used, cf. [Wolff et al. 2011b]:

$$\boldsymbol{\varepsilon}'_{cp} = \Lambda \frac{\boldsymbol{\sigma}^* - \mathbf{X}^*_{cp}}{\|\boldsymbol{\sigma}^* - \mathbf{X}^*_{cp}\|}.$$

The plastic multiplier is determined by the yield condition (2.19).

2.1.9 Remark (Calculation of the Plastic Multiplier)
From (2.17) *and the yield condition* (2.19) *we obtain*

$$(\boldsymbol{\sigma} - \mathbf{X}_{cp}) : \boldsymbol{\varepsilon}'_{cp} = (\boldsymbol{\sigma} - \mathbf{X}_{cp}) : \Lambda(\boldsymbol{\sigma}^* - \mathbf{X}_{cp}) = \frac{2}{3}(R_0 + R)^2\Lambda,$$

$$(\boldsymbol{\sigma} - \mathbf{X}_{cp}) : \boldsymbol{\varepsilon}'_{cp} = \frac{1}{\Lambda}\boldsymbol{\varepsilon}'_{cp} : \boldsymbol{\varepsilon}'_{cp}$$

and therefore

$$\Lambda = \frac{3}{2(R_0 + R)} s'_{cp}, \tag{2.20}$$

where the accumulated plastic strain is given via:

$$s_{cp}(\mathbf{x},t) := \int_0^t \left(\frac{2}{3}\boldsymbol{\varepsilon}'_{cp}(\mathbf{x},\tau) : \boldsymbol{\varepsilon}'_{cp}(\mathbf{x},\tau)\right)^{\frac{1}{2}} \mathrm{d}\tau. \tag{2.21}$$

2.1.10 Example (Alternative Formula for the Plastic Multiplier)
For sake of simplicity we neglect kinematic and isotropic hardening and TRIP for a moment and get the following formula for the plastic multiplier Λ *in case of perfect plasticity. Differentiating the yield condition and using the yield condition itself as well as the relation* $\boldsymbol{\sigma}^* = 2\mu\,(\boldsymbol{\varepsilon}(\mathbf{u}) - \boldsymbol{\varepsilon}_{trip} - \boldsymbol{\varepsilon}_{cp})$ *for the stress deviator, we get*

$$\Lambda = \begin{cases} \frac{1}{R_0}\boldsymbol{\sigma}^* : \boldsymbol{\varepsilon}(\mathbf{u}'), & \text{if} \quad F(\boldsymbol{\sigma}, R_0) = 0 \quad \text{and} \quad \boldsymbol{\sigma}^* : \boldsymbol{\varepsilon}(\mathbf{u}') > 0 \\ 0 & \text{else} \end{cases} \tag{2.22}$$

A similar formula can be derived in this manner in the case of kinematic and isotropic hardening and TRIP.

2.1.11 Remark (Application: Material Behaviour of Steel)
In the case of perfect plasticity, the behaviour of the material is largely idealised. Therefore, the model of perfect plasticity is not well describing the material behaviour of steel because kinematic and isotropic hardening have an important influence.

2.1.6 Transformation-induced Plasticity (TRIP)

Phase transformations under non-vanishing deviatoric (non-isotropic) stress yield a permanent deviatoric deformation which cannot be described by classical plasticity at the macroscopic level (cf. [Leblond et al. 1986a,b]). We assume that TRIP has no yield condition (cf. the references in [Wolff et al. 2008c] for details and for alternatives) and that its evolution can be described by

$$\boldsymbol{\varepsilon}'_{trip} = b\left(\boldsymbol{\sigma}^* - \mathbf{X}^*_{trip}\right), \tag{2.23}$$

where $b \geq 0$ depends essentially on the phase evolution and $\mathbf{X}_{trip}$ – back-stress associated with TRIP. More details about TRIP will be given in Section 2.5.

2.1.7 Initial and Boundary Conditions

We assume the following initial conditions f.a. $\mathbf{x}$ in Ω:

$$\mathbf{u}(\mathbf{x},0) = \mathbf{u}_0(\mathbf{x}), \qquad \mathbf{u}'(\mathbf{x},0) = \mathbf{u}_1(\mathbf{x}), \qquad \theta(\mathbf{x},0) = \theta_0(\mathbf{x}), \tag{2.24}$$

(2.25) $\quad \boldsymbol{\varepsilon}_{trip}(\mathbf{x}, 0) = \mathbf{0}, \qquad \boldsymbol{\varepsilon}_{cp}(\mathbf{x}, 0) = \mathbf{0}, \qquad \boldsymbol{\xi}(\mathbf{x}, 0) = \mathbf{0},$

(2.26) $\quad \mathbf{p}(\mathbf{x}, 0) = \mathbf{p}_0(\mathbf{x})$

with

(2.27) $\quad \sum_{i=1}^{m} p_{0i} = 1, \qquad p_{0i} \geq 0 \qquad \text{for} \qquad i = 1, \ldots, m.$

In certain situations it might be useful to apply similar initial conditions like $\boldsymbol{\xi}(\mathbf{x}, 0) = \boldsymbol{\xi}_0(\mathbf{x})$. Furthermore, boundary conditions for $\mathbf{u}$ and θ have to be added. We assume mixed boundary conditions for $\mathbf{u}$

(2.28) $\quad \mathbf{u} = \mathbf{0}$ f.a. points in Γ_1, $\qquad \boldsymbol{\sigma} \cdot \boldsymbol{\nu}_{\Gamma_2} = \mathbf{0}$ f.a. points in Γ_2,

where Γ_1 and Γ_2 are mutually disjoint parts of the boundary of Ω s.t. meas$(\Gamma_1) > 0$ and Γ_1 is closed and $\boldsymbol{\nu}_{\Gamma_2}$ is the outward unit normal vector on Γ_2 (this case without any boundary load is important e.g. in heat treatment processes, where no load is applied on the surface). Let the temperature fulfill the Robin condition

(2.29) $$-\lambda_\theta \frac{\partial \theta}{\partial \boldsymbol{\nu}_\Gamma} = \delta \left(\theta - \theta_\Gamma\right) \text{ on } \Gamma_T,$$

where λ_θ – heat conductivity, δ – heat-exchange coefficient, θ_Γ – temperature of the surrounding medium and $\boldsymbol{\nu}_\Gamma$ – outward unit normal vector to the boundary $\Gamma := \partial\Omega$.

2.1.12 Remark (Boundary Conditions)
For $\Gamma_1 = \emptyset$ we have a pure traction stress problem and $\Gamma_2 = \emptyset$ indicates a pure displacement problem. Of course, this has an influence on the solvability of the full problem. The corresponding Neumann problem, e.g. has no unique solution, cf. [Ciarlet 1988].
Different boundary conditions for θ, e.g. mixed boundary conditions are also reasonable and manageable, cf. Problem ($\mathbf{P}_{NUM}$).
Using the Stefan-Boltzmann law for thermal radiation gives the boundary condition

$$-\lambda_\theta \frac{\partial \theta}{\partial \boldsymbol{\nu}_\Gamma} = \delta \left(\theta^4 - \theta_\Gamma^4\right) \text{ on } \Gamma_T,$$

which is also reasonable from an application point of view, but more difficult to treat mathematically (concerning regularity of θ).
As mentioned above, we focus on quenching situations. If necessary, one can use the non-homogeneous boundary conditions

$$\mathbf{u} = \mathbf{b}_1 \quad \text{f.a. points in } \Gamma_1, \qquad \boldsymbol{\sigma} \cdot \boldsymbol{\nu}_{\Gamma_2} = \mathbf{b}_2 \quad \text{f.a. points in } \Gamma_2,$$

instead of (2.28). Thus, we allow a mechanical input due to a given displacement $\mathbf{b}_1$ and/or surface stress $\mathbf{b}_2$. In [Duvaut and Lions 1976] is problem is solved via homogenisation of the boundary values.

2.2 Dissipation Inequality

Discussing the constitutive model developed in the previous section in a thermodynamical framework (cf. e.g. [Haupt 2000; Lemaitre and Chaboche 1990]), we assume that the free energy Ψ, the stress tensor $\boldsymbol{\sigma}$, the entropy η and the heat-flow vector $\mathbf{q}$ are given in the following form:

$$(2.30) \qquad \Psi = \tilde{\Psi}(\boldsymbol{\varepsilon}_{te}, \boldsymbol{\varepsilon}_{cp}, \boldsymbol{\varepsilon}_{trip}, \theta, \nabla\theta, \boldsymbol{\xi}, \mathbf{p}), \qquad \boldsymbol{\sigma} = \tilde{\boldsymbol{\sigma}}(\boldsymbol{\varepsilon}_{te}, \boldsymbol{\varepsilon}_{cp}, \boldsymbol{\varepsilon}_{trip}, \theta, \nabla\theta, \boldsymbol{\xi}, \mathbf{p}),$$

$$(2.31) \qquad \eta = \tilde{\eta}(\boldsymbol{\varepsilon}_{te}, \boldsymbol{\varepsilon}_{cp}, \boldsymbol{\varepsilon}_{trip}, \theta, \nabla\theta, \boldsymbol{\xi}, \mathbf{p}), \qquad \mathbf{q} = \tilde{\mathbf{q}}(\boldsymbol{\varepsilon}_{te}, \boldsymbol{\varepsilon}_{cp}, \boldsymbol{\varepsilon}_{trip}, \theta, \nabla\theta, \boldsymbol{\xi}, \mathbf{p})$$

$\boldsymbol{\xi} = (\xi_1, \ldots, \xi_n)$, ξ_j – scalars or tensors denote additional internal variables besides $\boldsymbol{\varepsilon}_{cp}$, $\boldsymbol{\varepsilon}_{trip}$ and $\mathbf{p}$ (cf. Subsection 2.6.3 or [Wolff et al. 2008c] for examples of $\boldsymbol{\xi}$ and the discussion of internal variables in [Coleman and Gurtin 1967]). The internal variables have to fulfill the evolution equations

$$(2.32) \qquad \boldsymbol{\varepsilon}'_{cp} = \tilde{\boldsymbol{\varepsilon}}_{cp}(\boldsymbol{\sigma}, \boldsymbol{\varepsilon}_{cp}, \boldsymbol{\varepsilon}_{trip}, \theta, \boldsymbol{\xi}, \boldsymbol{\xi}', \mathbf{p}), \qquad \boldsymbol{\varepsilon}'_{trip} = \tilde{\boldsymbol{\varepsilon}}_{trip}(\boldsymbol{\sigma}, \boldsymbol{\varepsilon}_{cp}, \boldsymbol{\varepsilon}_{trip}, \theta, \boldsymbol{\xi}, \boldsymbol{\xi}', \mathbf{p}, \mathbf{p}'),$$

$$(2.33) \qquad \boldsymbol{\xi}' = \tilde{\boldsymbol{\xi}}(\boldsymbol{\varepsilon}_{cp}, \boldsymbol{\varepsilon}_{trip}, \boldsymbol{\varepsilon}'_{cp}, \boldsymbol{\varepsilon}'_{trip}, \theta, \boldsymbol{\xi}, \mathbf{p})$$

and the evolution equations for phase transformations (cf. [Wolff et al. 2008c]). Note that (2.17) and (2.23) are special cases of (2.32). Standard arguments (inserting (2.30) in (2.3), cf. [Wolff et al. 2008c]) lead to the potential relations

$$(2.34) \qquad \boldsymbol{\sigma} = \rho_0 \frac{\partial \tilde{\Psi}}{\partial \boldsymbol{\varepsilon}_{te}}, \qquad \eta = -\frac{\partial \tilde{\Psi}}{\partial \theta}, \qquad \frac{\partial \tilde{\Psi}}{\partial \nabla\theta} = 0$$

and to the dissipation inequality

$$(2.35) \qquad (\boldsymbol{\sigma} - \mathbf{X}_{cp}) : \boldsymbol{\varepsilon}'_{cp} + (\boldsymbol{\sigma} - \mathbf{X}_{trip}) : \boldsymbol{\varepsilon}'_{trip} - \rho_0 \frac{\partial \tilde{\Psi}}{\partial \boldsymbol{\xi}} \boldsymbol{\xi}' - \rho_0 \frac{\partial \tilde{\Psi}}{\partial \mathbf{p}} : \mathbf{p}' - \frac{1}{\theta} \mathbf{q} \cdot \nabla\theta \geq 0,$$

with the back-stress $\mathbf{X}_{cp}$ associated with (classical) plasticity and the back-stress $\mathbf{X}_{trip}$ associated with TRIP, defined by

$$(2.36) \qquad \mathbf{X}_{cp} := \rho_0 \frac{\partial \tilde{\Psi}}{\partial \boldsymbol{\varepsilon}_{cp}} \qquad \text{and} \qquad \mathbf{X}_{trip} := \rho_0 \frac{\partial \tilde{\Psi}}{\partial \boldsymbol{\varepsilon}_{trip}}.$$

Since the first terms in (2.35) are independent of $\nabla\theta$, one obtains the inequality of dissipation due to plasticity and TRIP and to phase transitions:

$$(2.37) \qquad (\boldsymbol{\sigma} - \mathbf{X}_{cp}) : \boldsymbol{\varepsilon}'_{cp} + (\boldsymbol{\sigma} - \mathbf{X}_{trip}) : \boldsymbol{\varepsilon}'_{trip} - \rho_0 \frac{\partial \tilde{\Psi}}{\partial \boldsymbol{\xi}} \boldsymbol{\xi}' - \rho_0 \frac{\partial \tilde{\Psi}}{\partial \mathbf{p}} : \mathbf{p}' \geq 0.$$

This inequality is a representation of restrictions for evolution equations for internal variables and for the free energy. It is sufficient condition for thermodynamic consistency of the presented model (cf. Section 2.7 and [Wolff et al. 2008c]). We require the heat-conduction inequality

$$\frac{1}{\theta}\mathbf{q}\cdot\nabla\theta \leq 0. \tag{2.38}$$

An extension of the classical Fourier law for the heat flow is

$$\tilde{\mathbf{q}}(\boldsymbol{\varepsilon}_{te},\boldsymbol{\varepsilon}_{cp},\boldsymbol{\varepsilon}_{trip},\theta,\nabla\theta,\boldsymbol{\xi},\mathbf{p}) := -\lambda_\theta(\boldsymbol{\varepsilon}_{te},\boldsymbol{\varepsilon}_{cp},\boldsymbol{\varepsilon}_{trip},\theta,\nabla\theta,\boldsymbol{\xi},\mathbf{p})\,\nabla\theta, \tag{2.39}$$

where $\lambda_\theta \geq 0$ is the heat conductivity. Hence, (2.38) is always fulfilled.

2.2.1 Remark (Alternative Heat Flux)
In [Bien 2003] the heat flux is postulated by a constitutive relation of the form

$$\mathbf{q} := -\kappa\,\nabla\theta - k\,\nabla\theta'$$

with the coefficient of thermal conductivity of the body κ *and some non-negative parameter* k*. This has direct consequences for the regularity of* θ*, cf. Subsection 5.3.8.*

2.3 Special Form of the Heat Equation

We assume, that the free energy Ψ and the entropy η depend on the same variables. Then, the relation (2.5) leads to the representation of the internal energy e

$$e = \tilde{e}(\boldsymbol{\varepsilon}_{te},\boldsymbol{\varepsilon}_{cp},\boldsymbol{\varepsilon}_{trip},\theta,\nabla\theta,\boldsymbol{\xi},\mathbf{p}). \tag{2.40}$$

Inserting (2.39) and (2.40) in (2.2), some standard calculations (cf. [Wolff et al. 2008c]) lead to

$$\begin{aligned}\rho_0 c_d\theta' - \operatorname{div}(\lambda_\theta\nabla\theta) &= (\boldsymbol{\sigma}-\mathbf{X}_{cp}) : \boldsymbol{\varepsilon}'_{cp} + (\boldsymbol{\sigma}-\mathbf{X}_{trip}) : \boldsymbol{\varepsilon}'_{trip} - \rho_0\frac{\partial\tilde{\Psi}}{\partial\boldsymbol{\xi}}\boldsymbol{\xi}' \\ &+ \theta\frac{\partial\tilde{\boldsymbol{\sigma}}}{\partial\theta} : \boldsymbol{\varepsilon}'_{te} + \theta\frac{\partial\mathbf{X}_{cp}}{\partial\theta} : \boldsymbol{\varepsilon}'_{cp} + \theta\frac{\partial\mathbf{X}_{trip}}{\partial\theta} : \boldsymbol{\varepsilon}'_{trip} + \rho_0\theta\frac{\partial^2\tilde{\Psi}}{\partial\theta\partial\boldsymbol{\xi}}\boldsymbol{\xi}' + \rho_0\frac{\partial\tilde{e}}{\partial\mathbf{p}}\mathbf{p}' + r\end{aligned} \tag{2.41}$$

with the specific heat (at constant deformation) defined by

$$c_d := \frac{\partial\tilde{e}}{\partial\theta} = -\theta\frac{\partial^2\tilde{\Psi}}{\partial\theta^2}. \tag{2.42}$$

The latent heats L_i ($i = 2, \dots, m$) are defined in this context by

$$L_i := \frac{\partial \tilde{e}}{\partial p_1} - \frac{\partial \tilde{e}}{\partial p_i}. \tag{2.43}$$

The latent heat L_i characterises the transformation of the initial first phase into the ith phase. Taking (2.8) into account, we get

$$-\rho_0 \frac{\partial \tilde{e}}{\partial \mathbf{p}} \mathbf{p}' = -\rho_0 \sum_{i=1}^{m} \frac{\partial \tilde{e}}{\partial p_i} p_i' = -\rho_0 \sum_{i=2}^{m} \left(\frac{\partial \tilde{e}}{\partial p_i} - \frac{\partial \tilde{e}}{\partial p_1} \right) p_i' = \rho_0 \sum_{i=1}^{m} L_i \gamma_i. \tag{2.44}$$

In practice, the latent heat L_i is usually approximated by $L_i := h_i - h_1$ (h_i – free enthalpy of the ith phase) (cf. [Wolff et al. 2008c]). Another possible approach combines the relation for the chemical potential difference, the temperature difference and the activation energy for the transformation[2].

2.3.1 Remark (Derivation of the Heat Equation)
The heat equation was derived in Section 2.3 in a standard way from the energy equation. A physical derivation of the energy equation or rather the heat equation via the conservation of energy in a brief and simplified approach can be found in [Larson and Thomée 2003] and [Suhr 2010].

2.4 General Macroscopic Approach for Phase Transitions in Steel

Steel is a polycrystalline material consisting mainly of iron and carbon. At a certain temperature, the crystal lattice becomes unstable and solid-solid phase transformations occur. The transformations of different phases do not interact – each phase has a temperature interval in which a transformation is possible.

Phase transformations are an important phenomenon in the material behaviour of steel and represent a major field of research. Isothermal diffusive transformations are well described by the Johnson-Mehl-Avrami-Kolmogorov kinetics and models for martensitic tranformation are based on the Koistinen-Marburger equation, but there are some essential open questions in modelling the non-isothermal transformation and phase transformations under stress. In [Wolff et al. 2007a] one can find a good description of a phenomenological model for phase transformations, but there are a lot of proposals for modifications and generalisations of phenomenological models for phase transformations,

[2]private communications, M. Wolff, 2011

cf. e.g. [Aeby-Gautier and Cailletaud 2004; Avrami 1939, 1940, 1941; Böhm et al. 2004; Burke 1965; Dreyer et al. 2008; Fischer et al. 1994; Hömberg 1995; Leblond et al. 1985, 1986a,b; Magee 1968; Mittemeijer and Sommer 2002; Wolff and Böhm 2002b; Wolff et al. 2003c, 2007a; Visintin 1987].

In this section we want to specialise $\boldsymbol{\gamma}$ in (2.6) in order to obtain applicable models for phase transformations for multi-phase simultaneous and consecutive reactions.

The following considerations are not only valid for phase transformations in steel, but also for general (chemical or other) reactions in coexisting mixtures. In the context of macroscopic modelling, we consider steel as a coexisting mixture of m ($m \geq 2$) phases (constituents), which may transform into another under appropriate conditions. Furthermore, we neglect any diffusion of these phases, assuming that they stay at their original places of formation.

We assume that the transformation of the ith phase into the jth phase (for $i \neq j$, $i, j = 1, \ldots, m$, abbreviation: $i \to j$) has the transformation rate $-a_{ij}$, i.e. for the transformation $i \to j$ the change of p_i in favour of p_j can be described by the transformation law

$$p_i' = -a_{ij}. \tag{2.45}$$

In accordance with (2.45) the growth of p_j at expense of p_i is expressed as

$$p_j' = a_{ij}. \tag{2.46}$$

As a consequence of (2.45) and (2.46), we set

$$a_{ij} \geq 0, \qquad i, j = 1, \ldots, m \qquad \text{and} \qquad a_{ii} := 0, \qquad i = 1, \ldots, m. \tag{2.47}$$

If there is no transformation $i \to j$, we set $a_{ij} = 0$. In general, the a_{ij} depend on the same variables like the γ_i in Equation (2.6). Based on the above considerations, the system (2.6) can be described by

$$p_i' = -\sum_{j=1}^{m} a_{ij} + \sum_{j=1}^{m} a_{ji}, \qquad i = 1, \ldots, m, \tag{2.48}$$

fulfilling the condition (2.47). Similarly to [Wolff et al. 2007a], we propose for phase transformations in steel:

2.4.1 Assumption

For $i, j = 1, 2, \ldots, m$ with $i \neq j$, there exists a limit value $\bar{p}_{ij} = \bar{p}_{ij}(\theta, \boldsymbol{\xi}, p_k)$ ($k \neq i, j$), s.t. the phase transformation $p_i \to p_j$ is only possible if $p_i > 0$ and $\bar{p}_{ij} - p_j > 0$. We set $\bar{p}_{ij} = 0$ if the phase transformation does not occur. Sometimes, $\bar{p}_{ij}$ is considered as

the equilibrium fraction (maximal possible fraction) $\bar{p}_i$ *of the* i*th phase (cf. [Wolff et al. 2007a]).*
For each phase j *there exist two limit temperatures* $\theta_{jf} = \theta_{jf}(\boldsymbol{\xi})$ *and* $\theta_{js} = \theta_{js}(\boldsymbol{\xi})$ *s.t. the transformation* $i \to j$ *takes place only, if* $\theta_{jf} \leq \theta < \theta_{js}$*.* θ_{jf} *and* θ_{js} *are called finish and start temperature, respectively. This last condition can be taken into account by a switch function* G_{ij}*, which is* 1 *if the condition is fulfilled and which is* 0 *otherwise (cf. [Wolff et al. 2007a]).*

The Heaviside function H is defined by

$$H(s) := 0 \quad \text{for} \quad s \leq 0 \quad \text{and} \quad H(s) = 1 \quad \text{for} \quad s > 0. \tag{2.49}$$

Taking Assumption 2.4.1 into account, we propose the following general model for phase transformations in steel

$$p_i' = -\sum_{j=1}^{m} a_{ij} H(p_i) H(\bar{p}_{ij} - p_j) G_{ij} + \sum_{j=1}^{m} a_{ji} H(p_j) H(\bar{p}_{ji} - p_i) G_{ji}, \quad i = 1, \dots, m. \tag{2.50}$$

If there are only two phases, i and j, and if the conditions are fulfilled, (2.50) reduces to (2.45) and (2.46). We specify the a_{ij} (cf. [Wolff et al. 2007a]):

$$a_{ij} := (e_{ij}(\theta, \boldsymbol{\xi}) + p_j)^{r_{ij}(\theta,\boldsymbol{\xi})} (\bar{p}_{ij} - p_j)^{s_{ij}(\theta,\boldsymbol{\xi})} g_{ij}(\theta, \boldsymbol{\xi}) h_{ij}(\theta') \quad \text{for} \quad i, j = 1, \dots, m. \tag{2.51}$$

The parameters e_{ij}, r_{ij}, s_{ij} and g_{ij} have to fulfill

$$e_{ij} \geq 0, \quad r_{ij} \geq 0, \quad s_{ij} \geq 0, \quad g_{ij} \geq 0, \quad h_{ij} \geq 0 \quad \text{f.a. admissible arguments.}$$

In accordance with (2.47), we set

$$e_{ii} = r_{ii} = s_{ii} = g_{ii} = 0 \text{ for } i = i, \dots, m.$$

For detailed discussion refer to [Wolff et al. 2007a].

2.4.2 Example (Leblond-Devaux Proposal)
A simple specialisation of (2.45) consists of

$$a_{ij} = \mu_{ij} p_i, \tag{2.52}$$

where the non-negative μ_{ij} *may depend on the same arguments as* a_{ij} *in (2.51). This means that the decomposition rate of* i *into* j *is proportional to the fraction of* i *available for decomposition. In case of only two present phases this leads to* $p_1' = -\mu_{12} p_1$ *(cf. [Wolff et al. 2007a]). Using the relation (2.48), we obtain for the forming phase (*$\bar{p}_{12} = 1$ *and* $\bar{p}_{21} = 0$*)*

$$p_2' = \mu_{12} (1 - p_2),$$

which is a special case of the Leblond-Devaux proposal (cf. [Wolff et al. 2007a]).

We specialise the general model to the Johnson-Mehl-Avrami-Kolmogorov kinetics for diffusive phase transformations, cf. [Wolff et al. 2007a; Fanfoni and Tomellini 1998].

2.4.3 Example (Johnson-Mehl-Avrami-Kolmogorov Kinetics)
At constant temperature the forming phase for a simple transformation of austenite into ferrite for a hypoeutectoid low alloyed steel grows according to the Johnson-Mehl-Avrami-Kolmogorov formula (or JMAK equation)

$$p_2(t) = \bar{p}_{12}\Big(1 - \exp\Big(-\Big(\frac{t}{\tau_{12}}\Big)^{n_{12}}\Big)\Big), \tag{2.53}$$

where $\tau_{12} > 0$, $n_{12} > 1$ are temperature(stress)-dependent material parameters, characterising the phase transformation (cf. [Wolff et al. 2007a]). The equilibrium phase fraction, i.e. the maximal possible phase fraction at the particular time (or temperature), where the transformation starts, is denoted by $\bar{p}$. Taking the time derivative (for constant temperature) and excluding the exponential and the explicit time dependence, we get from (2.53) the autonomous ODE

$$p_2'(t) = (\bar{p}_{12} - p_2(t))\,\frac{n_{12}}{\tau_{12}}\left(-\ln\left(1 - p_2(t)\bar{p}_{12}^{-1}\right)\right)^{1-\frac{1}{n_{12}}}, \qquad p_2(0) = 0.$$

Therefore, assuming the Johnson-Mehl-Avrami-Kolmogorov kinetics f.a. possible transformations $i \to j$, we define (cf. [Wolff et al. 2007a])

$$a_{ij} = (\bar{p}_{ij} - p_j(t))\,\frac{n_{ij}}{\tau_{ij}}\left(-\ln\left(1 - p_j(t)\bar{p}_{ij}^{-1}\right)\right)^{1-\frac{1}{n_{ij}}} \quad \text{for} \quad i,j = 1,\dots,m,\ i \neq j,$$

$$a_{ii} = 0 \quad \text{for} \quad i,j = 1,\dots,m.$$

2.4.4 Remark (Koistinen-Marburger Equation)
It is well known, that the martensitic transformation does not follow the Johnson-Mehl-Avrami-Kolmogorov kinetics (cf. [Wolff et al. 2007a]). Leblond and Devaux suggested applying the linear approach in Example 2.4.2 to martensitic transformation as well. Due to Koistinen and Marburger (cf. [Koistinen and Marburger 1959; Lemaitre 2001]) the martensite fraction forming from a given austenite fraction p_1 at the temperature θ less than the martensite start temperature reads as

$$\bar{p}_{12}(\theta) = p_1(t_{ms})\left(1 - \exp\left(-\frac{\theta_{ms} - \theta}{\theta_{m0}}\right)\right) \tag{2.54}$$

equivalent to

$$\bar{p}_{12}'(t) = (1 - \bar{p}_{12}(t))\frac{-\theta'(t)}{\theta_{m0}}, \qquad \bar{p}_{12}(0) = 0.$$

The 'limit value' of martensite in the multi-phase case is given by

$$\bar{p}_{1m}(t) := 1 - \sum_{j=2}^{m-1} p_j(t).$$

An alternative description for the martensitic transformation is the model of Yu (cf. [Yu 1997]):

$$\bar{p}(\theta) = \frac{\theta_{ms} - \theta}{\theta_{ms} - \theta - \beta(\theta_{mf} - \theta)},$$

where θ_{ms} – martensite start, θ_{mf} – finish temperature and the model parameters θ_{m_0}, β, μ are determined via optimisation in dilatometer experiments and/or Gleeble™ experiments, cf. [Hömberg et al. 2009; Wolff et al. 2007c] (all other material parameters are obtained in the same way). Typical values for the austenite-martensite transformation can be found in Table 8.1 and in [Acht et al. 2008a,b].

2.4.5 Remark (Application: Steel)
The effect of stress/strain-dependent transformation behaviour is negligible for steel 100Cr6 cf. [Dalgic and Löwisch 2006; Dalgic et al. 2006; Wolff et al. 2007b,c]. Phase transition models in context of steel are discussed in [Wolff and Böhm 2002a,b; Wolff et al. 2004, 2005a, 2006e,c].

An overview of different models for phase transitions in steel is given in [Böhm et al. 2003; Wolff et al. 2007c].

2.5 Transformation-induced Plasticity (TRIP)

Transformation-induced plasticity (TRIP) is an important factor in distortion of steel. The main characteristic is the occurrence of macroscopic permanent (plastic) non-isotropic deformation during phase transitions due to density differences of the individual phases although the applied (macroscopic) thermal and/or mechanical stress is below the yield strength of the weaker phase. The TRIP strain causes the changes of volume and shape of the crystal structure during the phase transformation, in which the softer phase which is involved in the transformation has to adapt its appearance to the environment of the harder phase. This results in a complicated state of inertial stresses in the environment of the transformation front, which often lead to locally plastic deformations without the effect of external stresses. Therefore TRIP cannot be explained by the classical plasticity at a macroscopic level. This effect appears in steel (also in other alloys) both

in continuous and in isothermal transformation at all levels of transformation. People worked on this phenomenon since the 1960s, but it is still the subject of intensive research. For detailed information on the phenomenon of TRIP and at its modelling we refer to e.g. [Ahrens 2003; Fischer et al. 1996; Fischer 1997; Fischer et al. 2000a; Dalgic and Löwisch 2006; Dalgic et al. 2006; Leblond and Devaux 1989a,b; Mielke 2007; Mitter 1987; Wolff and Böhm 2003; Wolff et al. 2003c, 2004, 2005a; Wolff and Böhm 2006a,b].
Two effects play an important role:

- In the ferritic and pearlitic transformation only the so-called Greenwood-Johnson effect is presumed. Density differences between existing and forming phase cause plastic deformation; deviatoric stresses lead to macroscopic permanent strain.
- Magee effect: change of atomic lattice structure during martensitic transformation.

TRIP is usually based on Greenwood/Johnson and developed by an improved modelling approach from Franitza/Mitter/Leblond (cf. e.g. [Antretter et al. 2002, 2004; Dalgic et al. 2007; Fischer et al. 2000b; Franz et al. 2004; Wolff and Böhm 2002a,b, 2003; Wolff et al. 2003c,b, 2004, 2009a]). In this ansatz we have to add a strain tensor, which is proportional to the stress deviator. For non-constant load this strain tensor can be represented as an integral, which takes the history of the acting deviatoric stress into account. A material parameter and a function which characterises the type of transformation occur as quantities to be determined.
Because this macroscopic phenomenon leads to a permanent non-isotropic deformation if phase transformations occur under non-vanishing deviatoric stress, the modelling and simulation of the bulk behaviour of steel must take these phenomena into account. The first proposal for TRIP in the multi-phase case is (Franitza-Mitter-Leblond ansatz):

$$\boldsymbol{\varepsilon}'_{trip} = \frac{3}{2}\left(\boldsymbol{\sigma}^* - \mathbf{X}^*_{trip}\right) \sum_{i=1}^{m} \kappa_i \frac{\partial \Phi_i}{\partial p_i}(p_i) \max\{p'_i, 0\}, \tag{2.55}$$

where κ_i – Greenwood-Johnson parameter (may depend on temperature and stress direction, cf. [Wolff et al. 2008c]) and Φ_i – saturation function of the ith phase satisfying

$$\kappa_i \geq 0, \qquad \Phi(0) = 0, \qquad \Phi(1) = 1, \qquad \frac{\partial \Phi_i}{\partial p_i}(p) \geq 0 \text{ f.a. } 0 < p < 1. \tag{2.56}$$

The function Φ_i describes the dependence of the transformed phase fraction p_i on the strain due to TRIP.

2.5.1 Example (Saturation Function)
There are various proposals for saturation functions in the literature (cf. [Wolff et al. 2003b]), partially based on experiments, partially derived from theoretical considerations:

$$\Phi(p) = p \qquad \textit{(Tanaka)}$$

$$\Phi(p) = p(2-p) \qquad \textit{(Desalos, Denis)}$$
$$\Phi(p) = p(1-\ln(p)) \qquad \textit{(Leblond)}$$
$$\Phi(p) = p(3-2\sqrt{p}) \qquad \textit{(Abrassart)}$$
$$\Phi(p) = \frac{p}{k-1}\left(k - p^{k-1}\right),\ p \in [0,1],\ k \geq 2 \qquad \textit{(Sjöström)}$$
$$\Phi(p) = \frac{1}{2}\left\{1 + \frac{\sin(k(2p-1))}{\sin(k)}\right\},\ p \in [0,1],\ k \in \left]0, \frac{\pi}{2}\right[\qquad \textit{(Wolff)}$$

In contrast to the extended model for TRIP with more than one forming phase according to (2.55), it is possible to use another concept, summarising all forming phases in one (cf. [Wolff et al. 2008c]). This approach leads to a model of the following structure

$$\boldsymbol{\varepsilon}'_{trip} = \frac{3}{2}\left(\boldsymbol{\sigma}^* - \mathbf{X}^*_{trip}\right)\kappa\Phi'\left(\sum_{i=1}^{m} p_i H(p'_i)\right)\sum_{i=1}^{m}\max\{p'_i, 0\} \tag{2.57}$$

with 'combined' κ and Φ (H – Heaviside function, cf. (2.49)). Obviously, the models (2.55) and (2.57) coincide in case of one forming phase and have the same basic structure (2.23). The scalar b is given by

$$b = \frac{3}{2}\sum_{i=1}^{m}\kappa_i\frac{\partial\Phi_i}{\partial p_i}(p_i)\max\{p'_i, 0\}$$

for the evolution equation (2.55) and

$$b = \frac{3}{2}\kappa\Phi'\left(\sum_{i=1}^{m} p_i H(p'_i)\right)\max\{p'_i, 0\}$$

for the evolution equation (2.57).

2.6 Constitutive Relation for the Free Energy*

In order to complete the model, it is necessary to make proposals for the free energy and the internal variables. It is important that these proposals lead to a thermodynamic consistent model and describe the real material behaviour sufficiently well (cf. [Wolff et al. 2008c] for details and discussion). We define the free energy density

$$\tilde{\Psi} = \tilde{\Psi}_{te} + \tilde{\Psi}_{in} \tag{2.58}$$

with its thermo-elastic part $\tilde{\Psi}_{te}$, which includes the density changes due to phase transformations and with its inelastic part $\tilde{\Psi}_{in}$, which represents the energy storage due to inelastic deformations, i.e. plasticity and TRIP as in [Wolff et al. 2008c]. In contrast to the strain, we do not split up the inelastic part of the free energy because of possible interaction between plasticity and TRIP.

2.6.1 Ansatz for the Thermo-elastic Part*

The extension of the standard approach of linear thermo-elasticity (cf. [Wolff et al. 2008c]) leads to

$$(2.59)\quad \tilde{\Psi}_{te}(\boldsymbol{\varepsilon}_{te}, \theta, \mathbf{p}) := \frac{1}{\rho}\Big\{\mu\boldsymbol{\varepsilon}^*_{te} : \boldsymbol{\varepsilon}^*_{te} + \frac{K}{2}\left(\mathrm{tr}(\boldsymbol{\varepsilon}_{te})\right)^2 - 3K_\alpha\left(\theta - \theta_0\right)\mathrm{tr}(\boldsymbol{\varepsilon}_{te}) - K\sum_{i=1}^{m}\left(\frac{\rho_0}{\rho_i(\theta_0)} - 1\right)p_i\,\mathrm{tr}(\boldsymbol{\varepsilon}_{te})\Big\} + \frac{1}{\rho_0}\sum_{i=1}^{m} C_i(\theta)p_i,$$

where C_i – calorimetric functions of ith phase (cf. [Wolff et al. 2008c]). The bulk entities μ, K and K_α generally depend on the temperature and the phase fractions. A linear mixture rule gives:

$$\mu(\theta, \mathbf{p}) := \sum_{i=1}^{m} \mu_i(\theta)p_i, \qquad K(\theta, \mathbf{p}) := \sum_{i=1}^{m} K_i(\theta)p_i, \qquad K_\alpha(\theta, \mathbf{p}) := \sum_{i=1}^{m} K_i(\theta)\alpha_i(\theta)p_i,$$

where $\mu_i > 0$ shear modulus, $K_i > 0$ compression modulus and α_i linear heat-dilatation coefficient of the ith phase.

2.6.2 Ansatz for the Inelastic Part*

The concrete structure of the inelastic part of the free energy essentially depends on the kind of hardening and energy storage and dissipation during hardening (cf. [Wolff et al. 2008c] for a general model). The inelastic part of the energy is defined as a non-negative quadratic form w.r.t. $\boldsymbol{\varepsilon}_{cp}$ and $\boldsymbol{\varepsilon}_{trip}$:

$$(2.60)\quad \tilde{\Psi}_{in}(\boldsymbol{\varepsilon}_{cp}, \boldsymbol{\varepsilon}_{trip}, s_{cp}, s_{trip}, \theta, \mathbf{p}) := \frac{1}{2\rho_0}\Big\{c_{cp}(\theta, \mathbf{p})\boldsymbol{\varepsilon}_{cp} : \boldsymbol{\varepsilon}_{cp} + 2c_{int}(\theta, \mathbf{p})\boldsymbol{\varepsilon}_{cp} : \boldsymbol{\varepsilon}_{trip} + c_{trip}(\theta, \mathbf{p})\boldsymbol{\varepsilon}_{trip} : \boldsymbol{\varepsilon}_{trip} - \gamma_{cp}(\theta, \mathbf{p})\left(s_{cp}\right)^2\Big\}.$$

As the energy has to be non-negative, the temperature and phase dependent coefficients have to fulfill:

$$c_{cp}, c_{trip}, \gamma_{cp} \geq 0, \quad c_{int}^2 \leq c_{cp}c_{int}, \quad c_{cp}(\theta, \mathbf{p}) = \sum_{i=1}^{m} c_{cpi}(\theta)p_i, \quad \gamma_{cp}(\theta, \mathbf{p}) = \sum_{i=1}^{m} \gamma_{cpi}(\theta)p_i,$$

where c_{cpi} and γ_{cpi} characterise the plastic behaviour of the ith phase.

2.6.1 Remark (Generalisation of the free contribution $\tilde{\Psi}_{in}$)
A slight generalisation of (2.60) in order to include the back-stresses $\mathbf{X}_{trip}$ and $\mathbf{X}_{cp}$ can be found in [Wolff et al. 2008c]. In [Wolff et al. 2008c] additional symmetric tensorial internal variables α_{cp} and α_{trip} of strain type are introduced in order to describe the energy dissipation during kinematic hardening and during evolution of TRIP back-stress. The scalar internal variable r_{cp} is of strain type and models the energy dissipation during isotropic hardening. We replace $\boldsymbol{\varepsilon}_{cp}$ by $\boldsymbol{\varepsilon}_{cp} - \alpha_{cp}$, $\boldsymbol{\varepsilon}_{trip}$ by $\boldsymbol{\varepsilon}_{trip} - \alpha_{trip}$ and $s_{cp} - r_{cp}$ and suppose

$$\alpha'_{cp} = \frac{a_{cp}}{c_{cp}}\mathbf{X}_{cp}s'_{cp}, \qquad \alpha'_{trip} = \frac{a_{trip}}{c_{trip}}\mathbf{X}_{trip}s'_{trip}, \qquad r'_{cp} = \frac{\beta_{cp}}{\gamma_{cp}}Rs'_{cp},$$

where a_{cp}, a_{trip}, β_{cp} are further coefficients (possible dependent on θ, $\mathbf{p}$ etc.) with $a_{cp}, a_{trip}, \beta_{cp} > 0$.

2.6.3 Evolution Equations for Internal Variables*

We proceed with the yield function F (cf. Equation (2.15)). Let the initial yield stress R_0 introduced as a linear mixture rule

$$R_0(\theta, \boldsymbol{\xi}, \mathbf{p}) := \sum_{i=1}^{m} R_{0,i}(\theta, \boldsymbol{\xi})p_i, \qquad R := -\rho_0 \frac{\partial \tilde{\Psi}_{in}}{\partial r_p}, \tag{2.61}$$

where $R_{0,i}$ – initial yield stress of the ith phase and R – increment of R_0 due to isotropic hardening (scalar having the character of a stress, cf. Subsection 2.1.5). Thus, the remaining dissipation inequality (Clausius-Planck-Inequality) reads as

$$(\boldsymbol{\sigma} - \mathbf{X}_{cp}) : \boldsymbol{\varepsilon}'_{cp} + (\boldsymbol{\sigma} - \mathbf{X}_{trip}) : \boldsymbol{\varepsilon}'_{trip} - Rs'_{cp} - \rho_0 \frac{\partial \tilde{\Psi}}{\partial p} p' \geq 0. \tag{2.62}$$

This inequality leads to substantial restrictions (cf. [Wolff et al. 2008c]). From (2.36), (2.60) and (2.61) we get the back-stress relations

$$\mathbf{X}_{cp} = c_p(\theta, \mathbf{p})\boldsymbol{\varepsilon}_{cp} + c_{int}(\theta, \mathbf{p})\boldsymbol{\varepsilon}_{trip}, \tag{2.63}$$

$$\mathbf{X}_{trip} = c_{int}(\theta, \mathbf{p})\boldsymbol{\varepsilon}_{cp} + c_{trip}(\theta, \mathbf{p})\boldsymbol{\varepsilon}_{trip}, \tag{2.64}$$

$$R = \gamma_{cp}s_{cp}. \tag{2.65}$$

We assume the following kind of hardening (cf. [Wolff et al. 2008c] and the discussion in [Håkansson et al. 2005]).

Isotropic Hardening*

For further investigations, it is very useful, to get more information about the isotropic hardening variable R. Equation (2.61) implies an integral equation for the isotropic hardening (in the situation of Remark 2.6.1)

$$R(t) = \gamma_{cp}\Big(s_{cp}(t) - \int_0^t \frac{\beta_{cp}}{\gamma_{cp}} R s'_{cp}(s)\,\mathrm{d}s\Big) \tag{2.66}$$

as well as the linear ODE (differentiating Equation (2.66))

$$R' = \gamma s'_p - \Big(\beta s'_p - \frac{\gamma'_{cp}}{\gamma_{cp}}\Big) R. \tag{2.67}$$

Clearly, the solution of the ODE (2.67) (for the initial value $R(0) = 0$) reads as

$$R(t) = \gamma_{cp} \int_0^t s'_{cp}(s) \exp\Big(- \int_0^s \beta_{cp} s'_{cp}(\tau)\,\mathrm{d}\tau\Big)\,\mathrm{d}s.$$

Moreover, we get the following estimate

$$0 \le R = R(t) \le \frac{\gamma_{cp}}{\min(\beta_{cp})}\Big(1 - \exp\big(- \min(\beta_{cp}) s_{cp}\big)\Big).$$

We oppress the general spatial dependence of the involved functions. In general, the parameters β_{cp} and γ_{cp} depend on the temperature and the phase fractions. For constant γ_{cp} and β_{cp} we have that R is a slope of s_{cp}

$$R = \frac{\gamma_{cp}}{\beta_{cp}}\Big(1 - \exp\big(- \beta_{cp} s_{cp}\big)\Big).$$

The curve R has its decline γ_{cp} and its saturation value is $\frac{\gamma_{cp}}{\beta_{cp}}$. Besides this, R is an increasing function of s_{cp}, as one expects in isotropic hardening (cf. [Wolff et al. 2008c]).

$$0 \le R \le \frac{\gamma_{cp}}{\beta_{cp}}, \qquad 0 \le R' \le \gamma_{cp} s'_{cp}(t). \tag{2.68}$$

2.6.2 Remark (Ramberg-Osgood Model)
The isotropic part is, due to Ramberg-Osgood (cf. [Suhr 2010])

$$R = c_{iso}(\theta) s_p^n$$

with a temperature-dependent parameter $c_{iso} \ge 0$ and a parameter $n \in]0, 1]$.

2.6.3 Remark (Alternative Ansatz for Isotropic Hardening)
In [Macherauch 1992] an experimental based ansatz is suggested. R is a monotone decreasing function as a function of temperature (and monotone increasing as a function of time, if we assume a cooling scenario) given by

$$R = R_G + R_0^*\left[1 - \left(\frac{\theta}{\theta_0}\right)^n\right]^m,$$

where θ – temperature, θ_0 – initial temperature and R_G, R_0, n, m – appropriate material parameters.

Kinematic Hardening*

Using (2.63) and (2.64) in the situation of Remark 2.6.1, we obtain equations for the back-stresses

$$\text{(2.69)} \quad \mathbf{X}_{cp}(t) = c_{cp}\left(\boldsymbol{\varepsilon}_{cp} - \int_0^t \frac{a_{cp}}{c_{cp}}\mathbf{X}_{cp}s'_{cp}\,\mathrm{d}s\right) + c_{int}\left(\boldsymbol{\varepsilon}_{trip} - \int_0^t \frac{a_{trip}}{c_{trip}}\mathbf{X}_{trip}s'_{trip}\,\mathrm{d}s\right),$$

$$\text{(2.70)} \quad \mathbf{X}_{trip}(t) = c_{int}\left(\boldsymbol{\varepsilon}_{cp} - \int_0^t \frac{a_{cp}}{c_{cp}}\mathbf{X}_{cp}s'_{cp}\,\mathrm{d}s\right) + c_{trip}\left(\boldsymbol{\varepsilon}_{trip} - \int_0^t \frac{a_{trip}}{c_{trip}}\mathbf{X}_{trip}s'_{trip}\,\mathrm{d}s\right).$$

These relations may be understood as generalisations of the well-known Armstrong-Frederick equations in plasticity (cf. [Lemaitre and Chaboche 1990; Haupt 2000; Jiang and Kurath 1996]). For a constant coefficient c_{cp} and without TRIP, from (2.69) follows the classical Armstrong-Frederick equation for non-linear hardening from 1966

$$\text{(2.71)} \qquad \mathbf{X}'_{cp} = c_{cp}\boldsymbol{\varepsilon}'_{cp} - a_{cp}\mathbf{X}_{cp}s'_{cp}.$$

In a similar manner, for constant c_{trip} and without classical plasticity, from (2.70) follows an analogon for TRIP

$$\mathbf{X}'_{trip} = c_{trip}\boldsymbol{\varepsilon}'_{trip} - a_{trip}\mathbf{X}_{trip}s'_{trip}.$$

For a given evolution of θ, $\mathbf{p}$, $\boldsymbol{\varepsilon}_{cp}$ and $\boldsymbol{\varepsilon}_{trip}$ (and therefore s_{cp} and s_{trip}) the Equations (2.69) and (2.70) are a coupled system of Volterra integral equations with a unique solution ($\mathbf{X}_{cp}$, $\mathbf{X}_{trip}$) (in the class of continuous functions under suitable conditions). Due to (2.11) we obtain the following relations

$$\mathrm{tr}(\mathbf{X}_{cp}) = 0, \qquad \mathrm{tr}(\mathbf{X}_{trip}) = 0, \qquad \mathrm{tr}(\alpha_{cp}) = 0, \qquad \mathrm{tr}(\alpha_{trip}) = 0.$$

It is well-known, that in the case of purely classical plasticity the Armstrong-Frederick equation (2.71) leads to a bounded back-stress for given $\boldsymbol{\varepsilon}_{cp}$ (saturation effect, cf. [Wolff et al. 2008c]). A similar result for the case of coupled backstresses $\mathbf{X}_{cp}$ and $\mathbf{X}_{trip}$ is not obvious. For constant c_{cp}, c_{int}, c_{trip}, the differentiation of (2.69) and (2.70) yields the following coupled system of ODEs

$$\mathbf{X}'_{cp} = c_{cp}\boldsymbol{\varepsilon}'_{cp} - a_{cp}\mathbf{X}_{cp}s'_{cp} + c_{int}\boldsymbol{\varepsilon}'_{trip} - \frac{c_{int}a_{trip}}{c_{trip}}\mathbf{X}_{trip}s'_{trip},$$

$$\mathbf{X}'_{trip} = c_{int}\boldsymbol{\varepsilon}'_{cp} - \frac{c_{int}a_{cp}}{c_{cp}}\mathbf{X}_{cp}s'_{cp} + c_{trip}\boldsymbol{\varepsilon}'_{trip} - a_{trip}\mathbf{X}_{trip}s'_{trip}.$$

This linear system of ODEs has a unique solution (for given evolution of $\boldsymbol{\varepsilon}_{cp}$ and $\boldsymbol{\varepsilon}_{trip}$). Using the results from the theory of ODEs, we obtain the following results:
Under the assumptions

$$c_{cp}, c_{int}, c_{trip}, a_{cp}, a_{trip} = \text{ const.}, \quad a_{cp}, a_{trip} > 0 \quad \text{and} \quad c_{int}^2 < c_{cp}\, c_{trip}$$

the back-stresses are bounded

$$|\mathbf{X}_{cp}(\mathbf{x}, t)| \leq c < \infty, \qquad |\mathbf{X}_{trip}(\mathbf{x}, t)| \leq c < \infty \qquad \text{f.a. } (\mathbf{x}, t) \in \Omega_T$$

and

$$\|\boldsymbol{\sigma}^* - \mathbf{X}^*_{cp}\| \leq c, \qquad \|\boldsymbol{\sigma}^*\| \leq c, \qquad \|\boldsymbol{\sigma}^* - \mathbf{X}^*_{trip}\| \leq c, \qquad \|\boldsymbol{\varepsilon}^*_{te}\| \leq c.$$

These results can be obtained in the case of general a_{cp}, a_{trip}. Non-constant c_{cp}, c_{int}, c_{trip} are also possible, but lead to difficulties. The global boundedness of $\mathbf{X}_{cp}$, $\mathbf{X}_{trip}$, R, $\boldsymbol{\sigma}^*$ and $\boldsymbol{\varepsilon}^*_{te}$ (uniformly w.r.t. $\boldsymbol{\varepsilon}_{cp}$, $\boldsymbol{\varepsilon}_{trip}$, s_{cp} and s_{trip}) is an important consequence of the general nonlinear hardening (cf. assumptions on the free energy, saturation property). In contrast to this, the case $a_{cp} = a_{trip} = \beta_{cp} = 0$ leads to a linear relation between $\mathbf{X}_{cp}$, $\mathbf{X}_{trip}$, $\boldsymbol{\varepsilon}_{cp}$, $\boldsymbol{\varepsilon}_{trip}$, R and s_{cp}. Thus, the model has an unbounded growth of the hardening variables for unbounded growing strains in this case. Moreover, the stress deviator might be growing unbounded, too.

2.6.4 Remark (Prager Model)
Moreover, the kinematic part of hardening was described by Prager in 1949 with the help of the following (simple) model for linear hardening:

$$\mathbf{X}'_{cp} = \frac{2}{3} c_{kin}(\theta)\, \boldsymbol{\varepsilon}'_{cp},$$

where $c_{kin} \geq 0$.

2.6.5 Remark (Generalisation of the Armstrong-Frederick Equations in Plasticity)
In [Suhr 2010] the following model is considered:

$$\mathbf{X}'_{cp} = \frac{2}{3}\hat{c}\, \boldsymbol{\varepsilon}'_{cp} - \hat{b}\, \mathbf{X}_{cp}\, s'_{cp} + \frac{2}{3} c_{int}\, \boldsymbol{\varepsilon}'_{trip},$$

$$\mathbf{X}'_{trip} = \frac{2}{3} c_{int}\, \boldsymbol{\varepsilon}'_{cp} - c_{int} \frac{\hat{b}}{\hat{c}} \mathbf{X}_{cp}\, s'_{cp} + \frac{2}{3} c_{trip}\, \boldsymbol{\varepsilon}'_{trip},$$
$$s'_{cp} = \sqrt{\frac{2}{3}} \gamma_{cp}, \quad R = \sum p_i\, R_i, \quad R'_i = \hat{\gamma}_{cpi} s'_{cp} - \hat{\beta}_{cpi} R_i s'_{cp},$$

where $c_{int}^2 \leq c_{trip}\hat{c}$, $\hat{\beta}_{cpi} = \hat{\beta}_{cpi}(\theta)$, $\hat{b} = \hat{b}(\theta)$ *and* $\hat{c}, \hat{\gamma}_{cpi}, c_{int}, c_{trip} = \text{const}.$

2.6.6 Remark (Generalisation of the Ramberg-Osgood and Prager Model in Plasticity)
Moreover, in [Suhr 2010] the following model is explained:

$$\mathbf{X}'_{cp} = \frac{2}{3} c_{cp}(\theta)\, \boldsymbol{\varepsilon}'_{cp} + \frac{2}{3} \frac{\mathrm{d}}{\mathrm{d}t} c_{cp}(\theta)\, \boldsymbol{\varepsilon}_{cp} + \frac{2}{3} c_{int}\, \boldsymbol{\varepsilon}'_{trip},$$
$$\mathbf{X}'_{trip} = \frac{2}{3} c_{int}\, \boldsymbol{\varepsilon}'_{cp} + \frac{2}{3} c_{up}\, \boldsymbol{\varepsilon}'_{trip},$$
$$s'_{cp} = \sqrt{\frac{2}{3}} \gamma_{cp}, \quad R = \sum p_i\, R_i, \quad R_i = c_i(\theta)\, (s_{cp})^{m_i(\theta)},$$

where $c_{int}^2 \leq c_{up} c_{cp}$, $c_{int}, c_{trip} = \text{const}.$ *and for fixed* s_{cp} *holds* $c_{cp}(\theta) = \frac{\mathrm{d}}{\mathrm{d}t} R(\theta, s_{cp})$.

2.6.7 Remark (Chaboche Model)
Chaboche suggested 1976 the ansatz to split up the back-stress as $\mathbf{X} = \sum_{i=1}^m \mathbf{X}_i$ *and describe the evolution for each partial back-stress* $\mathbf{X}_i$, $i = 1, \ldots, m$, *cf. [Chaboche 2008].*

2.6.8 Remark (Extensions to Visco-Plasticity)
A few remarks on extensions to visco-elasticity and visco-plasticity, creep and relaxation phenomena, which play an important role in the material behaviour of metallic materials, are mentioned in [Wolff et al. 2008c].
Perzyna developed in 1964 an visco-plastic approach to model rate-dependent effects of visco-plastic material behaviour (cf. [Perzyna 1971]):

$$\boldsymbol{\varepsilon}'_{cp} = \Lambda_{vp} \left(\boldsymbol{\sigma}^* - \mathbf{X}^*_{cp} \right), \tag{2.72}$$

where

$$\Lambda_{vp} = \Lambda_{vp}(F) > 0 \quad \text{for} \quad F(\boldsymbol{\sigma}, \mathbf{X}_{cp}, R_0, R) > 0,$$
$$\Lambda_{vp}(F) = 0 \quad \text{for} \quad F(\boldsymbol{\sigma}, \mathbf{X}_{cp}, R_0, R) \leq 0$$

holds for the visco-plastic multiplier

$$\Lambda_{vp} = \frac{2}{3\eta} \left\langle \frac{F(\boldsymbol{\sigma}, \mathbf{X}_{cp}, R_0, R)}{D} \right\rangle^m, \tag{2.73}$$

where $\eta > 0$, $m > 0$, $D > 0$, *cf. [Wolff et al. 2008c].*

(Classical) plasticity can be seen as a limit case of visco-plasticity for $\eta \to 0$, *cf. [Duvaut and Lions 1976; Anzellotti and Luckhaus 1987].*

2.7 Thermodynamic Consistency*

The macroscopic behaviour of material bodies is not only determined by a mechanical process. Among various effects, thermo-mechanical energy transformations play an important role. They change the temperature and influence the mechanical behaviour of the material and vice versa.
The modelling of thermo-mechanical material behaviour has to observe the fundamental law of mechanics. It also has to adhere to the natural laws of thermodynamics and the principle of irreversibility.
From the point of view of thermo-mechanics the Clausius-Duhem inequality is a condition which every solution of the basic equations has to fulfill, if it is to represent a physically possible process at all (cf. [Haupt 2000] for the following definition).

2.7.1 Definition (Thermodynamic Consistency)
The temporal course of the independent variables, combined with the material response determined by means of the constitutive equations, is called a thermo-mechanical process. A set of constitutive equations, fulfilling the dissipation inequality f.a. imaginable thermo-mechanical processes, is called thermo-mechanically consistent.

In order to focus on the mathematical investigation of the proposed model, we do not deal extensively with thermodynamic considerations and investigations on thermodynamic consistency here.
The introduced model is thermodynamically consistent (for suitable conditions: free energy must be non-negative, initial yield stress must be positive) in the sense that the intrinsic dissipation is non-negative. The detailed investigation of thermodynamic consistency can be found in [Wolff et al. 2008c].

2.8 Non-Dimensionalisation*

The first and probably most important step in the (numerical) analysis of a system of differential equations is non-dimensionalisation. It has several important issues:

(1) It identifies the ratio of dimensional parameters, which control the solution behaviour.

(2) Terms in the equations are dimensionless. Therefore it its possible to compare their sizes and to identify the important terms in the equations and their interaction in different regimes in a systematic manner, giving an insight into the structure of solutions and the important physical mechanisms. Moreover, proper scaling shows its advantage in numerical simulations, as the results are directly comparable.

In particular, negligible terms can be identified leading to simplification in many circumstances.

(3) It allows estimates of the effects of additional features to the original model through the new dimensional parameters associated with the additional terms. This allows measurement of the effect of the physical features in the model.

(4) Finally, it can reduce the number of parameters occurring in the problem by forming the non-dimensional parameters.

It is well-known that quenching of metallic workpieces may lead to distortion – this distortion depends on the parameters of the employed model, e.g. heat exchange coefficient, geometry, initial and surrounding temperature, heat capacity, elastic parameters and hardening parameters. The underlying model of thermo-elasto-plastic material behaviour is quite complex; even in case of partial linearisation. Thus, there is a practical interest in finding dimensionless numbers in order to deduce the model and get to know, which parameters are needed from the physical point of view.

Although dimensional analysis is not frequently used in solid mechanics, we focus on a special problem (see [Melnik 2001] for an ansatz and [Wolff et al. 2008b; Şimşir et al. 2009; Şimşir et al. 2010; Wolff et al. 2009b] for details). Here is not the space to re-formulate the full system, therefore we consider an exemplary problem of thermo-elasto-plasticity without phase transitions and TRIP. Let Ω be a cylinder with length $L > 0$ and diameter $D > 0$ as well as $T > 0$ for fixed time. We assume all material parameters to be constant. Furthermore, we suppress external sources. We introduce a dimensionless displacement vector $\mathbf{w}$ and a dimensionless temperature ϑ by

$$w_1 := \frac{u_1}{D}, \qquad w_2 := \frac{u_2}{D}, \qquad w_3 := \frac{u_3}{L}, \qquad \vartheta := \frac{\theta - \theta_\Gamma}{\theta_0 - \theta_\Gamma}.$$

Moreover, we introduce a linear operator acting on $\mathbb{R}^3$ via

$$\mathbf{u} = \mathbf{A}\mathbf{w}, \qquad \mathbf{A}\mathbf{w} := \left(Dw_1, Dw_2, Lw_3\right)^T.$$

Therefore we get $\vartheta(\mathbf{x}, 0) = 1$, $\mathbf{x} \in \Omega$ and $-\kappa \nabla \vartheta \nu = \delta \vartheta$ on $\partial\Omega$. The Lamé coefficients μ and λ and the compression modulus K may be expressed through the Young modulus E and the Poisson number ν, cf. Remark 2.13. Hence, the thermo-elasto-plastic system consisting of equations (2.1), (2.12) and (2.41) reads as in the new form:

$$\frac{\rho_0(1+\nu)}{E}\frac{\partial^2}{\partial t^2}(\mathbf{A}\mathbf{w}) - \operatorname{Div}\left(\boldsymbol{\varepsilon}(\mathbf{A}\mathbf{w}) + \frac{\nu}{1-2\nu}\operatorname{tr}\left(\boldsymbol{\varepsilon}(\mathbf{A}\mathbf{w})\right)\mathbf{I}\right.$$
$$\left. - \frac{1+\nu}{1-2\nu}\alpha\left(\theta_0 - \theta_\Gamma\right)\vartheta\,\mathbf{I}\right) = -\operatorname{Div}(\boldsymbol{\varepsilon}_{cp})$$

$$\rho c_e \frac{\partial \vartheta}{\partial t} - \operatorname{div}(\kappa \nabla \vartheta) + \frac{E\alpha\theta_0}{(1-2\nu)(\theta_0 - \theta_\Gamma)} \operatorname{Div}(\mathbf{A}\mathbf{w}') = \frac{\sigma_0}{\theta_0 - \theta_\Gamma} s'_{cp}$$

Using the approach of Buckingham (cf. [Hutter and Jöhnk 2004]) we find a complete and independent set of dimensionless numbers corresponding to the above material parameters – four independent dimensions (length, time, mass, temperature), the important Fourier number $F_0 := \frac{T\kappa}{\rho_0 c_e D^2}$ and the Biot number $Bi := \frac{D\delta}{\kappa}$ reduce the number of parameters.
Using a coordinate transformation $\mathbf{W}(\boldsymbol{\xi}, \tau) := \mathbf{w}(\mathbf{x}, t)$, $\Theta(\boldsymbol{\xi}, \tau) := \vartheta(\mathbf{x}, t)$ etc. (dimensionless form w.r.t. space and time) with

$$\xi_1 := \frac{x_1}{D}, \qquad \xi_2 := \frac{x_2}{D}, \qquad \xi_3 := \frac{x_3}{L}, \qquad \tau := \frac{t}{T},$$

we can transform the variational formulation to a reference domain

$$\Omega := \left\{ \mathbf{x} \in \mathbb{R}^3 : 0 < x_1^2 + x_2^2 < \frac{D^2}{4}, -\frac{L}{2} < x_3 < \frac{L}{2} \right\}.$$

2.9 The Complete Model of Material Behaviour

The complex material behaviour of steel including phase transformations, plasticity and TRIP consists of

- linear-momentum equation (2.1),
- heat-conduction equation (2.41),
- phase transformation laws (2.6),
- decomposition of strain tensor (2.4) and (2.10),
- material law of linear thermo-elasticity with phase transformation (2.12),
- equation for TRIP strain (2.23),
- equation for plastic strain and the flow rule (2.17),
- general constraint for the system (2.15)
- relation for plastic multiplier (2.18) and (2.19),
- equations for back-stresses and for increment of initial yield stress (2.36), (2.61),
- evolution equations for internal variables (cf. Subsection 2.6.3),
- boundary and initial equations (2.24) – (2.29).

The complex system of coupled differential equations, completed by the constraint for plasticity, that arises from the complete macroscopic model describing the material behaviour is summerised in the next chapter.

Chapter 3

The Fully Coupled Problem of Thermo-Elasto-Plasticity with Phase Transitions

In this chapter we present a complex (macroscopic) model of the material behaviour of steel including specific phenomena like stress-dependent phase transitions, TRIP and the possible interaction with classical plasticity developed in the preceding chapter for small deformations.
In Section 3.1, a mathematical model is formulated, which is suitable for further investigations and numerical simulations. Some (basic) assumptions and modifications of the original problem are discussed in Section 3.2. Due to the great difficulties that arise when dealing with the fully coupled model in order to show an existence and uniqueness result, we consider (and investigate) three modified settings (a Steklov regularisation, a visco-elastic regularisation and a quasi-static model) in this thesis, which are summarised in Section 3.3. References and comments are given in Section 3.4.

3.1 The Original Problem

Modelling the relevant interactions between temperature, mechanical behaviour and phase transitions leads to an initial boundary value problem (IBVP) for a system of coupled nonlinear partial and ordinary differential equations and inequalities for the time and space-dependent temperature, displacement and phase fractions.

Summarising all the model equations of Chapter 2 needed to describe the evolution of displacement, temperature and phase fractions leads to the following IBVP:

Problem (P)
Find the displacement $\mathbf{u} : \overline{\Omega}_T \to \mathbb{R}^3$*, the temperature* $\theta : \overline{\Omega}_T \to \mathbb{R}$ *and the phase fractions* $\mathbf{p} : \overline{\Omega}_T \to \mathbb{R}^m$ *s.t.*

$$\rho_0 \frac{\partial^2 \mathbf{u}}{\partial t^2} - 2\,\mathrm{Div}\left(\mu \boldsymbol{\varepsilon}(\mathbf{u})\right) - \mathrm{grad}\left(\lambda\,\mathrm{div}(\mathbf{u})\right) + 3\,\mathrm{grad}\left(K_\alpha(\theta - \theta_0)\right)$$

$$+\,\mathrm{grad}\left(K \sum_{i=1}^{m} \left(\frac{\rho_0}{\rho_i(\theta_0)} - 1\right) p_i\right) + 2\,\mathrm{Div}\left(\mu \boldsymbol{\varepsilon}_{trip}\right) + 2\,\mathrm{Div}\left(\mu \boldsymbol{\varepsilon}_{cp}\right) = \mathbf{f} \quad \text{in} \quad \Omega_T$$

$$\rho_0 c_e \frac{\partial \theta}{\partial t} - \mathrm{div}\left(\lambda_\theta \nabla \theta\right) = \left(\boldsymbol{\sigma} - \mathbf{X}_{trip}\right) : \boldsymbol{\varepsilon}'_{trip} + \left(\boldsymbol{\sigma} - \mathbf{X}_{cp}\right) : \boldsymbol{\varepsilon}'_{cp}$$

$$+\theta \frac{\partial \boldsymbol{\sigma}}{\partial \theta} : \boldsymbol{\varepsilon}'_{te} + \rho_0 \sum_{i=2}^{m} L_i p'_i + r \quad \text{in} \quad \Omega_T$$

$$\frac{\partial \mathbf{p}}{\partial t} = \boldsymbol{\gamma}(\mathbf{p}, \theta, \theta', \mathrm{tr}(\boldsymbol{\sigma}), \boldsymbol{\sigma}^* : \boldsymbol{\sigma}^*) \quad \text{in} \quad \Omega_T$$

$$\boldsymbol{\varepsilon}'_{cp} = \Lambda\left(\boldsymbol{\sigma}^* - \mathbf{X}_{cp}\right), \quad \Lambda \geq 0 \text{ for } F = 0 \text{ and } \Lambda = 0 \text{ for } F < 0 \quad \text{in} \quad \Omega_T$$

$$\boldsymbol{\varepsilon}'_{trip} = \frac{3}{2}\left(\boldsymbol{\sigma}^* - \mathbf{X}_{trip}\right) \sum_{i=1}^{m} \kappa_i \frac{\partial \Phi_i}{\partial p_i}(p_i) \max\left\{p'_i, 0\right\} \quad \text{in} \quad \Omega_T$$

including the relations

$$\boldsymbol{\varepsilon} := \frac{1}{2}\left(\nabla \mathbf{u} + \nabla \mathbf{u}^T\right), \quad \boldsymbol{\varepsilon} = \boldsymbol{\varepsilon}_{te} + \boldsymbol{\varepsilon}_{trip} + \boldsymbol{\varepsilon}_{cp},$$

$$F := \sqrt{\frac{3}{2}\left(\boldsymbol{\sigma}^* - \mathbf{X}^*_{cp}\right) : \left(\boldsymbol{\sigma}^* - \mathbf{X}^*_{cp}\right)} - (R_0 + R) \leq 0,$$

$$R(t) = \gamma_{cp}\left(s_{cp}(t) - \int_0^t \frac{\beta_{cp}}{\gamma_{cp}} R s'_{cp}(s)\,\mathrm{ds}\right), t \in S,$$

$$\mathbf{X}_{cp}(t) = c_{cp}\left(\boldsymbol{\varepsilon}_{cp} - \int_0^t \frac{a_{cp}}{c_{cp}} \mathbf{X}_{cp} s'_{cp}\,\mathrm{ds}\right) + c_{int}\left(\boldsymbol{\varepsilon}_{trip} - \int_0^t \frac{a_{trip}}{c_{trip}} \mathbf{X}_{trip} s'_{trip}\,\mathrm{ds}\right), t \in S,$$

$$\mathbf{X}_{trip}(t) = c_{int}\left(\boldsymbol{\varepsilon}_{cp} - \int_0^t \frac{a_{cp}}{c_{cp}} \mathbf{X}_{cp} s'_{cp}\,\mathrm{ds}\right) + c_{trip}\left(\boldsymbol{\varepsilon}_{trip} - \int_0^t \frac{a_{trip}}{c_{trip}} \mathbf{X}_{trip} s'_{trip}\,\mathrm{ds}\right), t \in S$$

as well as initial values $\mathbf{u}(0) = \mathbf{u}_0$, $\mathbf{u}'(0) = \mathbf{u}_1$, $\theta(0) = \theta_0$, $\boldsymbol{\varepsilon}_{trip}(0) = \mathbf{0}$, $\boldsymbol{\varepsilon}_{cp}(0) = \mathbf{0}$, $\boldsymbol{\xi}(0) = \mathbf{0}$, $\mathbf{p}(0) = \mathbf{p}_0$ *in* Ω *and boundary values* $\mathbf{u} = \mathbf{0}$ *f.a. points in* Γ_1, $\boldsymbol{\sigma} \cdot \boldsymbol{\nu}_{\Gamma_2} = \mathbf{0}$ *f.a. points in* Γ_2 *and* $-\lambda_\theta \frac{\partial \theta}{\partial \nu_\Gamma} = \delta\,(\theta - \theta_\Gamma)$ *on* Γ_T.

3.2 Modifications of the Original Problem

In order to investigate Problem (**P**) we make some general (modelling) assumptions and discuss an equivalent formulation for the strains due to transformation-induced and classical plasticity in the next subsections.

3.2.1 General Modelling Assumptions

In the following we make some general (modelling) assumptions:

3.2.1 Assumption (General Simplifications)
As in the theory of linear thermo-elasticity for small deformations commonly accepted (cf. e.g. [Haupt 2000]), we obtain (after linearisation in θ, $\theta \approx \theta_0$, cf. [Dautray and Lions 1992; Jiang and Racke 2000]) the following approximation of the thermo-mechanical dissipation

$$\theta \frac{\partial \boldsymbol{\sigma}}{\partial \theta} : \boldsymbol{\varepsilon}'_{te} = -3\, K_\alpha\, \theta_0\, \operatorname{div}(\mathbf{u}'), \tag{3.1}$$

where θ_0 denotes a (constant) reference temperature close to the temperature. Moreover, we neglect kinematic hardening, i.e. we assume that there are no back-stresses and no accumulated strains, which means

$$\mathbf{X}_{trip} = \mathbf{0} \quad \text{and} \quad \mathbf{X}_{cp} = \mathbf{0}, \qquad s_{trip} = 0 \quad \text{and} \quad s_{cp} = 0. \tag{3.2}$$

Furthermore, we consider constant material parameters, i.e.

$$\rho_0, \rho_i, \mu_i, K_i, \alpha_i, R_{0i}, c_e, \lambda_\theta, \kappa_i, L_i, \theta_0, \theta_\Gamma \in \mathbb{R}^+,\ \delta \in \mathbb{R}_0^+, \qquad i = 1, \ldots, m. \tag{3.3}$$

The Assumption 3.2.1 can be modified in the following manner (cf. the following remarks) so that the solution theory which is used in the sequel is applicable and yields the unique existence of a weak solution to the corresponding initial-boundary value problem.

3.2.2 Remark (Material Parameters)
In practice, some of the material parameters depend on the temperature and on the phase fractions. In order to keep constant coefficients for first investigations of the material behaviour, sometimes one uses their averages over the temperatures between

θ_0 and θ_Γ (cf. [Şimşir et al. 2009]). Instead of (3.3) *f.a. material parameters it would be possible (from a mathematical point of view) to consider:*

$$\theta_0 \in \mathbb{R}^+,\ \rho_0 \in \mathbb{R}^+,\ \rho_i(\theta_0) \in \mathbb{R}^+,\ c_e,\ \mu_i,\ K_i,\ \alpha_i,\ \lambda_\theta,\ \kappa_i,\ R_{0i},\ \delta,\ L_i \in C(\mathbb{R}) \cap L^\infty(\mathbb{R})$$

f.a. $i = 1, \ldots, m$ with

$$\begin{aligned}
&\exists \mu_{i0}, \mu_{i1} > 0 && \forall s \in \mathbb{R}: && 0 < \mu_{i0} \le \mu_i(s) \le \mu_{i1} < \infty,\\
&\exists K_{i0}, K_{i1} > 0 && \forall s \in \mathbb{R}: && 0 < K_{i0} \le K_i(s) \le K_{i1} < \infty,\\
&\exists \alpha_{i0}, \alpha_{i1} > 0 && \forall s \in \mathbb{R}: && -\infty < \alpha_{i0} \le \alpha_i(s) \le \alpha_{i1} < \infty,\\
&\exists R^0_{0i}, R^1_{0i} > 0 && \forall s \in \mathbb{R}: && 0 < R^0_{0i} \le R_{0i}(s) \le R^1_{0i} < \infty,\\
&\exists \kappa_{i1} > 0 && \forall s \in \mathbb{R}: && 0 \le \kappa_i(s) \le \kappa_{i1} < \infty,\\
&\exists \lambda_{\theta 0}, \lambda_{\theta 1} > 0 && \forall s \in \mathbb{R}: && 0 < \lambda_{\theta 0} \le \lambda_\theta(s) \le \lambda_{\theta 1} < \infty,\\
&\exists \delta_1 > 0 && \forall s \in \mathbb{R}: && 0 \le \delta(s) \le \delta_1 < \infty,\\
&\exists c_{e0}, c_{e1} > 0 && \forall s \in \mathbb{R}: && 0 < c_{e0} \le c_e(s) \le c_{e1} < \infty,\\
&\exists L_{i0}, L_{i1} > 0 && \forall s \in \mathbb{R}: && -\infty < L_{i0} \le L_i(s) \le L_{i1} < \infty.
\end{aligned}$$

We remark that the investigation of the parameter-dependence is not the main objective of this thesis. We try to consider all essential information of the material behaviour in the model but also keep it as simple as possible. Therefore we use constant material parameters in the following.

3.2.3 Remark (Thermo-mechanical Dissipation)
The reference temperature in (3.1) *could be the initial temperature (we use this in our setting although this is questionable in the context of heat treatment processes) but it is also possible (and reasonable) to use an average of the temperature between θ_0 and θ_Γ. Using the thermo-mechanical dissipation*

$$\theta \frac{\partial \boldsymbol{\sigma}}{\partial \theta} : \boldsymbol{\varepsilon}'_{te} = -3\, K_\alpha\, \theta\, \mathrm{div}(\mathbf{u}') \tag{3.4}$$

instead of the approximation in (3.1) *would be possible within the fixed-point argumentation, cf. [Chelminski et al. 2007] for a similar problem resp. argumentation. We remark that* (3.4) *is the exact representation for the thermo-mechanical dissipation (in the special case considered here) if no back-stresses are considered.*
Alternative approaches resp. further simplifications are e.g.

$$\theta \frac{\partial \boldsymbol{\sigma}}{\partial \theta} : \boldsymbol{\varepsilon}'_{te} = -3\, K_\alpha\, f(\theta)\, \mathrm{div}(\mathbf{u}')$$

instead of (3.1), *cf. [Chelminski and Racke 2006] or*

$$\theta \frac{\partial \boldsymbol{\sigma}}{\partial \theta} : \boldsymbol{\varepsilon}'_{te} = -3\, K_\alpha\, f(\mathbf{p})\ \mathrm{div}(\mathbf{u}')$$

instead of (3.1), *cf. [Hömberg and Khludnev 2006a,b] for an appropriate switch function*

$$f(\varsigma) := \begin{cases} 1 & \text{for } \varsigma < K \\ [0,1] & \text{for } \varsigma = K \\ 0 & \text{for } \varsigma > K \end{cases}$$

(K – upper bound of θ resp. $\mathbf{p}$ *due to physical argumentation) in order to have control for the function θ resp.* $\mathbf{p}$.

3.2.4 Remark (Back-stresses and Accumulated Strains)
The equations for the back-stresses are ODEs in t (time) with a parameter $\mathbf{x}$ *(space variable) like the evolution equations for the TRIP strain. Therefore it would be possible to use the same methods as in Section 5.3.4 in order to incorporate kinematic hardening.*

3.2.2 Equivalent Formulation of the TRIP Model

Consider the following problem:

Problem ($\mathbf{P}_{\boldsymbol{\varepsilon}_{trip}}$)
Find the strain $\boldsymbol{\varepsilon}_{trip} : \overline{\Omega}_T \to \mathbb{R}^{3\times 3}_{sym}$ *s.t.*

(3.5) $$\boldsymbol{\varepsilon}'_{trip}(\mathbf{x},t) = \frac{3}{2}\boldsymbol{\sigma}^*(\mathbf{x},t)\sum_{i=1}^{m} \kappa_i \frac{\partial \Phi_i}{\partial p_i}(p_i(\mathbf{x},t))\max\{p_i'(\mathbf{x},t),0\}, \qquad (\mathbf{x},t)\in\Omega_T$$

(3.6) $$\boldsymbol{\varepsilon}_{trip}(\mathbf{x},0) = \mathbf{0}, \qquad \mathbf{x}\in\Omega$$

Obviously, we get from (2.12),(2.10)

(3.7) $$\boldsymbol{\sigma}^* = 2\mu\boldsymbol{\varepsilon}^*_{te} = 2\mu(\boldsymbol{\varepsilon}^*(\mathbf{u}) - \boldsymbol{\varepsilon}_{trip} - \boldsymbol{\varepsilon}_{cp}), \qquad \mathrm{tr}(\boldsymbol{\sigma}) = 2\mu\,\mathrm{tr}(\boldsymbol{\varepsilon}(\mathbf{u})) = 2\mu\,\mathrm{div}(\mathbf{u}).$$

Now, the problem (3.5), (3.6) can be formulated as an equivalent initial value problem:

(3.8) $$\boldsymbol{\varepsilon}'_{trip}(t) = b(t)(\boldsymbol{\varepsilon}^*(\mathbf{u}(t)) - \boldsymbol{\varepsilon}_{cp}(t)) - b(t)\boldsymbol{\varepsilon}_{trip}(t), \qquad t\in S$$

(3.9) $$\boldsymbol{\varepsilon}_{trip}(0) = \mathbf{0}$$

where

$$b(t) = 3\mu\sum_{i=1}^{m}\kappa_i\frac{\partial\Phi_i}{\partial p_i}(p_i(t))\max\left\{\frac{\partial p_i}{\partial t}(t),0\right\}, \qquad t\in S.$$

3.2.3 Equivalent Formulation of the Flow Rule

For further mathematical investigation it is convenient to reformulate the described model, using a variational inequality (or rather a differential inclusion) in order to characterise the plastic deformation. We look at the following (original) problem:

Problem $(\mathbf{P}_{\boldsymbol{\varepsilon}_{cp}})$
Find the strain $\boldsymbol{\varepsilon}_{cp} : \overline{\Omega}_T \to \mathbb{R}^{3\times 3}_{sym}$ *with* $\mathrm{tr}(\boldsymbol{\varepsilon}_{cp}) = 0$, *s.t.*

$$\boldsymbol{\varepsilon}'_{cp}(\mathbf{x}, t) = \Lambda\,\boldsymbol{\sigma}^*(\mathbf{x}, t), \qquad (\mathbf{x}, t) \in \Omega_T \tag{3.10}$$
$$\boldsymbol{\varepsilon}_{cp}(\mathbf{x}, 0) = \mathbf{0}, \qquad \mathbf{x} \in \Omega \tag{3.11}$$

where the plastic multiplier Λ *has to fulfill*

$$\Lambda = 0, \qquad \textit{if} \qquad F(\boldsymbol{\sigma}, R_0, R) < 0 \tag{3.12}$$
$$\Lambda \geq 0, \qquad \textit{if} \qquad F(\boldsymbol{\sigma}, R_0, R) = 0 \tag{3.13}$$

f.a. $\boldsymbol{\sigma} \in \mathbb{R}^{3\times 3}_{sym}$ *with* $F(\boldsymbol{\sigma}, R_0, R) \leq 0$.

Taking (3.13), (3.10) and the yield function (2.15) into account, we reformulate (3.10) as

$$\boldsymbol{\varepsilon}'_{cp} = \frac{2}{3}\Lambda(R_0 + R)\frac{\partial F}{\partial \boldsymbol{\sigma}}(\boldsymbol{\sigma}, R_0, R) = \tilde{\Lambda}\frac{\partial F}{\partial \boldsymbol{\sigma}}(\boldsymbol{\sigma}, R_0, R) \tag{3.14}$$

with a new multiplier $\tilde{\Lambda}$. This reformulation is often called 'normality rule', cf. [Lubliner 2006; Maugin 1992].

We define $F : \mathbb{R}^{3\times 3} \to \mathbb{R}$ via

$$F(\boldsymbol{\sigma}) := \sqrt{\frac{3}{2}\boldsymbol{\sigma}^* : \boldsymbol{\sigma}^*} - (R_0 + R) \tag{3.15}$$

for given constants $R_0, R \in \mathbb{R}^+$. The set of all admissible $\boldsymbol{\sigma}$ is convex (cf. [Han and Reddy 1999b]). We define

$$\begin{aligned} \mathbf{K}_F &:= \Big\{\boldsymbol{\tau} \in \mathbb{R}^{3\times 3}_{\mathrm{sym}}, \mathrm{tr}(\boldsymbol{\tau}) = 0 : F(\boldsymbol{\tau}) \leq 0\Big\}, \\ \mathbf{K} &:= \Big\{\boldsymbol{\tau} \in \mathbf{H}_\sigma : \boldsymbol{\tau}(\mathbf{x}) \in \mathbf{K}_F \text{ f.a.a. } \mathbf{x} \in \Omega\Big\}. \end{aligned} \tag{3.16}$$

The constraint in ($\mathbf{P}_{\boldsymbol{\varepsilon}_{cp}}$) is non-linear. An adequate tool for dealing with non-linear constraints for equations is a variational inequality (cf. e.g. [Han and Reddy 1999b]). Based on the identity

$$\mathbf{A}^* : \mathbf{B}^* = \mathbf{A}^* : \mathbf{B} = \mathbf{A} : \mathbf{B} - \frac{1}{3}\operatorname{tr}(\mathbf{A})\operatorname{tr}(\mathbf{B}) \qquad \text{for} \qquad \mathbf{A}, \mathbf{B} \in \mathbb{R}^9$$

the relations (2.15) and (3.10) – (3.13) are equivalent to the variational inequality

$$(3.17) \qquad \boldsymbol{\varepsilon}'_{cp} : \left(\boldsymbol{\tau} - \boldsymbol{\sigma}\right) \leq 0 \qquad \text{in} \qquad \Omega_T$$

f.a. $\boldsymbol{\tau} \in \mathbb{R}^9$ with $\boldsymbol{\tau} = \boldsymbol{\tau}^T$ and $F(\boldsymbol{\tau}) \leq 0$, where $\boldsymbol{\sigma}$ has to fulfill (2.15).

3.2.5 Remark (Equivalence between normality rule and variational inequality)
The normality rule (3.14) *implies the variational inequality* (3.17). *The Equation* (3.17) *is obviously fulfilled for* $\Lambda = 0$. *For* $\Lambda > 0$ *it follows* $F(\boldsymbol{\sigma}) = 0$ *and therefore* $\frac{\partial F}{\partial \sigma}(\boldsymbol{\sigma}) \neq 0$. *Thus,* $\boldsymbol{\varepsilon}'_{cp}$ *and* $\frac{\partial F}{\partial \sigma}$ *are parallel and have the same orientation. Because of the convexity of* $\mathbf{K}_F$ *it follows* $\boldsymbol{\varepsilon}'_{cp} : \left(\boldsymbol{\tau}^* - \boldsymbol{\sigma}^*\right) \leq 0$ *in* Ω_T *f.a.* $\boldsymbol{\tau} \in \mathbb{R}^9$ *with* $\boldsymbol{\tau} = \boldsymbol{\tau}^T$ *and* $F(\boldsymbol{\tau}) \leq 0$, *which is equivalent to* (3.17).
On the other hand, the variational inequality (3.17) *implies the normality rule* (3.14). *If* $F(\boldsymbol{\sigma}) < 0$, *then it follows* $\boldsymbol{\sigma}^* \in \operatorname{int}(\mathbf{K}_F)$ *and* (3.17) *implies* $\boldsymbol{\varepsilon}'_{cp} = 0 = \tilde{\Lambda}\frac{\partial F}{\partial \sigma}$ *with* $\tilde{\Lambda} = 0$, *compatible with* (3.13). *Let* $F(\boldsymbol{\sigma}) = 0$. *Hence,* $\boldsymbol{\sigma}^* \in \partial\mathbf{K}_F$. *Putting* $\boldsymbol{\tau} = \boldsymbol{\sigma} + \mathbf{v}$ *with* $\mathbf{v} : \frac{\partial F}{\partial \sigma} \leq 0$ *s.t.* $F(\boldsymbol{\sigma}) \leq 0$ *in* (3.17) *we have* $\boldsymbol{\varepsilon}'_{cp} : \mathbf{v} \leq 0$ *f.a.* $\mathbf{v}$ *with* $\mathbf{v} : \frac{\partial F}{\partial \sigma} \leq 0$. *Therefore,* $\boldsymbol{\varepsilon}'_{cp}$ *and* $\frac{\partial F}{\partial \sigma}$ *are parallel and have the same orientation. Thus,* (3.14) *is valid.*

As a result of the re-formulation via a variational inequality, the plastic multiplier is excluded. If $\boldsymbol{\sigma}$ fulfills (2.15) with a strong inequality, than (3.17) leads to

$$\boldsymbol{\varepsilon}'_{cp} = \mathbf{0},$$

i.e. there is no plastic deformation.

If no plastic deformation occurs, the model 'thermo-elasticity with phase transitions and TRIP without classical plasticity' (cf. e.g. [Boettcher 2007; Kern 2011]) is directly applicable.

In order to prepare further mathematical investigations, we want to reformulate the variational inequality as a differential inclusion. This approach is also used in [Amassad et al. 2001] for instance. Mathematical problems of perfect plasticity are discussed in [Duvaut and Lions 1976; Ebobisse and Reddy 2004].

There are at least two possibilities when dealing with the variational inequality (3.17). Either one can exclude $\boldsymbol{\varepsilon}'_{cp}$ via $\boldsymbol{\sigma}^*$ or vice versa.

(1) Eliminating $\boldsymbol{\sigma}$ in the variational inequality: Using (2.10) and (3.7) we rewrite (3.17). This leads to to a new variational inequality for $\boldsymbol{\varepsilon}^*_{te}$. For convenience we denote $\boldsymbol{\eta} := \boldsymbol{\varepsilon}^*_{te}$. Thus, we look for a function $\boldsymbol{\eta} : \Omega_T \to \mathbb{R}^9$, s.t.

$$\operatorname{tr}(\boldsymbol{\eta}) = 0, \quad F(2\mu\boldsymbol{\eta}) \leq 0,$$
$$\boldsymbol{\eta}' : (\boldsymbol{\tau} - \boldsymbol{\eta}) \geq (\varepsilon^*(\mathbf{u}') - \boldsymbol{\varepsilon}'_{trip}) : (\boldsymbol{\tau} - \boldsymbol{\eta})$$

f.a. $\boldsymbol{\tau} \in \mathbb{R}^{3\times3}_{\text{sym}}$ with $\boldsymbol{\tau} = \boldsymbol{\tau}^T$, $\operatorname{tr}(\boldsymbol{\tau}) = 0$ and $F(2\mu\boldsymbol{\tau}) \leq 0$. Based on (2.10) and (3.7), $\boldsymbol{\varepsilon}_{te}$ is given by $\boldsymbol{\varepsilon}_{te} = \boldsymbol{\eta} + \frac{1}{3}\operatorname{tr}(\boldsymbol{\varepsilon}(\mathbf{u}))\,\mathbf{I}$.
Moreover, based on (2.10), we have

$$\boldsymbol{\varepsilon}_{cp} = \boldsymbol{\varepsilon}(\mathbf{u})^* - \boldsymbol{\varepsilon}_{trip} - \boldsymbol{\eta}, \qquad \boldsymbol{\varepsilon}'_{cp} = \boldsymbol{\varepsilon}(\mathbf{u}')^* - \boldsymbol{\varepsilon}'_{trip} - \boldsymbol{\eta}'.$$

(2) Eliminating $\boldsymbol{\varepsilon}_{cp}$ in the variational inequality: This approach is similar to the idea in [Duvaut and Lions 1976]. In our situation we get from (2.10) and (3.7)

$$\boldsymbol{\varepsilon}'_{cp} = -\Big(\frac{1}{2\mu}\boldsymbol{\sigma}^*\Big)' + \boldsymbol{\varepsilon}^*(\mathbf{u}') - \boldsymbol{\varepsilon}'_{trip}.$$

Therefore, the inequality (3.17) reads as

$$\Big(\frac{1}{2\mu}\boldsymbol{\sigma}^*\Big)' : (\boldsymbol{\tau} - \boldsymbol{\sigma}) - \boldsymbol{\varepsilon}^*(\mathbf{u}') : (\boldsymbol{\tau} - \boldsymbol{\sigma}) + \boldsymbol{\varepsilon}'_{trip} : (\boldsymbol{\tau} - \boldsymbol{\sigma}) \geq 0 \tag{3.18}$$

f.a. $\boldsymbol{\tau} \in \mathbb{R}^9$ with $\boldsymbol{\tau} = \boldsymbol{\tau}^T$ and $F(\boldsymbol{\tau}) \leq 0$, where $\boldsymbol{\sigma}$ has to be symmetric and to fulfill the constraint $F(\boldsymbol{\sigma}) \leq 0$. Note, that only in the special case without influence of $\boldsymbol{\varepsilon}_{cp}$ on $\boldsymbol{\varepsilon}_{trip}$, one can exclude $\boldsymbol{\varepsilon}'_{trip}$ from (3.18) without any return of $\boldsymbol{\varepsilon}_{cp}$ after substituting $\boldsymbol{\varepsilon}'_{trip}$ in (3.18). But, in the case of hardening, one re-imports $\boldsymbol{\varepsilon}_{cp}$ via the back-stress relations. That is why this approach seems to be inconvenient.

3.2.6 Remark (Weak Formulation)
We define a weak formulation of the variational inequality. That means, we look for $\boldsymbol{\eta} \in \mathfrak{H}_\sigma$ *with* $\boldsymbol{\eta}(t) \in \mathbf{K}$ f.a.a. $t \in S$ *s.t.*

$$\int_\Omega \boldsymbol{\eta}'(t) : (\boldsymbol{\sigma} - \boldsymbol{\eta}(t))\,\mathrm{d}\mathbf{x} \geq \int_\Omega \Big(\boldsymbol{\varepsilon}^*(\mathbf{u}'(t)) - \boldsymbol{\varepsilon}'_{trip}\Big) : (\boldsymbol{\sigma} - \boldsymbol{\eta}(t))\,\mathrm{d}\mathbf{x} \tag{3.19}$$

f.a. $\boldsymbol{\sigma} \in \mathbf{H}_\sigma$ *with* $\operatorname{tr}(\boldsymbol{\sigma}) = 0$, $\boldsymbol{\sigma} = \boldsymbol{\sigma}^T$ *and* $F(2\mu\boldsymbol{\sigma}) \leq 0$.
The idea to tackle this problem is to use solution methods for solving parabolic variational inequalities, cf. [Naumann 1984].

Let $\chi_K : \mathbf{H}_\sigma \to \mathbb{R} \cup \{+\infty\}$ be the indicator function on $\mathbf{K}$, i.e. $\chi_K(\mathbf{u}) = 0$ if $\mathbf{u} \in \mathbf{K}$ and $\chi_K(\mathbf{u}) = +\infty$ if $\mathbf{u} \notin \mathbf{K}$. The variational inequality (3.17) can be rewritten as a differential inclusion $\boldsymbol{\varepsilon}'_{cp} \in \partial\chi_K(\boldsymbol{\sigma})$, cf. [Růžička 2004; Zeidler 1985] for details. Using (3.18) and this differential inclusion (2μ is a positive multiplier), the Problem ($\mathbf{P}_{\boldsymbol{\varepsilon}_{cp}}$) is equivalent to

Problem ($\mathbf{P}'_{\varepsilon_{cp}}$)
Find the stress deviator $\boldsymbol{\sigma}^* : \overline{\Omega}_T \to \mathbb{R}^{3\times 3}_{sym}$*, s.t.*

$$(\boldsymbol{\sigma}^*(t))' + \partial\chi_{\mathbf{K}}(\boldsymbol{\sigma}^*(t)) \ni 2\mu\Big(\boldsymbol{\varepsilon}^*(\mathbf{u}'(t)) - \boldsymbol{\varepsilon}'_{trip}(t)\Big) \text{ f.a.a. } t \in S \tag{3.20}$$

$$\boldsymbol{\sigma}^*(0) = \boldsymbol{\sigma}^*_0 := 2\mu\boldsymbol{\varepsilon}^*(\mathbf{u}_0). \tag{3.21}$$

3.2.7 Remark (Time-dependent **K**)
Due to the general time-dependence of R_0 *and* R*, the set of admissible stresses varies in time, when considering a time-dependent process (cf. e.g. [Han and Reddy 1999b]). In the application problem we are concerned with cooling or quenching processes. Therefore, we assume a growing yield radius with the time. We define* $F : \mathbb{R}^{3\times 3} \times \mathbb{R} \times \mathbb{R}^3 \to \mathbb{R}$ *via*

$$\begin{aligned} F(\boldsymbol{\sigma}, t, \mathbf{x}) &:= \sqrt{\frac{3}{2}\boldsymbol{\sigma}^* : \boldsymbol{\sigma}^*} - (R_0 + R(t, \mathbf{x})), \\ \mathbf{K}_F(t, \mathbf{x}) &:= \Big\{\boldsymbol{\tau} \in \mathbb{R}^{3\times 3}_{sym}, \operatorname{tr}(\boldsymbol{\tau}) = 0 : F(\boldsymbol{\tau}, t, \mathbf{x}) \le 0\Big\}, \\ \mathbf{K}(t) &:= \Big\{\boldsymbol{\tau} \in \mathbf{H}_\sigma : \boldsymbol{\tau}(\mathbf{x}) \in \mathbf{K}_F(t, \mathbf{x}) \ \textit{f.a.a. } \mathbf{x} \in \Omega\Big\}, \end{aligned}$$

where R *is defined, e.g., as in Equation* (2.66) *or* (2.67).

3.2.8 Remark (Parameter-dependent **K**)
In [Babadjian et al. 2011; Chelminski and Racke 2006] the yield function depends explicitly on the temperature. We assume in our application problem cooling or quenching processes and therefore assume a decreasing temperature, i.e. a growing yield radius. We define $F : \mathbb{R}^{3\times 3} \times \mathbb{R} \to \mathbb{R}$ *via*

$$\begin{aligned} F(\boldsymbol{\tau}, \theta) &:= \sqrt{\frac{3}{2}\boldsymbol{\sigma}^* : \boldsymbol{\sigma}^*} - (R_0 + R(\theta)), \\ \mathbf{K}_F(\theta) &:= \Big\{\boldsymbol{\tau} \in \mathbb{R}^{3\times 3}_{sym}, \operatorname{tr}(\boldsymbol{\tau}) = 0 : F(\boldsymbol{\tau}, \theta) \le 0\Big\}, \\ \mathbf{K}(\theta) &:= \Big\{\boldsymbol{\tau} \in \mathbf{H}_\sigma : \boldsymbol{\tau}(\mathbf{x}) \in \mathbf{K}_F(\theta) \ \textit{f.a.a. } \mathbf{x} \in \Omega\Big\}, \end{aligned}$$

where R *is defined, e.g., as in Remark 2.6.2 or in Remark 2.6.3 and* θ *is a given parameter, e.g. the temperature.*

3.3 The Investigated Problems

The main idea is to prove existence and uniqueness of a solution of the corresponding IBVP for the coupled problem of linear thermo-elasto-plasticity including phase transitions and TRIP in steel under appropriate conditions. But even in the case of the simplified prerequisites in Assumption 3.2.1, great difficulties arise when dealing with this complex system of coupled differential equations, completed by the constraint for plasticity.
We have not been able to deal with the fully coupled problem, summarised as Problem $(\mathbf{P})$. It has not been possible to show an existence and uniqueness result for the whole problem with the help of the used means and methods presented in Chapter 4. Therefore we look at three slightly different modifications of the original problem, which we discuss in the next three subsections.
Three different (alternative) settings are considered: In the first one, a Steklov regularisation of the fully coupled problem is employed, cf. Subsection 3.3.1. In the second one, a visco-elastic regularisation of the fully coupled problem is studied (cf. Subsection 3.3.2) and in the third setting, a quasi-static model for the displacement is considered (cf. Subsection 3.3.3).
The strategy in the two regularised settings would be

(1) to establish existence and uniqueness of a solution $(\mathbf{u}_h, \theta_h, \mathbf{p}_h)$ to the regularised problem for any h,

(2) to show that the solution $(\mathbf{u}_h, \theta_h, \mathbf{p}_h)$ is bounded, independent of h, in appropriate spaces,

(3) to use the above boundedness and the compactness properties of the spaces concerned to deduce the existence of subsequences that converge to limit $(\mathbf{u}, \theta, \mathbf{p})$ and

(4) to verify that the limit of the subsequences in fact solves the original problem.

Unfortunately, only the first step is obtained in the two given regularised settings.
Nevertheless, it remains a great challenge to prove the (unique) existence of a weak solution of each problem, i.e. of the two regularised settings (which is the first step) and of the quasi-static setting, cf. Chapters 5 – 7.

3.3.1 Formulation of Problem $(\mathbf{P}_A)$

'Regularisation via averaging' (or more precise 'Steklov regularisation' in this particular case) is the first setting that we consider in our mathematical investigation. There are various methods of averaging (e.g. the method of volume averaging, cf. [Whitaker 1999]) available in order to regularise the fully coupled problem. The idea is to improve the regularity of the involved functions (resp. to use smooth functions) in order to solve the problem. In this setting we use an integral average, in particular the 'Steklov average'. 'Steklov regularisation' means that we replace the time derivative of $\mathbf{u}$ in the heat equation and in the differential inclusion (resp. variational inequality arising from the flow rule) by the difference quotient (cf. Appendix B.1)

$$\mathcal{D}_h\mathbf{u}(t) := \begin{cases} \frac{1}{h}\big(\mathbf{u}(t+h) - \mathbf{u}(t)\big) & \text{for} \quad t \in [0, T-h] \\ 0 & \text{for} \quad t \in]T-h, T] \end{cases}, \qquad h > 0$$

and we replace θ in the relation for the stress tensor by its Steklov average (cf. Appendix B.2)

$$\mathcal{S}_h\theta(t) := \begin{cases} \frac{1}{h}\int_t^{t+h}\theta(s)\,\mathrm{d}s & \text{for} \quad t \in [0, T-h] \\ 0 & \text{for} \quad t \in]T-h, T] \end{cases}, \qquad h > 0.$$

The reason for this approach is basically the solvability of the problem. The main two justifications for this approach are that in the numerical realisation of such problems also difference quotients are used instead of the derivatives. Also for the application or modelling point of view, it makes sense to work with averages, because changes of quantities cannot be measured exactly at a particular point at a certain time.
From the point of view of numerics this regularisation is no loss – for the point of view of the analysis it is indeed a small loss, but there still is the chance to show the passage to the limit of the comprehensive model to the original one (which is not fulfilled in this thesis).

Problem $(\mathbf{P}_A)$
Find the displacement field $\mathbf{u} : \overline{\Omega}_T \to \mathbb{R}^3$, *the temperature* $\theta : \overline{\Omega}_T \to \mathbb{R}$ *and the phase fractions* $\mathbf{p} : \overline{\Omega}_T \to \mathbb{R}^m$ *s.t.*

$$\begin{aligned}
&\rho_0\frac{\partial^2\mathbf{u}}{\partial t^2} - 2\,\mathrm{Div}\,(\mu\boldsymbol{\varepsilon}(\mathbf{u})) - \mathrm{grad}\,(\lambda\,\mathrm{div}(\mathbf{u})) + 3\,\mathrm{grad}\,(K_\alpha(\mathcal{S}_h(\theta) - \theta_0)) \\
&+\mathrm{grad}\left(K\sum_{i=1}^{m}\left(\frac{\rho_0}{\rho_i(\theta_0)} - 1\right)p_i\right) + 2\,\mathrm{Div}\,(\mu\boldsymbol{\varepsilon}_{trip}) + 2\,\mathrm{Div}\,(\mu\boldsymbol{\varepsilon}_{cp}) = \mathbf{f} \quad \textit{in} \quad \Omega_T \\
&\rho_0 c_e\frac{\partial\theta}{\partial t} - \mathrm{div}\,(\lambda_\theta\,\nabla\theta) + 3K_\alpha\theta_0\,\mathrm{div}\,(\mathcal{D}_h(\mathbf{u})) = \boldsymbol{\sigma} : \boldsymbol{\varepsilon}'_{trip} + \boldsymbol{\sigma} : \boldsymbol{\varepsilon}'_{cp} \\
&\qquad\qquad + \rho_0\sum_{i=2}^{m}L_i p_i' + r \quad \textit{in} \quad \Omega_T \\
&\frac{\partial\mathbf{p}}{\partial t} = \boldsymbol{\gamma}(\mathbf{p}, \theta, \theta', \mathrm{tr}(\boldsymbol{\sigma}), \boldsymbol{\sigma}^* : \boldsymbol{\sigma}^*) \quad \textit{in} \quad \Omega_T
\end{aligned}$$

$$\boldsymbol{\varepsilon}_{cp} = \boldsymbol{\varepsilon}^*(\mathcal{S}_h(\mathbf{u})) - \boldsymbol{\varepsilon}_{trip} - \frac{1}{2\mu}\boldsymbol{\sigma}^* \quad \textit{with}$$

$$(\boldsymbol{\sigma}^*)' + \partial\chi_K(\boldsymbol{\sigma}^*) \ni 2\mu\left(\boldsymbol{\varepsilon}^*(\mathcal{D}_h(\mathbf{u})) - \boldsymbol{\varepsilon}'_{trip}\right) \quad \textit{in} \quad \Omega_T$$

$$\boldsymbol{\varepsilon}'_{trip} = 3\mu\,(\boldsymbol{\varepsilon}^*(\mathbf{u}) - \boldsymbol{\varepsilon}_{cp} - \boldsymbol{\varepsilon}_{trip}) \sum_{i=1}^{m} \kappa_i \frac{\partial\Phi_i}{\partial p_i}(p_i)\max\{p_i', 0\} \quad \textit{in} \quad \Omega_T$$

including conditions (2.4), (2.10), (2.11), (3.15), (3.16) *as well as initial and boundary values* (2.24) – (2.29), (3.21).

3.3.2 Formulation of Problem ($\mathbf{P}_{VE}$)

In the 'visco-elastic regularisation' we consider a slightly different medium with a small viscosity (or rather damping force in the context of friction). We add the regularising term $h\boldsymbol{\varepsilon}\left(\frac{\partial u}{\partial t}\right)$ in the stress tensor $\boldsymbol{\sigma}$. Again, the passage to the limit in the regularisation parameter is not subject of this thesis.

Problem ($\mathbf{P}_{VE}$)
Find the displacement $\mathbf{u} : \overline{\Omega}_T \to \mathbb{R}^3$, *the temperature* $\theta : \overline{\Omega}_T \to \mathbb{R}$ *and the phase fractions* $\mathbf{p} : \overline{\Omega}_T \to \mathbb{R}^m$ *s.t.*

$$\rho_0 \frac{\partial^2\mathbf{u}}{\partial t^2} - 2\,\mathrm{Div}\,(\mu\boldsymbol{\varepsilon}(\mathbf{u})) - \mathrm{grad}\,(\lambda\,\mathrm{div}(\mathbf{u})) - \mathrm{Div}\left(h\boldsymbol{\varepsilon}\left(\frac{\partial\mathbf{u}}{\partial t}\right)\right)$$

$$+3\,\mathrm{grad}\,(K_\alpha\,(\theta-\theta_0)) + \mathrm{grad}\left(K\sum_{i=1}^{m}\left(\frac{\rho_0}{\rho_i(\theta_0)} - 1\right)p_i\right)$$

$$+2\,\mathrm{Div}\,(\mu\boldsymbol{\varepsilon}_{trip}) + 2\,\mathrm{Div}\,(\mu\boldsymbol{\varepsilon}_{cp}) = \mathbf{f} \quad \textit{in} \quad \Omega_T$$

$$\rho_0 c_e \frac{\partial\theta}{\partial t} - \mathrm{div}(\lambda_\theta\,\nabla\,\theta) + 3K_\alpha\theta_0\,\mathrm{div}\left(\frac{\partial\mathbf{u}}{\partial t}\right) = \boldsymbol{\sigma} : \left(\boldsymbol{\varepsilon}'_{trip} + \boldsymbol{\varepsilon}'_{cp}\right) + \rho_0\sum_{i=2}^{m} L_i p_i' + r \quad \textit{in} \quad \Omega_T$$

$$\frac{\partial\mathbf{p}}{\partial t} = \boldsymbol{\gamma}(\mathbf{p}, \theta, \theta', \mathrm{tr}(\boldsymbol{\sigma}), \boldsymbol{\sigma}^* : \boldsymbol{\sigma}^*) \quad \textit{in} \quad \Omega_T$$

$$\boldsymbol{\varepsilon}_{cp} = \boldsymbol{\varepsilon}^*(\mathbf{u}) - \boldsymbol{\varepsilon}_{trip} - \frac{1}{2\mu}\boldsymbol{\sigma}^* \quad \textit{with} \quad (\boldsymbol{\sigma}^*)' + \partial\chi_K(\boldsymbol{\sigma}^*) \ni 2\mu\left(\boldsymbol{\varepsilon}^*(\mathbf{u}') - \boldsymbol{\varepsilon}'_{trip}\right) \quad \textit{in} \quad \Omega_T$$

$$\boldsymbol{\varepsilon}'_{trip} = 3\mu\,(\boldsymbol{\varepsilon}^*(\mathbf{u}) - \boldsymbol{\varepsilon}_{cp} - \boldsymbol{\varepsilon}_{trip}) \sum_{i=1}^{m} \kappa_i \frac{\partial\Phi_i}{\partial p_i}(p_i)\max\{p_i', 0\} \quad \textit{in} \quad \Omega_T$$

including conditions (2.4), (2.10), (2.11), (3.15), (3.16) *as well as initial values* (2.24) – (2.27), (3.21) *and boundary values* (2.29) *and*

(3.22) $\mathbf{u} = \mathbf{0}$ *f.a. points in* Γ_1, $\left(\boldsymbol{\sigma} + h\boldsymbol{\varepsilon}(\mathbf{u}')\right) \cdot \nu_{\Gamma_2} = \mathbf{0}$ *f.a. points in* Γ_2,

where $\boldsymbol{\sigma}$ *is defined as in Equation* (2.12).

3.3.1 Remark (Visco-Elastic Model)
In Problem ($\mathbf{P}_{VE}$) *we follow a (mathematical) approach to solve the fully coupled problem* ($\mathbf{P}$). *In order to describe visco-elastic material behaviour, i.e. a material with a (small) viscosity h, we must add an additional dissipation term* $h\,|\boldsymbol{\varepsilon}(\mathbf{u}')|^2$ *to the right-hand side of the heat equation. The (mathematical) analysis of this thermo-visco-elastic material behaviour leads to local (in time) solutions (using a fixed-point argument, the Galerkin method and a non-linear Gronwall inequality in order to obtain the necessary a-priori estimates, cf. [Bonetti and Bonfanti 2005, 2008; Roubíček 2005] for details). But in this work we do not cover visco-elastic material behaviour, because steel is not a visco-elastic material.*

3.3.3 Formulation of Problem ($\mathbf{P}_{QS}$)

In this approach the displacement $\mathbf{u}$ (and therefore the stress $\boldsymbol{\sigma}$) is governed by the quasi-static momentum balance, i.e. the hyperbolic equation turns into an elliptic equation for the balance of momentum. The background to this setting is that in most cases it is sensible to neglect the inertial term $\frac{\partial^2 \mathrm{u}}{\partial t^2}$ because we are usually not discussing situations, where stresses appear and vanish abruptly (cf. comparative simulations in [Suhr 2010] and a similar approach in [Kern 2011]). Of course, it is worth discussing if it makes sense to use the dissipation term $\operatorname{div}\left(\frac{\partial \mathrm{u}}{\partial t}\right)$ in this context.

Problem ($\mathbf{P}_{QS}$)
Find the displacement $\mathbf{u} : \overline{\Omega}_T \to \mathbb{R}^3$, *the temperature* $\theta : \overline{\Omega}_T \to \mathbb{R}$ *and the phase fractions* $\mathbf{p} : \overline{\Omega}_T \to \mathbb{R}^m$ *s.t.*

$$-2\,\mathrm{Div}\,(\mu\boldsymbol{\varepsilon}(\mathbf{u})) - \mathrm{grad}\,(\lambda\,\mathrm{div}(\mathbf{u})) + 3\,\mathrm{grad}\,(K_\alpha(\theta - \theta_0))$$
$$+\,\mathrm{grad}\left(K\sum_{i=1}^{m}\left(\frac{\rho_0}{\rho_i(\theta_0)} - 1\right)p_i\right) + 2\,\mathrm{Div}\,(\mu\boldsymbol{\varepsilon}_{trip}) + 2\,\mathrm{Div}\,(\mu\boldsymbol{\varepsilon}_{cp}) = \mathbf{f} \quad in \quad \Omega_T$$

$$\rho_0 c_e \frac{\partial \theta}{\partial t} - \operatorname{div}(\lambda_\theta \nabla \theta) + 3K_\alpha \theta_0 \operatorname{div}\left(\frac{\partial \mathbf{u}}{\partial t}\right) = \boldsymbol{\sigma} : \left(\boldsymbol{\varepsilon}'_{trip} + \boldsymbol{\varepsilon}'_{cp}\right) + \rho_0 \sum_{i=2}^{m} L_i p'_i + r \quad in \ \Omega_T$$

$$\frac{\partial \mathbf{p}}{\partial t} = \boldsymbol{\gamma}(\mathbf{p}, \theta, \theta', \operatorname{tr}(\boldsymbol{\sigma}), \boldsymbol{\sigma}^* : \boldsymbol{\sigma}^*) \quad in \ \Omega_T$$

$$\boldsymbol{\varepsilon}_{cp} = \boldsymbol{\varepsilon}^*(\mathbf{u}) - \boldsymbol{\varepsilon}_{trip} - \frac{1}{2\mu}\boldsymbol{\sigma}^* \quad with \quad (\boldsymbol{\sigma}^*)' + \partial\chi_K(\boldsymbol{\sigma}^*) \ni 2\mu\left(\boldsymbol{\varepsilon}^*(\mathbf{u}') - \boldsymbol{\varepsilon}'_{trip}\right) \quad in \ \Omega_T$$

$$\boldsymbol{\varepsilon}'_{trip} = 3\mu\,(\boldsymbol{\varepsilon}^*(\mathbf{u}) - \boldsymbol{\varepsilon}_{cp} - \boldsymbol{\varepsilon}_{trip}) \sum_{i=1}^{m} \kappa_i \frac{\partial \Phi_i}{\partial p_i}(p_i) \max\{p'_i, 0\} \quad in \ \Omega_T$$

including conditions (2.4), (2.10), (2.11), (3.15), (3.16) *as well as initial and boundary values* $(2.24)_3$, (2.25) – (2.29) *and* (3.21).

3.4 References and Comments

We remark that the solution of each regularised problem depends on the regularisation parameter h. The passage to the limit $h \to 0$ is a-priori not obtained.

In this section we give a summary of the literature that exist associated with Problem (**P**) and the corresponding regularised Problems ($\mathbf{P}_A$), ($\mathbf{P}_{VE}$) and ($\mathbf{P}_{QS}$). Of course, this list is not complete. In this section we concentrate on the mechanical background – for the mathematical background or related works, cf. Section 4.5.

Associated with this topic there exists a huge amount of literature.

Material Behaviour of Steel The term 'Distortion Engineering' is introduced (or relevant) in [Hoffmann et al. 2002; Frerichs et al. 2007a,b, 2009; Thoben et al. 2002; Şimşir et al. 2009].
Steel and metallic workpieces are introduced in [Berns and Theisen 2006; Bleck 2001; Bleck et al. 2003; Dahl 1993; Fonseca 1996; Horstmann 1992; Liedtke 2005a,b; Rose and Hougardy 1972; Schröter et al. 1995; Wegst and Wegst 2004] and the material behaviour is discussed in [Berthelot 1999; Denis et al. 2001; Helm 1998; Ilschner and Singer 2005; Lemaitre 2001; Macherauch 1992; Müller 1972; Seidel 1999; Stüwe 1978; Wever and Rose 1954].
In [Kohtz 1994; Liedtke 2005a,b; Wolff et al. 2000] heat treatment processes are

investigated, especially quenching processes are looked at in [Inoue et al. 1989; Pietzsch 2000; Shi et al. 2004; Sjöström 1985; Wolff et al. 2003d].

Continuum Mechanics For general modelling in continuum mechanics there exists a vast amount of literature, e.g. purely engineering [Altenbach and Altenbach 1994; Betten 1993; Bowen 1976; Chandrasekharaiah and Debnath 1994; Dill 2007; Feynman et al. 1991; Haupt 2000; Parisch 2003; Šilhavý 1997; Temam and Miranville 2000; Wilmanski 1998; Wladimirow 1972; Ziegler 1998], with a strong mathematical flavour: [Dautray and Lions 1990; Duvaut and Lions 1976; Eck et al. 2008; Ladyženskaya 1985; Zeidler 1988] or with special emphasis on thermo-mechanics: [Levitas 1998], solid mechanics [Lemaitre and Chaboche 1990], heat transfer [Baehr and Stephan 2006], elasticity [Ciarlet 1988; Landau and Lifschitz 1989; Lurie 2005], thermo-elasticity [Nowacki 1986], plasticity [Burth and Brocks 1992; Burghahn et al. 1996; Chaboche 2008; Chakrabarty 1987; Dunne and Petrinic 2005; Han and Reddy 1999b; Hashiguchi 2005; Haupt 1977; Hill 1950; Khan and Huang 1995; Maugin 1992; Mendelson 1968; Reckling 1967; Salencon 1974; Sawczuk 1989; Simo and Hughes 1998], elasto-plasticity [Dachkovski and Böhm 2004c] etc.

Phase Transitions and TRIP Phase transitions are investigated in [Avrami 1939, 1940, 1941; Caballero et al. 2001]. The kinetics of phase transitions are discussed in [Burke 1965] and (slightly more general) the theory of kinetic transformation is the topic in [Cahn 1956; Christian 1965; Inoue et al. 1981; Johnson and Mehl 1939; Larsson and Mangard 1995]. Moreover, phase transitions, especially in the context of the material behaviour of steel, are modelled in [Böhm et al. 2003, 2004; Dachkovski et al. 2003; Dachkovski and Böhm 2005; Dalgic and Löwisch 2004; Denis et al. 2002; Denis 1997; Denis et al. 1999, 1992; Fischer et al. 2000a, 2003; Medeiros Fonseca 1996; Koistinen and Marburger 1959; Leblond and Devaux 1984; Leblond et al. 1985, 1986a,b; Miokovic et al. 2004; Mittemeijer and Sommer 2002; Mittemeijer 1992; Oberste-Brandenburg 1999; Porter and Easterling 1992; Scheil 1929; Verdi and Visintin 1987; Videau et al. 1994; Wolff and Böhm 2005]. Together with phase transitions in steel, the phenomenon of TRIP has to be taken into account, which is explained and/or investigated in [Ahrens et al. 2000, 2002; Ahrens 2003; Besserdich 1993; Dalgic and Löwisch 2006; Dalgic et al. 2006; Fischer et al. 1996; Fischer 1997; Fischer et al. 1998; Garcia de Andres et al. 1998; Hougardy and Yamazaki 1986; Hunkel et al. 1999; Inoue and Tanaka 2006; Leblond and Devaux 1989a,b; Mitter 1987; Mahnken et al. 2009; Schmidt et al. 2003; Tanaka et al. 2003; Taleb and Sidoroff 2003; Turteltaub and Suiker 2005; Tanaka and Sato 1985]. In this context also the parameter identification of phase transition models via dilatometer and Gleeble™ experiments (cf. Figure 1.7) play a role and are mentioned

in [Hömberg et al. 2009; Wolff et al. 2003a; Wolff and Suhr 2003; Wolff et al. 2006b, 2007b,d,c, 2010b].
In [Wolff et al. 2007a,c] some macroscopic models for phase transformations in steel were tested and evaluated. In addition, a general phenomenological model for phase transformations in steel was presented and evaluated.

Coupled Models For coupled models there exists literature in a smaller scale. Problems of thermo-elasticity are treated in [Weinmann 2009], modelling of elasto-plastic problems can be found in [Lubarda 2002; Palmov 1998].
In connection with phase transitions a thermoplastic problem is discussed in [Dachkovski and Böhm 2004a] and the coupling with an elasto-plastic problem can be found in [Dachkovski and Böhm 2004b]. In [Inoue and Wang 1985] stress, temperature and phase transitions are coupled. TRIP and plasticity is considered in [Taleb and Petit 2006].
Coupled models of thermo-elasto-plasticity with phase transitions and TRIP are discussed in [Wolff et al. 2005b, 2011c; Wolff and Böhm 2002a,b, 2003] for small deformations. An important contribution to the modelling of thermo-elasto-plasticity with phase transitions and TRIP of the material behaviour of steel and the basis of this thesis are [Wolff et al. 2008c, 2011b,a].
Coupled models that neglect the classical plasticity are discussed in e.g. [Wolff et al. 2003c,b, 2004, 2005a, 2006e,c; Wolff and Böhm 2006a,b,b; Wolff et al. 2008a, 2007a; Wolff 2008a,b]. For such models dimensional analysis (cf. [Wolff et al. 2008b]), carbon diffusion (e.g. [Wolff et al. 2006d,a]) and creep [Wolff and Böhm 2010; Bökenheide et al. 2011] are investigated.

Chapter 4

Mathematical Preliminaries

The main objective of this thesis is the investigation of coupled models for the material behaviour of steel, which describe the phase transformations, transformation-induced and classical plasticity in addition to the temperature and the deformation. Especially those models have been insufficiently examined in a strict mathematical and numerical context so far (we mention some existing results in the literature in Section 4.5). Problem (**P**), which is summarised in Section 3.1, integrates the complex physical behaviour of steel materials in more general models of thermo-elasto-plasticity. Since we mentioned the main difficulties that arise when dealing with the fully coupled problem in the preceding chapter, we restrict ourselves to the mathematical analysis of the regularised problems ($\mathbf{P}_A$), ($\mathbf{P}_{VE}$) and ($\mathbf{P}_{QS}$), as summarised in Section 3.3. The focus of this work is the proof of existence and uniqueness results for the weak solvability of these mathematical problems of the linear thermo-elasto-plasticity including phase transitions and TRIP under suitable conditions.

In this chapter we make the necessary preparations for the investigation of the regularised problems ($\mathbf{P}_A$), ($\mathbf{P}_{VE}$) and ($\mathbf{P}_{QS}$) in the next three chapters. The mathematical background for the analysis of the regularised problems is discussed in Section 4.1. In Section 4.2 the function spaces used within this work are collected and in Section 4.3 the general (mathematical) assumptions required for the existence and uniqueness results are presented. Section 4.4 gives an overview of the main results. References and related works are given in Section 4.5.

4.1 Mathematical Context

This section defines the research context of the mathematical problem (**P**) resp. the regularised problems ($\mathbf{P}_A$), ($\mathbf{P}_{VE}$) and ($\mathbf{P}_{QS}$) and discusses the means and methods used for proving the existence and uniqueness results in this work.

Modelling the relevant interactions between temperature, mechanical behaviour and phase transitions in Chapter 2 leads to an initial boundary value problem for a complex system of coupled nonlinear partial and ordinary differential equations and inequalities (regarding to classical plasticity) for the time and space-dependent temperature, displacement and phase fractions (cf. Chapter 3 for the formulation of the mathematical problem(s)). Figure 4.1 visualises the structure of the Problem (**P**) (resp. the regularised problems ($\mathbf{P}_A$), ($\mathbf{P}_{VE}$) and ($\mathbf{P}_{QS}$)).

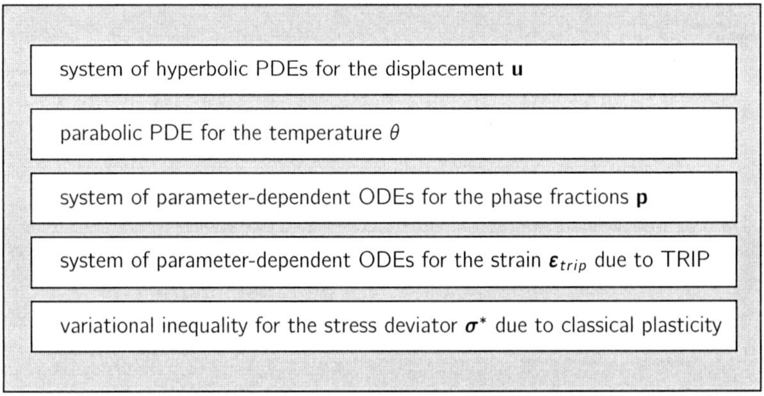

Figure 4.1: Summary of the structure of the 'Mathematical Problems'.

Obviously, it is not trivial to prove the (unique) existence of a weak solution of the fully coupled problem (**P**). Even in case of simplified assumptions (cf. Assumption 3.2.1), there arise great difficulties, when trying to solve this coupled system of equations consisting of a system of hyperbolic PDEs for the displacement $\mathbf{u}$, a parabolic PDE for the temperature θ, a system of parameter-dependent ODEs for the phase fractions $\mathbf{p}$, a system of parameter-dependent ODEs for the strain $\boldsymbol{\varepsilon}_{trip}$ due to TRIP and a (parabolic) variational inequality for the strain $\boldsymbol{\varepsilon}_{cp}$ resp. for the stress deviator $\boldsymbol{\sigma}^*$ due to classical plasticity (cf. Chapter 3; in particular Section 3.1 for the original problem).
Therefore we introduced three slightly modified problems of the Problem (**P**) (cf. Section 3.3), which will be investigated in the next three chapters.
Table 4.1 gives an overview of the considered problems and Table 4.2 summarises the considered subproblems mentioned in Figure 4.1.

Name	Description	Page
($\mathbf{P}$)	Fully Coupled Problem	48
($\mathbf{P}_A$)	Regularisation via Averaging, i.e. Steklov Regularisation	57
($\mathbf{P}_{VE}$)	Visco-Elastic Regularisation	58
($\mathbf{P}_{QS}$)	Quasi-Static Problem	59
($\mathbf{P}_{NUM}$)	Simulated Problem	197

Table 4.1: Overview of the 'Mathematical Problems' introduced in Chapter 3.

Name	Description	Page
($\mathbf{P}_{\mathrm{p}}$)	Subproblem of Phase Transitions	95
($\mathbf{P}_{\varepsilon_{trip}}$)	Subproblem of TRIP	51
($\mathbf{P}_{\varepsilon_{cp}}$)	Subproblem of Classical Plasticity	52
($\mathbf{P}_{\theta}$)	Subproblem of Heat Equation	114
($\mathbf{P}_{\mathrm{u}}$)	Subproblem of Displacement	102

Table 4.2: Overview of all 'Mathematical Subproblems'.

The mathematical analysis follows the methodological approach of the general treatment of (abstract) evolution equations in [Emmrich 2004; Gajewski et al. 1974; Renardy and Rogers 1996; Wloka 1987; Zeidler 1990b] (cf. [Brézis 1971; Dautray and Lions 1992; Evans 1998; Gilbarg and Trudinger 2001; Lions and Magenes 1973a,b,c; Larson and Thomée 2003; Roubíček 2005; Showalter 1997] for further information).
We use 'variational methods' (or 'energy methods'; in particular the 'Galerkin Method') in the standard 'L^2-setting'[1] for solving the evolution equations[2] and follow the ideas in [Barbu 1976; Brézis 1973; Showalter 1997; Zeidler 1985] to solve the evolution variational inequality.
Finally, the fixed-point theorems of Banach and Schauder (cf. e.g. [Zeidler 1986]) are applied in order to prove the existence and uniqueness results for the parameter-

[1]This approach is motivated by well-known results for systems of hyperbolic equations and variational inequalities in this case. Moreover, we do not consider a L^p-setting for $p \neq 2$.
[2]Rothe's method could be an alternative to the Galerkin method, but we do not use it in this context.

dependent ODEs and for the coupled system. These methods, i.e. the Galerkin Method in combination with a fixed-point argumentation are generally used for mathematical investigations of problems in mechanics of solids, cf. e.g. [Eck et al. 2005; Chelminski et al. 2007]. Therefore (and because more general right-hand sides than for other methods (v.i.) are allowed) we also use the Galerkin method and various fixed-point arguments. Of course, there are other methods for dealing with evolution equations available, e.g. the 'Fourier Method or Method of Diagonalisation', integral transformation methods like the 'Fourier and Laplace Transforms' or the 'Method of Semi-groups' (cf. [Dautray and Lions 1992]), but we do not use them in this work[3].

The complications for the mathematical investigation of the fully coupled problem stem particularly from the coupling terms of the displacement (resp. elasticity) equation and the heat equation, i.e. the thermal part of the stress tensor in the equation for the displacement and the dissipation term in the heat equation, and the material laws for transformation-induced and classical plasticity create difficulties. Moreover, we consider (general) mixed Dirichlet-Neumann boundary conditions for the displacement and Robin boundary conditions for the temperature. This choice of boundary conditions reproduces the physical reality quite well, but it creates additional difficulties for the mathematical analysis of the whole problem (cf. discussion in [Kern 2011], where (only) pure Dirichlet conditions are used).

The difficulties are caused by a lack of regularity, i.e. the regularity of the temperature or the displacement, which is needed in our approach in order to solve the linearised (single) equations for heat, displacement, phase fractions, stress and strains is not obtained by the solution of these equations. Moreover, the second time derivatives of the strains due to transformation-induced and classical plasticity does not necessarily exist. For example, in order to show $\boldsymbol{\varepsilon}_{cp} \in \mathfrak{H}_{\sigma}$, which is necessary in our approach, we need $\mathbf{u} \in \mathfrak{V}_{\mathrm{u}}$, but solving the displacement equation with $\boldsymbol{\varepsilon}_{cp} \in \mathfrak{H}_{\sigma}$ just gives us $\mathbf{u} \in \mathcal{V}_{\mathrm{u}}^{\infty} \cap \mathfrak{H}_{\mathrm{u}}$ (cf. Section 4.2 for the definition of function spaces). Moreover, $\mathbf{u} \in \mathfrak{V}_{\mathrm{u}}$ is also necessary in our approach to show $\theta \in \mathfrak{H}_{\theta}$, which influences the regularity of $\mathbf{u}$ and $\mathbf{p}$ and so on.

We collect some ideas in order to approach this problem:

- The same 'trick' or idea as in linear thermo-elasticity (cf. [Boettcher 2007; Gawinecki 1986] or [Ebenfeld 2002; Khalifa 2003; Zheng 1995] for the coupling of hyperbolic-parabolic equations) where a simultaneous Galerkin approach for the coupled system for $(\mathbf{u}, \theta)$ is used in order to neutralise the coupling terms does not work in our setting. First of all, the first a-priori estimate for the limit process in the heat equation is not sufficient and secondly $\mathbf{u}' \in \mathcal{V}_{\mathrm{u}}$ is necessary in order to treat the variational inequality for the strain due to classical plasticity. Therefore

[3] The semi-group method is applied for evolution equations in thermo-plasticity in [Racke 1990, 1992; Jiang and Racke 2000].

we follow a different approach here.

- Additional investigation of the regularity with the help of the lifting property, differentiation of the Galerkin equations or testing the Galerkin equations with special bases are not leading to the desired results. We refer to the discussion in [Boettcher 2007; Gawinecki 1986], where (zero) Dirichlet boundary conditions, constant parameters, additional regularity of initial data and of the right-hand sides as well as a regularisation of the TRIP strain and the use of a hyperbolic heat equation (cf. [Müller 1972]) are tested.

All in all, proving existence and uniqueness of a (global-in-time) weak solution of the fully coupled problem (**P**) with general (mixed) boundary values is, in this setting and with the help of the used techniques (L^2-setting, Galerkin method and fixed-point argumentation via Banach and Schauder), not obvious or even possible.

An alternative approach is the use of regularisations, cf. the discussion in [Boettcher 2007], in order to handle the problem with the help of the techniques mentioned before. We have chosen three different possibilities to (slightly) change the (physical) model. We will investigate the modified problems presented in Section 3.3 in the following subsections.

The problem of the regularisations is that either additional conditions (some parameters have to be small) are needed or the estimates are not independent of the regularisation parameter, which means that the limit process (of the regularisation parameter) is generally not possible. The idea is to regularise the problem in such a way as to obtain a problem with a solution in 'nice' spaces and then to show that the regularised solution converges in some sense to functions that solve the original problem. The main results of this thesis are as follows:

- We show existence and uniqueness of a weak solution of Problem ($\mathbf{P}_A$) with the help of Banach's fixed-point argumentation under suitable assumptions. The existence result can also be verified with the help of the Schauder fixed-point theorem. Additional regularity of the weak solution can be shown in the 'Banach'-setting under (more restrictive) assumptions and therefore the passage to the limit in the regularisation parameter is possible.
- The Problem ($\mathbf{P}_{VE}$) is treated in the same way as Problem ($\mathbf{P}_A$).
- We show existence and uniqueness of a weak solution of Problem ($\mathbf{P}_{QS}$) with the help of Banach's fixed-point argumentation under suitable assumptions.
  Whereas the problems ($\mathbf{P}_A$) and ($\mathbf{P}_{VE}$) are two different regularisations of the original problem (**P**), the Problem ($\mathbf{P}_{QS}$) is of a different type. Therefore, we do not go into details regarding regularity results and there is no passage to the limit in the regularisation parameter.

The character of the saturation function in the TRIP-strain, the dissipation term in the heat equation, the function for isotropic hardening in evolution inequality for the (classical) plastic strain and the right-hand sides of the evolution equations for the phase fractions may vary in the different settings – we will remark the necessary restrictions at the particular point.

4.2 Function Spaces

The (standard) definitions of function spaces on space-time domains and a collection of a few results that are important for the existence and uniqueness results in this chapter are given in Appendix A. We use the following notation for function spaces required for the weak formulation of the investigated problems:

Spaces regarding $\mathbf{u}$:

$$
\begin{aligned}
\mathbf{H}_u &:= [L^2(\Omega)]^3, & \mathbf{V}_u &:= \left\{\mathbf{u} \in [W^{1,2}(\Omega)]^3 : \mathbf{u}|_{\Gamma_1} = \mathbf{0}\right\},\\
\mathbf{W}_u &:= [W^{2,2}(\Omega)]^3 \cap \mathbf{V}_u, & \mathcal{H}_u &:= L^2(S;\mathbf{H}_u),\\
\mathcal{V}_u &:= L^2(S;\mathbf{V}_u), & \mathcal{W}_u &:= L^2(S;\mathbf{W}_u),\\
\mathcal{H}_u^\infty &:= L^\infty(S;\mathbf{H}_u), & \mathcal{V}_u^\infty &:= L^\infty(S;\mathbf{V}_u),\\
\mathcal{W}_u^\infty &:= L^\infty(S;\mathbf{W}_u), & \mathfrak{H}_u &:= W^{1,2}(S;\mathbf{H}_u),\\
\mathfrak{V}_u &:= W^{1,2}(S;\mathbf{V}_u), & \mathfrak{W}_u &:= W^{1,2}(S;\mathbf{W}_u),\\
\mathcal{U}_u &:= \{\mathbf{u} \in \mathcal{V}_u : \mathbf{u}' \in \mathcal{H}_u, \mathbf{u}'' \in \mathcal{V}_u^*\}, & \mathfrak{U}_u &:= \{\mathbf{u} \in \mathcal{W}_u : \mathbf{u}' \in \mathcal{V}_u, \mathbf{u}'' \in \mathcal{H}_u\}.
\end{aligned}
$$

Spaces regarding θ:

$$
\begin{aligned}
H_\theta &:= L^2(\Omega), & V_\theta &:= W^{1,2}(\Omega), & W_\theta &:= W^{2,2}(\Omega) \cap V_\theta,\\
\mathcal{H}_\theta &:= L^2(S;H_\theta), & \mathcal{V}_\theta &:= L^2(S;V_\theta), & \mathcal{W}_\theta &:= L^2(S;W_\theta),\\
\mathcal{H}_\theta^\infty &:= L^\infty(S;H_\theta), & \mathcal{V}_\theta^\infty &:= L^\infty(S;V_\theta), & \mathcal{W}_\theta^\infty &:= L^\infty(S;W_\theta),\\
\mathfrak{H}_\theta &:= W^{1,2}(S;H_\theta), & \mathfrak{V}_\theta &:= W^{1,2}(S;V_\theta), & \mathfrak{W}_\theta &:= W^{1,2}(S;W_\theta),\\
\mathcal{U}_\theta &:= W^{1,2}(S;V_\theta,V_\theta^*), & \mathfrak{U}_\theta &:= W^{1,2}(S;W_\theta,H_\theta^*).
\end{aligned}
$$

Spaces regarding $\mathbf{p}$:

$$
\begin{aligned}
\mathbf{H}_\mathrm{p} &:= [L^2(\Omega)]^m, & \mathbf{V}_\mathrm{p} &:= [W^{1,2}(\Omega)]^m, & \mathbf{X}_\mathrm{p} &:= [L^\infty(\Omega)]^m,\\
\mathcal{H}_\mathrm{p} &:= L^2(S;\mathbf{H}_\mathrm{p}), & \mathcal{V}_\mathrm{p} &:= L^2(S;\mathbf{V}_\mathrm{p}), & \mathcal{X}_\mathrm{p} &:= L^2(S;\mathbf{X}_\mathrm{p}),\\
\mathcal{H}_\mathrm{p}^\infty &:= L^\infty(S;\mathbf{H}_\mathrm{p}), & \mathcal{V}_\mathrm{p}^\infty &:= L^\infty(S;\mathbf{V}_\mathrm{p}), & \mathcal{X}_\mathrm{p}^\infty &:= [L^\infty(\Omega_T)]^m,\\
\mathfrak{H}_\mathrm{p} &:= W^{1,2}(S;\mathbf{H}_\mathrm{p}), & \mathfrak{V}_\mathrm{p} &:= W^{1,2}(S;\mathbf{V}_\mathrm{p}), & \mathfrak{X}_\mathrm{p}^\infty &:= [W^{1,\infty}(\Omega_T)]^m.
\end{aligned}
$$

Spaces regarding $\boldsymbol{\sigma}$, $\boldsymbol{\varepsilon}$, $\boldsymbol{\varepsilon}_{te}$, $\boldsymbol{\varepsilon}_{trip}$ and $\boldsymbol{\varepsilon}_{cp}$:

$$
\begin{aligned}
\mathbf{H}_\sigma &:= [L^2(\Omega)]^9, & \mathbf{V}_\sigma &:= [W^{1,2}(\Omega)]^9, & \mathbf{X}_\sigma &:= [L^\infty(\Omega)]^9,\\
\mathcal{H}_\sigma &:= L^2(S;\mathbf{H}_\sigma), & \mathcal{V}_\sigma &:= L^2(S;\mathbf{V}_\sigma), & \mathcal{X}_\sigma &:= L^2(S;\mathbf{X}_\sigma),\\
\mathcal{H}_\sigma^\infty &:= L^\infty(S;\mathbf{H}_\sigma), & \mathcal{V}_\sigma^\infty &:= L^\infty(S;\mathbf{V}_\sigma), & \mathcal{X}_\sigma^\infty &:= [L^\infty(\Omega_T)]^9,\\
\mathfrak{H}_\sigma &:= W^{1,2}(S;\mathbf{H}_\sigma), & \mathfrak{V}_\sigma &:= W^{1,2}(S;\mathbf{V}_\sigma), & \mathfrak{X}_\sigma &:= W^{1,2}(S;\mathbf{X}_\sigma).
\end{aligned}
$$

Table 4.3: Overview of function spaces.

4.3 General Mathematical Assumptions

We use the following assumptions throughout this chapter:

4.3.1 Assumption
Let $\Omega \subset \mathbb{R}^3$ be a bounded domain with Lipschitz boundary and $\Omega_T := \Omega\times]0,T[$ the corresponding time cylinder with $T \in \mathbb{R}^+$. Let Γ_1 be a closed subset of the boundary with positive surface measure and $\Gamma_2 := \partial\Omega \setminus \Gamma_1$. Moreover, let all material parameters be real constants with

$$\rho_0, \rho_i, \mu_i, K_i, \alpha_i, R_{0i}, c_e, \lambda_\theta, \kappa_i, L_i > 0,\ \delta \geq 0, \qquad i = 1, \ldots, m.$$

In addition, let

(A1) $\mathbf{u}_0 \in \mathbf{V}_\mathrm{u}$, $\mathbf{u}_1 \in \mathbf{H}_\mathrm{u}$ *and* $\mathbf{f} \in \mathcal{H}_\mathrm{u}$.

(A2) $\theta_0 \in H_\theta$, $r \in \mathcal{H}_\theta$, *and* $\theta_\Gamma \in L^2(\Gamma_T)$.

(A3) $F = F(\boldsymbol{\sigma})$ *be a convex function* $\mathbb{R}^9 \to \mathbb{R}$, *which is differentiable in* $\mathbb{R}^9 \setminus \{\mathbf{0}\}$ *and fulfills* $F(\mathbf{0}) < 0$. *Moreover,* $\frac{\partial F}{\partial \sigma}(\boldsymbol{\tau}) \neq 0$ *f.a.* $\boldsymbol{\tau} \in \mathbb{R}^9$ *with* $\boldsymbol{\tau} = \boldsymbol{\tau}^T$ *and* $F(\boldsymbol{\tau}) = 0$. *In addition, let* $\boldsymbol{\sigma}^*(0) \in \mathbf{K}$, *i.e.* $\mathbf{u}_0 \in \mathbf{V}_\mathrm{u}$ *with* $\|\mathbf{u}_0\|_{\mathrm{V_u}} \leq \frac{R_0+R}{\sqrt{6\mu}}$. *In this context we define*

$$\mathbf{K}_F := \left\{\boldsymbol{\tau} \in \mathbb{R}^{3\times 3}_{sym}, \operatorname{tr}(\boldsymbol{\tau}) = 0 : F(\boldsymbol{\tau}; R, R_0) \leq 0\right\},$$
$$\mathbf{K} := \left\{\boldsymbol{\tau} \in \mathbf{H}_\sigma : \boldsymbol{\tau}(\mathbf{x}) \in \mathbf{K}_F \text{ f.a.a. } \mathbf{x} \in \Omega\right\}.$$

Furthermore,

- *let* $R_0, R \in \mathbb{R}^+$ *(i.e.* $\mathbf{K}$ *is constant),*
- *or let* $R_0 \in \mathbb{R}^+$ *and* $R : \Omega_T \to \mathbb{R}^+$ *be a Lipschitz continuous and bounded function with* $R' \in L^2(\Omega_T)$ *(i.e.* $\mathbf{K}$ *is time-dependent),*
- *or let* $R_0 \in \mathbb{R}^+$ *and* $R : \mathbb{R} \to \mathbb{R}^+$ *be Lipschitz continuous, bounded and monotonically increasing function (i.e.* $\mathbf{K}$ *is parameter-dependent).*

(A4) $\Phi_i \in C^{0,1}([0,1]) \cap C^1(]0,1[)$ *with* $\Phi_i(0) = 0$, $\Phi_i(1) = 1$ *and*

$$0 \leq \frac{\partial \Phi_i}{\partial p}(p) \leq M_\Phi < \infty$$

f.a. $0 \leq p \leq 1$ *and f.a.* $i = 1, \dots, m$.

(A5) Let $\boldsymbol{\gamma} : \mathbb{R}^m \times \mathbb{R} \times \mathbb{R}^M \to \mathbb{R}^m$ *be a Lebesgue measurable and bounded function, i.e. there exist a non-negative function* $h \in C(\mathbb{R}^m)$ *and two constants* $c_1, c_2 \geq 0$ *such that*

$$\|\boldsymbol{\gamma}(\mathbf{p}, \theta, \boldsymbol{\xi})\|_\infty \leq c_1 + c_2 h(\mathbf{p}) \quad \text{f.a.} \quad \mathbf{p} \in \mathbb{R}^m, \quad \theta \in \mathbb{R}, \quad \boldsymbol{\xi} \in \mathbb{R}^M.$$

Moreover, let the function $\boldsymbol{\gamma}$ *be Lipschitz continuous in the following sense: There exist constants* $L_\mathrm{p} > 0$, $L_\theta > 0$ *and* $L_\xi > 0$ *s.t.*

$$\begin{aligned}
\|\boldsymbol{\gamma}(\mathbf{p}, \theta, \boldsymbol{\xi}) - \boldsymbol{\gamma}(\mathbf{q}, \theta, \boldsymbol{\xi})\|_\infty &\leq L_\mathrm{p}\|\mathbf{p} - \mathbf{q}\|_\infty && \text{f.a. } \mathbf{p}, \mathbf{q} \in \mathbb{R}^m, \theta \in \mathbb{R}, \boldsymbol{\xi} \in \mathbb{R}^M, \\
\|\boldsymbol{\gamma}(\mathbf{p}, \theta, \boldsymbol{\xi}) - \boldsymbol{\gamma}(\mathbf{p}, \vartheta, \boldsymbol{\xi})\|_\infty &\leq L_\theta|\theta - \vartheta| && \text{f.a. } \mathbf{p} \in \mathbb{R}^m, \theta, \vartheta \in \mathbb{R}, \boldsymbol{\xi} \in \mathbb{R}^M, \\
\|\boldsymbol{\gamma}(\mathbf{p}, \theta, \boldsymbol{\xi}) - \boldsymbol{\gamma}(\mathbf{p}, \theta, \boldsymbol{\zeta})\|_\infty &\leq L_\xi\|\boldsymbol{\xi} - \boldsymbol{\zeta}\|_\infty && \text{f.a. } \mathbf{p} \in \mathbb{R}^m, \theta \in \mathbb{R}, \boldsymbol{\xi}, \boldsymbol{\zeta} \in \mathbb{R}^M.
\end{aligned}$$

(A6) Furthermore, let $\mathbf{p}_0 \in \mathbf{X}_\mathrm{p}$.

Some additional assumptions are needed as well in the next chapter:

4.3.2 Assumption
Let $\Omega \subset \mathbb{R}^3$ be a bounded $\mathcal{C}^2$-domain. In addition, let

(A1) $\mathbf{u}_0 \in \mathbf{W}_{\mathrm{u}}$, $\mathbf{u}_1 \in \mathbf{V}_{\mathrm{u}}$ *and* $\mathbf{f} \in \mathcal{V}_{\mathrm{u}}$.

(A2) $\theta_0 \in H_\theta$, $r \in \mathcal{H}_\theta$ *and* $\theta_\Gamma \in W^{\frac{1}{4},2}(S; W^{\frac{1}{2},2}(\partial\Omega))$.

(A2') $\theta_0 \in V_\theta$, $r \in \mathcal{V}_\theta$ *and* $\theta_\Gamma \in W^{\frac{1}{4},2}(S; W^{\frac{1}{2},2}(\partial\Omega))$.

(A3) $\gamma_i, \frac{\partial \gamma_i}{\partial p_k}, \frac{\partial \gamma_i}{\partial \theta} : \mathbb{R}^m \times \mathbb{R} \to \mathbb{R}$ *be Lebesgue measurable and bounded functions for* $i, k = 1, \dots, m$.

(A3') $\frac{\partial \gamma_i}{\partial p_k}, \frac{\partial \gamma_i}{\partial \theta}, \frac{\partial^2 \gamma_i}{\partial p_j \partial p_k}, \frac{\partial^2 \gamma_i}{\partial \theta \partial p_k}, \frac{\partial^2 \gamma_i}{\partial p_k^2}, \frac{\partial^2 \gamma_i}{\partial \theta^2} : \mathbb{R}^m \times \mathbb{R} \to \mathbb{R}$ *be Lebesgue measurable and bounded functions for* $i, k = 1, \dots, m$.

4.4 Overview of the Main Results

In this section we collect some results of the auxiliary problems which are necessary for the proofs in Chapters 5, 6 and 7 (cf. Table 4.4). See Table 4.2 on page 65 for an overview of the investigated PDE problems and subproblems. Table 4.5 gives an overview of the proven main results.

Prob.	**Existence and Uniqueness**	**Continuity w.r.t. Parameter(s)**	**Regularity**	**Additional Properties**
$(\mathbf{P}_{\mathrm{p}})$	Lem. 5.3.8	Lem. 5.3.10	Lem. 5.6.1, 5.6.2 and Rem. 5.6.3	Sec. 5.7.1: Appl. to PT in Steel
$(\mathbf{P}_{\boldsymbol{\varepsilon}_{trip}})$	Lem. 5.3.13	Lem. 5.3.15	Lem. 5.6.8	
$(\mathbf{P}_{\boldsymbol{\varepsilon}_{cp}})$	Lem. 5.3.3	Lem. 5.3.6	Lem. 5.6.7	Sec. 5.7.2, 5.7.3
$(\mathbf{P}_\theta)$	Lem. 5.3.34 Lem. 5.4.5 Lem. 6.3.14	Lem. 5.3.35 Lem. 6.3.15	Lem. 5.3.36 Lem. 5.4.6 Lem. 6.3.16	Sec. 5.7.4
$(\mathbf{P}_{\mathrm{u}})$	Lem. 5.3.25, Lem. 6.3.9, Lem. 7.3.3	Lem. 5.3.26, Lem. 6.3.11, Lem. 7.3.4	Lem. 5.6.11, Lem. 6.4.8	

Table 4.4: Overview of the auxiliary results.

	Fully Coupled Problem		
	Regularisation via Averaging (Prob. ($\mathbf{P}_A$), cf. Chapter 5)	*Visco-elastic Regularisation* (Prob. ($\mathbf{P}_{VE}$), cf. Chapter 6)	*Quasi-static Problem* (Prob. ($\mathbf{P}_{QS}$), cf. Chapter 7)
Existence	Thm. 5.2.1: via Banach FPT without intrinsic dissipation Thm. 5.2.3: via Banach FPT and Schauder FPT without PT depending on θ'	Thm. 6.2.1: via Banach FPT without intrinsic dissipation	Thm. 7.2.1: via Banach FPT without intrinsic dissipation
Uniqueness	Thm. 5.2.1: via Banach FPT without intrinsic dissipation Rem. 5.2.4: via Banach FPT and Schauder FPT without PT depending on θ' and without intrinsic dissipation	Thm. 6.2.1: via Banach FPT without intrinsic dissipation	Thm. 7.2.1: via Banach FPT without intrinsic dissipation
Regularity	Thm. 5.5.2: via Banach FPT without intrinsic dissipation, PT only depending on θ, saturation function equals identity, zero Dirichlet BC for $\mathbf{u}$ and zero Dirichlet or Neumann BC for θ	Rem. 6.2.2: via Banach FPT without intrinsic dissipation, PT only depending on θ, saturation function equals identity, zero Dirichlet BC for $\mathbf{u}$ and zero Dirichlet or Neumann BC for θ	
Passage to the Limit in the Regularisation Parameter	Rem. 5.5.3: via Banach FPT without intrinsic dissipation, PT only depending on θ, saturation function equals identity, zero Dirichlet BC for $\mathbf{u}$ and zero Dirichlet or Neumann BC for θ	Rem. 6.2.2: via Banach FPT without intrinsic dissipation, PT only depending on θ, saturation function equals identity, zero Dirichlet BC for $\mathbf{u}$ and zero Dirichlet or Neumann BC for θ	

Table 4.5: Overview of the main results. (**K** may be constant, time- or parameter-dependent in all settings.)

4.5 References and Comments

In this section we collect some references that exist associated with the fully coupled Problem (**P**) and the corresponding regularised problems ($\mathbf{P}_A$), ($\mathbf{P}_{VE}$) and ($\mathbf{P}_{QS}$), cf. Sections 2.9 and 3.3. Of course, this list is not exhaustive. In contrast to Section 3.4 we concentrate in this section on related works which deal with the mathematical analysis of such problems. For literature that is looking at numerical analysis and simulation we refer to Section 8.4.

Problem of Elasticity In the literature there are extensive descriptions to deal with the mathematical problem of linear elasticity, cf. e.g. [Bacuta and Bramble 2003; Belishev and Lasiecka 2002; Brown and Mitrea 2009; Chang and Choe 2005; Ciarlet 1988; Ciarlet and Nečas 1985; Hughes et al. 1976; Herzog et al. 2011b; Hayasida and Wada 1999; Ito 1990a,b; Larsson and Mangard 1995; Lanza de Cristoforis and Valent 1982; Marsden and Hughes 1983; Mitrea and Monniaux 2010; Nečas and Štípl 1976; Valent 1988; Washizu 1968].

Problem (Elasticity)
Find the displacement field $\mathbf{u} : \overline{\Omega}_T \to \mathbb{R}^3$ *and the stress field* $\boldsymbol{\sigma} : \overline{\Omega}_T \to \mathbb{R}^{3\times3}_{sym}$ *s.t.*

$$\rho_0 \frac{\partial^2 \mathbf{u}}{\partial t^2} - 2\,\mathrm{Div}(\mu\boldsymbol{\varepsilon}(\mathbf{u})) - \mathrm{grad}(\lambda\,\mathrm{div}(\mathbf{u})) = \mathbf{f} \quad \text{in} \quad \Omega_T$$

$$\mathbf{u}(0) = \mathbf{u}_0,\ \mathbf{u}'(0) = \mathbf{u}_1 \quad \text{in} \quad \Omega$$

including boundary conditions (2.28).

Existence and uniqueness results for the stationary problem of linear elasticity can be found in [Duvaut and Lions 1976; Nečas and Hlavácek 1981; Marsden and Hughes 1983; Ciarlet 1988; Valent 1988; Zeidler 1988; Boettcher 2007]. The instationary problem is investigated e.g. in [Boettcher 2007; Duvaut and Lions 1976; Zeidler 1988] via energy methods. Hyperbolic equations in the context of elasticity are treated in [Dafermos and Hrusa 1985]. Regularity results, like $W^{1,p}$-regularity and the fact that we cannot expect $W^{1,\infty}$-regularity for the presented setting are discussed in [Herzog et al. 2011a,b].

Problems of Viscoelasticity are treated in [Awbi et al. 2000; Duvaut and Lions 1976; Bellout and Nečas 1994; Barral et al. 2007; Le Tallec 1990].

Problem of Thermo-Elasticity Extensive literature in order to investigate the mathematical problem of (linear) thermo-elasticity can be found in [Ames and Payne 1994; Bien 2003, 1996; Bloom 1975; Boettcher 2007; Elliott and Qi 1994; Gawinecki 2002, 1986, 1992; Giorgi and Pata 2001; Gawinecki et al. 2007; Jiang 1992; Jiang and Racke 2000, 1990; Kaliev and Mugafarov 2003; Lord and Shulman 1967; Marin 1995; Pawlow and Zochowski 2004; Ponce and Racke 1990; Pawlow and Zochowski 2002; Pawlow and Zajaczkowski 2005a,b; Racke 1990; Rieger 1998; Rivera and Racke 1998; Yoshikawa et al. 2007].

Problem (Thermo-Elasticity)
Find the displacement field $\mathbf{u} : \overline{\Omega}_T \to \mathbb{R}^3$, *the stress field* $\boldsymbol{\sigma} : \overline{\Omega}_T \to \mathbb{R}^{3\times 3}_{sym}$ *and the temperature* $\theta : \overline{\Omega}_T \to \mathbb{R}$ *s.t.*

$$\rho_0 \frac{\partial^2 \mathbf{u}}{\partial t^2} - 2\,\mathrm{Div}(\mu \boldsymbol{\varepsilon}(\mathbf{u})) - \mathrm{grad}(\lambda\,\mathrm{div}(\mathbf{u})) + 3\,\mathrm{grad}(K_\alpha(\theta - \theta_0)) = \mathbf{f} \quad \text{in} \quad \Omega_T$$

$$\rho_0 c_e \frac{\partial \theta}{\partial t} - \mathrm{div}(\lambda_\theta \nabla \theta) + 3K_\alpha \theta\,\mathrm{div}\left(\frac{\partial \mathbf{u}}{\partial t}\right) = r \quad \text{in} \quad \Omega_T$$

including initial and boundary values (2.24), (2.28) *and* (2.29).

Energy methods are employed in [Boettcher 2007; Gawinecki 1986; Dautray and Lions 1992]. Using a simultaneous Galerkin approach and the pure Dirichlet boundary condition

$$\mathbf{u} = 0 \quad \text{on} \quad \Gamma_T$$

instead of (2.28), integration by parts yields

$$3\int_\Omega K_\alpha\,\theta_0\,\mathrm{div}(\mathbf{u}'(\tau))\varphi(\tau)\,\mathrm{d}\mathbf{x} = -3\int_\Omega K_\alpha\,\theta_0\,\mathbf{u}'(\tau)\nabla\varphi(\tau)\,\mathrm{d}\mathbf{x} \qquad \text{and}$$

$$-3\int_\Omega K_\alpha\,\theta_0\,(\theta(t) - \theta_0)\,\mathrm{div}(\mathbf{u}'(t))\,\mathrm{d}\mathbf{x} = 3\int_\Omega \mathrm{grad}(K_\alpha\,\theta_0\,\theta(t))\,\mathbf{u}'(t)\,\mathrm{d}\mathbf{x}.$$

Therefore, the critical coupling terms vanish for constant coefficients and there is no need of additional regularity results (cf. the discussion in [Boettcher 2007]). Although these boundary conditions are convenient for the mathematical analysis, they do not represent well the physical reality, cf. the remarks in [Kern 2011]. The same 'reduction trick' is employed in [Jiang and Racke 1990, 2000] for a thermo-elastic system without (external) source terms and for different boundary conditions:

$$\mathbf{u}|_{\Gamma_1} = \mathbf{0},\ \boldsymbol{\sigma}\cdot\nu|_{\Gamma_2} = \mathbf{0} \quad \text{or} \quad \mathbf{u}|_{\partial\Omega} = \mathbf{0} \quad \text{and} \quad \theta|_{\partial\Omega} = 0 \quad \text{or} \quad \frac{\partial\theta}{\partial\nu}|_{\partial\Omega} = 0.$$

Then, for the (weak) solution one obtains $\mathbf{u} \in \mathfrak{W}_u^\infty \cap \mathcal{V}_u^\infty$ and $\theta \in \mathcal{H}_\theta^\infty \cap \mathcal{V}_\theta$ and additional regularity (time differentiation of the Galerkin equations and testing with the leading time derivatives) yields $\mathbf{u}'' \in \mathcal{H}_u^\infty$, $\mathbf{u}' \in \mathcal{V}_u^\infty$ and $\theta' \in \mathcal{H}_\theta^\infty \cap \mathcal{V}_\theta$.

Semi-group methods are used in [Jiang and Racke 2000, 1990; Ponce and Racke 1990; Racke 1992, 1990] in order to solve problems of the following type:

$$\begin{aligned} \mathbf{u}'' - \mu\Delta\mathbf{u} - (\mu+\lambda)\,\nabla(\operatorname{div}(\mathbf{u})) + \gamma\,\nabla\,\theta &= \mathbf{f}_1(\nabla\,\mathbf{u}, \nabla^2\,\mathbf{u}, \theta, \nabla\,\theta) && \text{in} \quad \Omega_T \\ \delta\theta' - \kappa\Delta\theta + \gamma\operatorname{div}(\mathbf{u}') &= f_2(\nabla\,\mathbf{u}, \nabla^2\,\mathbf{u}, \nabla\,\mathbf{u}', \theta, \nabla^2\,\theta) && \text{in} \quad \Omega_T \end{aligned}$$

with initial and boundary conditions $\mathbf{u}|_{\partial\Omega} = 0$, $\theta|_{\partial\Omega} = 0$, $\mathbf{u}(t=0) = \mathbf{u}_0$, $\mathbf{u}'(t=0) = \mathbf{u}_1$ and $\theta(t=0) = \theta_0$. The idea is to solve

$$\mathbf{v}' + \mathbf{A}\mathbf{v} = \mathbf{g}, \qquad \mathbf{v}(0) = \mathbf{v}_0 \tag{4.1}$$

in Hilbert space $\mathcal{H} := [H^1(\mathbb{R}^3)]^3 \times [L^2(\mathbb{R}^3)]^3 \times L^2(\mathbb{R}^3)$ where

$$\mathbf{v}(t) := \begin{pmatrix} \nabla\,\mathbf{u}(t) \\ \mathbf{u}'(t) \\ \theta(t) \end{pmatrix}, \quad \mathbf{v}_0 := \begin{pmatrix} \mathbf{u}_0 \\ \mathbf{u}_1 \\ \theta_0 \end{pmatrix}, \quad \mathbf{g}(t) := \begin{pmatrix} 0 \\ \mathbf{f}_1 \\ f_2 \end{pmatrix}, \quad \mathbf{A} := \begin{pmatrix} 0 & -\nabla & 0 \\ -\operatorname{div} & 0 & \gamma\,\nabla \\ 0 & \gamma\operatorname{div} & -\kappa\Delta \end{pmatrix},$$

$\mathrm{D}(\mathbf{A}) := \{\mathbf{v} \in \mathcal{H} : \mathbf{A}\mathbf{v} \in [L^2(\mathbb{R}^3)]^3\}$ and $\mathbf{A}$ generates a contraction semi-group. Semi-group methods are outlined in [Bénilan and Crandall 1991; Kato 1988; Lunardi 2004; Pazy 1983; Racke 1992; Yagi 2010]. A brief overview of semi-group theory is also given in [Dautray and Lions 1992; Renardy and Rogers 1996; Zeidler 1990a,b]

Problem of Thermo-Visco-Elasticity The problem of thermo-visco-elasticity is covered in [Amassad et al. 2002; Bonetti and Bonfanti 2003, 2005, 2008; Bartels and Roubíček 2009; Eck et al. 2005; Figueiredo and Trabucho 1995; Gawinecki 2003; Yoshikawa et al. 2009].

Problem (Thermo-Visco-Elasticity)
Find the displacement field $\mathbf{u} : \overline{\Omega}_T \to \mathbb{R}^3$, *the stress field* $\boldsymbol{\sigma} : \overline{\Omega}_T \to \mathbb{R}^{3\times 3}_{sym}$ *and the temperature* $\theta : \overline{\Omega}_T \to \mathbb{R}$ *s.t.*

$$\rho_0 \frac{\partial^2 \mathbf{u}}{\partial t^2} - 2\operatorname{Div}(\mu\boldsymbol{\varepsilon}(\mathbf{u})) - \nabla(\lambda\operatorname{div}(\mathbf{u})) + \operatorname{Div}(h\boldsymbol{\varepsilon}(\mathbf{u}')) + 3\nabla(K_\alpha(\theta-\theta_0)) = \mathbf{f} \quad \text{in} \quad \Omega_T$$

$$\rho_0 c_e \frac{\partial\theta}{\partial t} - \operatorname{div}(\lambda_\theta\,\nabla\,\theta) + 3K_\alpha\theta\operatorname{div}\left(\frac{\partial\mathbf{u}}{\partial t}\right) = r \quad \text{in} \quad \Omega_T$$

including initial and boundary values (2.24), (3.22) *and* (2.29).

The thermo-visco-elastic problem without additional dissipation is investigated in [Bonetti and Bonfanti 2003; Eck et al. 2005; Roubíček 2005] by using the Galerkin approximation and a fixed-point argumentation. [Roubíček 2005] proves the existence of a weak solution by Schauder's fixed-point technique involving a mapping from $W^{2,2}(S; W^{1,2}(\Omega)) \cap W^{1,\infty}(S; W^{2,2}(\Omega))$. A-priori estimate can be obtained by differentiation of the equation w.r.t. time and by testing it with by the acceleration $\mathbf{u}''$. Moreover, Green' formula, embedding inequalities, Gronwall's and Korn's inequality are used in the proof.
An additional viscous dissipation in the heat equation, given by $\mu\boldsymbol{\varepsilon}(\mathbf{u}') : \boldsymbol{\varepsilon}(\mathbf{u}')^2$ is considered in [Bonetti and Bonfanti 2005, 2008] and leads to local (in time) solutions. Again, the used methods are the simultaneous Galerkin method and the usage of a (general) Gronwall-Bihari inequality, cf. Remark 3.3.1.

Problem of Elasto-Plasticity Mathematical problems of plasticity are discussed in [Amassad et al. 2001, 1999a; Bensoussan and Frehse 1996; Babadjian et al. 2011; Chelminski 1999, 2001; Demyanov 2009; Djoko et al. 2007a,b; Ebobisse and Reddy 2004; Fuchs and Seregin 2000; Griesse and Meyer 2008; Gröger 1978c; Han and Reddy 1999b; Hashiguchi 2005; Hill 1950; Herzog et al. 2011a; Hömberg and Khludnev 2006a; Han and Reddy 2000, 1999a; Johnson 1976, 1978; Löbach 2007a,b; Frehse and Löbach 2008; Löbach 2008, 2010; Neff 2003, 2005; Nečas and Trávníček 1980; Showalter and Shi 1997, 1998; Temam 1985, 1986; Washizu 1968]. Mathematical modelling of plastic material behaviour (cf. references in Section 3.4) leads to the description with the help of a plastic flow rule via a (parabolic) variational inequality or equivalent via a differential inclusion (cf. e.g. [Wolff et al. 2008c]). This variational inequality can be solved via abstract results with the help of the Yosida approximation, cf. e.g. [Barbu 1976; Brézis 1973; Hlavácek et al. 1988; Roubíček 2005; Showalter 1997].
Problems in visco-plasticity are considered in [Alber and Chelminski 2002; Amassad and Fabre 2002, 2004; Amassad and Sofonea 1998; Nesenenko 2009].

Moreover, there is a huge amount of literature related to elasto-plasticity, cf. [Anzellotti and Luckhaus 1987; Alber 1994, 1998; Bulíček et al. 2009; Bumb and Knees 2009; Brokate and Khludnev 1998; Brokate and Krejci 1998b,a; Brokate 1998; Dal Maso et al. 2006; Gröger et al. 1979; Gröger 1978a,b, 1980; Knees 2005, 2006, 2008, 2009; Krejci and Sprekels 2006; Khludnev and Sokolowski 1997; Krejci et al. 2009; Lang et al. 2006a,b; Miersemann 1980; Martins et al. 2007; Moreau 1976, 1977; Neff and Chelminski 2005, 2007; Nečas and Hlavácek 1981; Neff and Knees 2008; Reddy 1992; Sofonea et al. 2004]. In addition, there are several works on inelastic deformation theory, cf. [Chelminski and Gwiazda 2000a,b, 2007; Chelminski 2002, 2003a,b, 1997, 1998; Chelminski and Naniewicz 2002; Gruber et al. 2010; Kaminski 2008, 2009a].

Problems of elasto-visco-plasticity are treated in [Fučik and Kufner 1978; Han and Sofonea 2002; Kaminski 2009b,c; Krejci et al. 2001; Riviére et al. 2004] in some works in context with the method of vanishing viscosity, cf. [Dal Maso et al. 2007, 2006, 2008; Duszek 1980].
The works of [Gröger 1978c; Johnson 1976; Temam 1985] are purely mathematical and deal with perfectly plastic problems.

Problem (Elasto-Plasticity)
Find the displacement field $\mathbf{u} : \overline{\Omega}_T \to \mathbb{R}^3$ *and the stress field* $\boldsymbol{\sigma} : \overline{\Omega}_T \to \mathbb{R}^{3\times 3}_{sym}$ *s.t.*

$$\begin{aligned} \rho_0 \frac{\partial^2 \mathbf{u}}{\partial t^2} - 2\,\mathrm{Div}(\mu \boldsymbol{\varepsilon}(\mathbf{u})) - \mathrm{grad}(\lambda\, \mathrm{div}(\mathbf{u})) + 2\,\mathrm{Div}(\mu \boldsymbol{\varepsilon}_{cp}) = \mathbf{f} \qquad & \textit{in} \quad \Omega_T \\ \boldsymbol{\varepsilon}'_{cp} = \Lambda \boldsymbol{\sigma}^*, \quad \Lambda \geq 0 \textit{ for } F = 0 \textit{ and } \Lambda = 0 \textit{ for } F < 0 \qquad & \textit{in} \quad \Omega_T \\ \mathbf{u}(0) = \mathbf{u}_0,\ \mathbf{u}'(0) = \mathbf{u}_1,\ \boldsymbol{\varepsilon}_{cp}(0) = \mathbf{0} \qquad & \textit{in} \quad \Omega \end{aligned}$$

including boundary conditions (2.28).

The problem can be rewritten as

$$\begin{aligned} (\mathbf{u}''(t), \mathbf{v}) + (\mathrm{Div}(\boldsymbol{\sigma}(t)), \mathbf{v}) = (\mathbf{f}(t), \mathbf{v}) \qquad & \forall\, \mathbf{v} \in \mathbf{V}_{\mathrm{u}} \\ \frac{\mathrm{d}}{\mathrm{d}t}\Big(\frac{1}{2\mu}\boldsymbol{\sigma}^*(t)\Big) : (\tau - \boldsymbol{\sigma}^*(t)) \geq (\boldsymbol{\varepsilon}^*(\mathbf{u}'(t)) - \boldsymbol{\varepsilon}'_{trip}(t)) : (\boldsymbol{\tau} - \boldsymbol{\sigma}^*(t)) \qquad & \forall\, \boldsymbol{\tau} \in \mathbf{K}. \end{aligned}$$

The idea of visco-plastic regularisation is used in [Anzellotti and Luckhaus 1987; Duvaut and Lions 1976].

In [Amassad et al. 2001] the problem of finding $y : \bar{S} \to H$ s.t.

$$Ay'(t) + \partial\Psi_{K\cap\Sigma(t)}(y(t)) \ni f(t) \text{ a.e. } t \in S, \qquad y(0) = y_0 \in H$$

where H is a real Hilbert space, $A : H \to H$ is a positive definite symmetric operator, K, Σ_0 are closed convex sets in H, $\chi : \bar{S} \to H$ is given, $\chi \in W^{1,\infty}(S; H)$, $\Sigma(t) = \Sigma_0 + \chi(t)$, $y_0 \in H$, $\Psi_{K\cap\Sigma(t)} : H \to]-\infty, +\infty]$ is the indicator function, $y_0 = \chi(0) \in \Sigma(0)$ and $f \in L^2(S; H)$ is discussed. Moreover, the existence of a unique solution $y \in W^{1,2}(S; H)$ has been proved.

The ansatz in [Alber and Chelminski 2002; Chelminski 2003a; Bensoussan and Frehse 1996; Kaminski 2008, 2009a] is to rewrite the problem as a system of parabolic equations, like (4.1) or in the case of elasto-plasticity as a system of parabolic inclusions and apply the theory of monotone operators in addition with fixed-point arguments.

Another possibility in order to solve the problem would be the usage of a fixed-point scheme like $T : \bar{\mathbf{u}} \in \mathcal{V}_{\mathrm{u}} \mapsto \boldsymbol{\sigma}^* \in \mathcal{H}_\sigma \mapsto \mathbf{u} \in \mathcal{V}_{\mathrm{u}}$.

Problem of Thermo-Elasto-Plasticity For the problem of thermo-elasto-plasticity there exists literature in a smaller scale, cf. [Duvaut and Lions 1976] for thermo-elasto-plasticity, [Bartczak 2011; Bartels and Roubíček 2008; Merouani and Messelmi 2010] for thermo-visco-plasticity and [Chelminski and Racke 2006; Gröger and Hünlich 1980] for mathematical problems in thermo-plasticity.

Problem (Thermo-Elasto-Plasticity)
Find the displacement field $\mathbf{u} : \overline{\Omega}_T \to \mathbb{R}^3$*, the stress field* $\boldsymbol{\sigma} : \overline{\Omega}_T \to \mathbb{R}^{3\times 3}_{sym}$ *and the temperature* $\theta : \overline{\Omega}_T \to \mathbb{R}$ *s.t.*

$$\rho_0 \frac{\partial^2 \mathbf{u}}{\partial t^2} - 2\,\mathrm{Div}(\mu \boldsymbol{\varepsilon}(\mathbf{u})) - \nabla(\lambda\,\mathrm{div}(\mathbf{u})) + 3\nabla(K_\alpha(\theta-\theta_0)) + 2\,\mathrm{Div}(\mu \boldsymbol{\varepsilon}_{cp}) = \mathbf{f} \quad in \quad \Omega_T$$

$$\rho_0 c_e \frac{\partial \theta}{\partial t} - \mathrm{div}(\lambda_\theta \nabla \theta) + 3K_\alpha \theta\,\mathrm{div}\left(\frac{\partial \mathbf{u}}{\partial t}\right) = \boldsymbol{\sigma} : \boldsymbol{\varepsilon}'_{cp} + r \quad in \quad \Omega_T$$

$$\boldsymbol{\varepsilon}'_{cp} = \Lambda \boldsymbol{\sigma}^*, \quad \Lambda \geq 0 \; for \; F = 0 \; and \; \Lambda = 0 \; for \; F < 0 \quad in \quad \Omega_T$$

including initial and boundary values (2.24), $(2.25)_2$, (2.28) *and* (2.29).

In [Chelminski and Racke 2006] thermo-plasticity with the Prandtl-Reuss flow rule and with a linear evolution equation for the kinematic hardening is studied. The yield function associated with the system under consideration depends explicitly on the temperature. To have a control on the temperature, the heat equation is slightly modified and it is proved that an approximation process, based on the Yosida approximation, converges to a global in time solution of the (modified) system of thermo-plasticity.

Problem of Thermo-Elasticity with Phase Transitions and TRIP In connection with phase transformations there exists very few mathematical literature. Especially coupled models for the material behaviour of steel, which describe phase transformations in addition to the temperature and the deformation, have been insufficiently investigated in a strict mathematical and numerical context so far. There are results in this direction which

only take into account the temperature and the phase transitions, cf. e.g. [Chelminski et al. 2007; Fernandes et al. 1985; Hömberg 1995, 1997; Hömberg and Khludnev 2006b; Hüßler 2007; Mielke 2007; Panizzi 2010]. The analysis of phase transitions is treated in [Fasano and Primicerio 1996; Fasano et al. 2007], and [Colli et al. 2004, 2007] connects the setting of phase transitions with differential inclusions.
Within the framework of a diploma thesis in the field of industrial mathematics [Boettcher 2007], the mathematical problem of linear thermo-elasticity taking into account phase transitions and TRIP was investigated. Under suitable conditions, existence and uniqueness results for the weak solvability of the corresponding initial boundary value problem for the equations of linear elasticity as well as for the equations of classical linear thermo-elasticity were given.
More references for the problem of thermo-elasticity with phase transitions and TRIP are [Boettcher 2007; Chelminski et al. 2007; Hömberg and Kern 2009; Hömberg and Khludnev 2006b; Kern 2011; Mielke 2007; Mainik and Mielke 2005].

Problem (Thermo-Elasticity with Phase Transitions and TRIP)
Find the displacement field $\mathbf{u} : \overline{\Omega}_T \to \mathbb{R}^3$*, the stress field* $\boldsymbol{\sigma} : \overline{\Omega}_T \to \mathbb{R}^{3\times 3}_{sym}$*, the temperature* $\theta : \overline{\Omega}_T \to \mathbb{R}$ *and the phase fractions* $\mathbf{p} : \overline{\Omega}_T \to \mathbb{R}^m$ *s.t.*

$$
\begin{aligned}
\rho_0 \frac{\partial^2 \mathbf{u}}{\partial t^2} - 2\,\mathrm{Div}(\mu \boldsymbol{\varepsilon}(\mathbf{u})) - \mathrm{grad}(\lambda\,\mathrm{div}(\mathbf{u})) + 3\,\mathrm{grad}(K_\alpha(\theta - \theta_0)) & \\
+ \mathrm{grad}(K \sum_{i=1}^{N} (\frac{\rho_0}{\rho_i(\theta_0)} - 1) p_i) + 2\,\mathrm{Div}(\mu \boldsymbol{\varepsilon}_{trip}) = \mathbf{f} & \quad in \quad \Omega_T \\
\rho_0 c_e \frac{\partial \theta}{\partial t} - \mathrm{div}(\lambda_\theta \nabla \theta) + 3K_\alpha \theta\,\mathrm{div}\left(\frac{\partial \mathbf{u}}{\partial t}\right) = \boldsymbol{\sigma} : \boldsymbol{\varepsilon}'_{trip} + \rho_0 \sum_{i=2}^{m} L_i p'_i + r & \quad in \quad \Omega_T \\
\frac{\partial \mathbf{p}}{\partial t} = \boldsymbol{\gamma}(\mathbf{p}, \theta) & \quad in \quad \Omega_T \\
\boldsymbol{\varepsilon}'_{trip} = \frac{3}{2}(\boldsymbol{\sigma}^* - \mathbf{X}_{trip}) \sum_{i=1}^{m} \kappa_i \frac{\partial \Phi_i}{\partial p_i}(p_i) \max\{p'_i, 0\} & \quad in \quad \Omega_T
\end{aligned}
$$

including initial and boundary values (2.24) – (2.29).

Problem of Thermo-Elasto-Plasticity with Phase Transitions and TRIP Because of the lack of literature, the task arose to integrate the complex physical behaviour of steel materials (especially the phase transformation, TRIP and classical plasticity) in more general models of thermo-elasticity.

Problem (Thermo-Elasto-Plasticity with Phase Transitions and TRIP)
Find the displacement field $\mathbf{u} : \overline{\Omega}_T \to \mathbb{R}^3$*, the stress field* $\boldsymbol{\sigma} : \overline{\Omega}_T \to \mathbb{R}^{3\times 3}_{sym}$*, the temperature* $\theta : \overline{\Omega}_T \to \mathbb{R}$ *and the phase fractions* $\mathbf{p} : \overline{\Omega}_T \to \mathbb{R}^m$ *s.t.*

$$\rho_0 \frac{\partial^2 \mathbf{u}}{\partial t^2} - 2\,\mathrm{Div}(\mu \boldsymbol{\varepsilon}(\mathbf{u})) - \mathrm{grad}(\lambda\,\mathrm{div}(\mathbf{u})) + 3\,\mathrm{grad}(K_\alpha(\theta - \theta_0))$$
$$+\,\mathrm{grad}(K \sum_{i=1}^{N} (\frac{\rho_0}{\rho_i(\theta_0)} - 1)p_i) + 2\,\mathrm{Div}(\mu \boldsymbol{\varepsilon}_{trip}) + 2\,\mathrm{Div}(\mu \boldsymbol{\varepsilon}_{trip}) = \mathbf{f} \quad in \quad \Omega_T$$
$$\rho_0 c_e \frac{\partial \theta}{\partial t} - \mathrm{div}(\lambda_\theta \nabla \theta) + 3K_\alpha \theta\,\mathrm{div}\left(\frac{\partial \mathbf{u}}{\partial t}\right) = \boldsymbol{\sigma} : (\boldsymbol{\varepsilon}'_{trip} + \boldsymbol{\varepsilon}'_{cp}) + \rho_0 \sum_{i=2}^{m} L_i p'_i + r \quad in \quad \Omega_T$$
$$\frac{\partial \mathbf{p}}{\partial t} = \boldsymbol{\gamma}(\mathbf{p}, \theta) \quad in \quad \Omega_T$$
$$\boldsymbol{\varepsilon}'_{cp} = \Lambda \boldsymbol{\sigma}^*, \quad \Lambda \geq 0 \text{ for } F = 0 \text{ and } \Lambda = 0 \text{ for } F < 0 \quad in \quad \Omega_T$$
$$\boldsymbol{\varepsilon}'_{trip} = \frac{3}{2} \boldsymbol{\sigma}^* \sum_{i=1}^{m} \kappa_i \frac{\partial \Phi_i}{\partial p_i}(p_i) \max\{p'_i, 0\} \quad in \quad \Omega_T$$

including initial and boundary values (2.24) – (2.29).

The quasi-static situation without phase transitions and TRIP is considered in [Bartczak 2011; Chelminski and Racke 2006]. The problem is rewritten as

$$\frac{1}{\rho}\,\mathrm{Div}(D(\boldsymbol{\varepsilon} - B\mathbf{z})) - c \nabla \theta = \mathbf{f}$$
$$\theta' = \kappa \Delta \theta - \gamma\,\mathrm{div}(\mathbf{u}')$$
$$\boldsymbol{\varepsilon}'_{cp} \in \partial \chi_{K(\theta)}(T - \alpha \boldsymbol{\varepsilon}_{cp})$$

in Ω_T and then, the theory of monotone operators is applied, cf. Appendix C.

The situation without mechanical dissipation is treated in [Suhr 2010]. The emphasis of this work is the numerical simulation including hardening. The idea for an analytical investigation in order to provide an existence and uniqueness result would be the following: Because there is no dissipation, the equations for $\mathbf{u}$ and (θ, p) are decoupled. We can prescribe a fixed temperature and use a fixed-point scheme (either Banach's or Schauder's fixed point theorem).

$$T_1 : \mathcal{H}_\theta \to \mathcal{W}_\theta \cap \mathfrak{H}_\theta \hookrightarrow \mathcal{H}_\theta, \qquad \bar{\theta} \mapsto \mathbf{p} \mapsto \theta,$$

$$T_2 : \mathfrak{V}_u \times \mathfrak{H}_\sigma \to \mathfrak{V}_u \times \mathfrak{H}_\sigma, \qquad (\bar{\mathbf{u}}, \bar{\boldsymbol{\varepsilon}}_{cp}) \mapsto \boldsymbol{\varepsilon}_{trip} \mapsto \boldsymbol{\sigma}^* \mapsto \boldsymbol{\varepsilon}_{cp} \mapsto \mathbf{u}.$$

The situation without temperature, phase transitions and TRIP is investigated in [Alber and Chelminski 2002; Kaminski 2008, 2009a]: Let Ω be an open, bounded domain with Lipschitz boundary and $g : \mathbb{R}^N \to \mathcal{P}(\mathbb{R}^N)$ maximal monotone with $0 \in g(0)$. Consider

$$\begin{aligned}
\mathbf{u}' &= \mathbf{v}, & \mathbf{u}(\mathbf{x},0) &= \mathbf{u}_0(\mathbf{x}),\\
\mathbf{v}' &= \frac{1}{\rho}\operatorname{Div}(D(\boldsymbol{\varepsilon} - B\mathbf{z})) + \frac{1}{\rho}\mathbf{f}, & \mathbf{v}(\mathbf{x},0) &= \mathbf{v}_0(\mathbf{x}),\\
\boldsymbol{\varepsilon}' &= \frac{1}{2}(\nabla \mathbf{v} + \nabla \mathbf{v}^T), & \boldsymbol{\varepsilon}(\mathbf{x},0) &= \boldsymbol{\varepsilon}_0(\mathbf{x}),\\
\mathbf{z}' &\in g(-\rho \nabla_z \Psi(\boldsymbol{\varepsilon}, \mathbf{z})), & \mathbf{z}(\mathbf{x},0) &= \mathbf{z}_0(\mathbf{x})
\end{aligned}$$

including boundary conditions $\mathbf{v}|_{\Gamma_1} = 0$ and $\rho \nabla_{\boldsymbol{\varepsilon}} \Psi(\boldsymbol{\varepsilon}, \mathbf{z})\mathbf{u}|_{\Gamma_2} = 0$.
The idea to look for a solution $(\mathbf{v}, \boldsymbol{\varepsilon}, \mathbf{z}) : \Omega_T \to \mathbf{W} := \mathbb{R}^3 \times \mathbb{R}^{3\times 3}_{sym} \times \mathbb{R}^N$ of this problem is to introduce an operator $A : L^2(\Omega, \mathbf{W}) \to \mathcal{P}(L^2(\Omega, \mathbf{W}))$, $D(A) := \{(\mathbf{v}, \boldsymbol{\varepsilon}, \mathbf{z}) \in \mathbf{V}_u \times \mathbf{H}_\sigma \times L^2(\Omega, \mathbb{R}^N) : A((\mathbf{v}, \boldsymbol{\varepsilon}, \mathbf{z})) \neq \emptyset\}$ and show that A is a maximal monotone operator regarding the scalar product

$$\langle (\mathbf{v}, \boldsymbol{\varepsilon}, \mathbf{z}), (\bar{\mathbf{v}}, \bar{\boldsymbol{\varepsilon}}, \bar{\mathbf{z}}) \rangle := \int_\Omega \rho \mathbf{v} \cdot \bar{\mathbf{v}} + (D(\boldsymbol{\varepsilon} - B\mathbf{z})) \cdot (\bar{\boldsymbol{\varepsilon}} - B\bar{\mathbf{z}}) + (L\mathbf{z}) \cdot \bar{\mathbf{z}} \, \mathrm{d}\mathbf{x}$$

in $L^2(\Omega, \mathbf{W})$ and that $\rho\Psi(\boldsymbol{\varepsilon}, \mathbf{z}) = \frac{1}{2}(D\boldsymbol{\varepsilon} - B\mathbf{z}) \cdot (\boldsymbol{\varepsilon} - B\mathbf{z}) + \frac{1}{2}L\mathbf{z} \cdot \mathbf{z}$ is quadratic and positive definite on $\mathbb{R}^{3\times 3}_{sym} \times \mathbb{R}^N$. Applying the general theory of monotone operators (the standard literature for such problems are e.g. [Barbu 1976; Brézis 1971; Showalter 1997; Zeidler 1985]) yields the existence of a solution.

In order to adapt this approach to our setting, we set $\mathbf{z} := (\boldsymbol{\varepsilon}_{trip}, \boldsymbol{\varepsilon}_{cp})$ (or possibly in case of hardening $\mathbf{z} := (\boldsymbol{\varepsilon}_{trip}, \boldsymbol{\varepsilon}_{cp}, \mathbf{X}_{trip}, \mathbf{X}_{cp})$) and

$$\begin{aligned}
\Psi = \frac{1}{\rho}(\mu \boldsymbol{\varepsilon}^*_{te} : \boldsymbol{\varepsilon}^*_{te} + \frac{K}{2}(\operatorname{tr}(\boldsymbol{\varepsilon}_{te}))^2 - 3K_\alpha(\theta - \theta_0)\operatorname{tr}(\boldsymbol{\varepsilon}_{te}) - K\sum(\frac{\rho_0}{\rho_i(\theta_0)} - 1)p_i \operatorname{tr}(\boldsymbol{\varepsilon}_{te}))\\
+ \frac{1}{\rho_0}\sum_i p_i + \frac{1}{2\rho_0} c_{cp}\boldsymbol{\varepsilon}_{cp} : \boldsymbol{\varepsilon}_{cp} + 2c_{int}\boldsymbol{\varepsilon}_{cp} : \boldsymbol{\varepsilon}_{trip} + c_{trip}\boldsymbol{\varepsilon}_{trip} : \boldsymbol{\varepsilon}_{trip}
\end{aligned}$$

and rewrite our system as

$$\begin{aligned}
\mathbf{u}' &= \mathbf{v}\\
\mathbf{v}' &= \frac{1}{\rho}\operatorname{Div}(\boldsymbol{\sigma}) + \frac{1}{\rho}\mathbf{f}
\end{aligned}$$

$$
\begin{aligned}
\boldsymbol{\varepsilon}' &= \frac{1}{2}(\nabla \mathbf{v} + \nabla \mathbf{v}^T) \\
\mathbf{p}' &= \boldsymbol{\gamma}(\mathbf{p}, \theta) \\
\boldsymbol{\varepsilon}'_{trip} &= b(\theta, \mathbf{p}, \mathbf{p}')\boldsymbol{\sigma}^* \\
\theta' &= \frac{\lambda_\theta}{c_e \rho}\Delta\theta + \frac{3K_\alpha}{c_e \rho}\operatorname{div}(\mathbf{v}) + \frac{\rho_0}{c_e \rho}\sum L_i p'_i + \frac{1}{c_e \rho} r \\
(\boldsymbol{\sigma}^*)' &\in \partial\chi_{K(\theta)} + \boldsymbol{\varepsilon}^*(\mathbf{v}) - b(\theta, \mathbf{p}, \mathbf{p}')\boldsymbol{\sigma}^*
\end{aligned}
$$

in Ω_T including initial and boundary conditions.
Unfortunately, the treatment of the fully coupled problem as a differential inclusion does not work. The phase transitions do not fit into this scheme, because the integrability conditions are not fulfilled, i.e. there is no representation as a potential of a conservative vector field for the evolution equations of the phase fractions.

Chapter 5

Analysis of Problem $(\mathbf{P}_A)$

In this chapter we investigate Problem $(\mathbf{P}_A)$ introduced in Subsection 3.3.1. We call this modification of the fully coupled problem 'Regularisation via Averaging' or 'Steklov Regularisation' because we replace the time derivatives of $\mathbf{u}$ in the heat equation and in the variational inequality for $\boldsymbol{\varepsilon}_{cp}$ (resp. $\boldsymbol{\sigma}^*$) by its difference quotient. Moreover, we replace θ in the stress tensor by its Steklov average.

In order to prove existence and uniqueness of a weak solution of the Problem $(\mathbf{P}_A)$ we use the analysed subproblems summarised in Tables 4.2 and 4.4. Subproblem means that we look at each evolution equation or rather inequality separately and treat the dependent variables as data. The subproblems will be extended sequentially by the other variables until the fully coupled problem is considered.
Moreover, we assume in this chapter $h > 0$ sufficiently small. Although the solution of the regularised problem depends on the regularisation parameter h, we omit this detail in the notation (unless otherwise expressly noted). The passage to the limit of the comprehensive model to the original one is not obtained a-priori.

The weak formulation of Problem $(\mathbf{P}_A)$ is given in Section 5.1. Section 5.2 collects the existence and uniqueness results for two different settings. First, the situation with thermal and without intrinsic dissipation is considered. Furthermore, we look at the situation with thermal and intrinsic dissipation. Sections 5.3 and 5.4 provide the associated proofs. In Sections 5.5 and 5.6 a regularity result for the first situation is proven. Section 5.7 provides further aspects and remarks.

5.1 Weak Formulation of the Problem

Now, we bring the problem into a weak formulation. This is not only useful for further mathematical investigations and numerical simulations, but also for dimensional analysis

(cf. Section 2.8). The weak formulation combines in a natural way the differential equations with the initial and boundary conditions (cf. Subsection 2.1.7) and reduces the formal assumptions on the smoothness of the involved functions.
The weak formulation is obtained by multiplying the PDEs with suitable test functions:

5.1.1 Definition (Weak Formulation of Problem ($\mathbf{P}_A$))
Under the Assumption 4.3.1 a quintuple $(\mathbf{u}, \theta, \mathbf{p}, \boldsymbol{\varepsilon}_{cp}, \boldsymbol{\varepsilon}_{trip}) \in \mathcal{U}_\mathrm{u} \times \mathcal{U}_\theta \times \mathfrak{X}_\mathrm{p}^\infty \times \mathfrak{H}_\sigma \times \mathfrak{H}_\sigma$ *is called a weak solution of the Problem* ($\mathbf{P}_A$), *if*

$$(5.1)\quad \left\langle \rho_0 \frac{\partial^2}{\partial t^2}\mathbf{u}(t), \mathbf{v}\right\rangle_{\mathrm{V}_\mathrm{u}^*\mathrm{V}_\mathrm{u}} + 2\int_\Omega \mu\, \boldsymbol{\varepsilon}(\mathbf{u}(t)) : \boldsymbol{\varepsilon}(\mathbf{v})\,\mathrm{d}\mathbf{x} + \int_\Omega \lambda\, \mathrm{div}(\mathbf{u}(t))\,\mathrm{div}(\mathbf{v})\,\mathrm{d}\mathbf{x}$$
$$= 3\int_\Omega K_\alpha\,(\mathcal{S}_h\theta(t) - \theta_0)\,\mathrm{div}(\mathbf{v})\,\mathrm{d}\mathbf{x} + \int_\Omega K \sum_{i=1}^m \left(\frac{\rho_0}{\rho_i(\theta_0)} - 1\right) p_i(t)\,\mathrm{div}(\mathbf{v})\,\mathrm{d}\mathbf{x}$$
$$+ 2\int_\Omega \mu\, \boldsymbol{\varepsilon}_{trip}(t) : \nabla\mathbf{v}\,\mathrm{d}\mathbf{x} + 2\int_\Omega \mu\, \boldsymbol{\varepsilon}_{cp}(t) : \nabla\mathbf{v}\,\mathrm{d}\mathbf{x} + \int_\Omega \mathbf{f}(t)\,\mathbf{v}\,\mathrm{d}\mathbf{x}$$

f.a. $\mathbf{v} \in \mathbf{V}_\mathrm{u}$, *f.a.a.* $t \in S$ *and* $\mathbf{u}(\mathbf{x}, 0) = \mathbf{u}_0(\mathbf{x})$ *a.e.*, $\mathbf{u}'(\mathbf{x}, 0) = \mathbf{u}_1(\mathbf{x})$ *a.e.*,

$$(5.2)\quad \left\langle \rho_0\, c_e \frac{\partial}{\partial t}\theta(t), \vartheta\right\rangle_{V_\theta^* V_\theta} + \int_\Omega \lambda_\theta\, \nabla\theta(t)\,\nabla\vartheta\,\mathrm{d}\mathbf{x} + \int_{\partial\Omega} \delta\,\theta(t)\,\vartheta\,\mathrm{d}\sigma_x$$
$$+ 3\int_\Omega K_\alpha\,\theta_0\,\mathrm{div}\left(\frac{\partial \mathcal{S}_h\mathbf{u}(t)}{\partial t}\right)\vartheta\,\mathrm{d}\mathbf{x} = \int_\Omega \boldsymbol{\sigma}(t) : \boldsymbol{\varepsilon}'_{trip}(t)\,\vartheta\,\mathrm{d}\mathbf{x} + \int_\Omega \boldsymbol{\sigma}(t) : \boldsymbol{\varepsilon}'_{cp}(t)\,\vartheta\,\mathrm{d}\mathbf{x}$$
$$+ \int_\Omega \rho_0 \sum_{i=2}^m L_i p'_i(t)\,\vartheta\,\mathrm{d}\mathbf{x} + \int_\Omega r(t)\,\vartheta\,\mathrm{d}\mathbf{x} + \int_{\partial\Omega} \delta\,\theta_\Gamma(t)\,\vartheta\,\mathrm{d}\sigma_x$$

f.a. $\vartheta \in V_\theta$, *f.a.a.* $t \in S$ *and* $\theta(\mathbf{x}, 0) = \theta_0(\mathbf{x})$ *a.e.*,

$$(5.3)\qquad \frac{\partial \mathbf{p}}{\partial t}(\mathbf{x}, t) = \boldsymbol{\gamma}(\mathbf{p}(\mathbf{x}, t), \theta(\mathbf{x}, t), \theta'(\mathbf{x}, t), \mathrm{tr}(\boldsymbol{\sigma}(\mathbf{x}, t)), \boldsymbol{\sigma}^*(\mathbf{x}, t) : \boldsymbol{\sigma}^*(\mathbf{x}, t))$$

f.a.a. $(\mathbf{x}, t) \in \Omega_T$ *and* $\mathbf{p}(\mathbf{x}, 0) = \mathbf{p}_0(\mathbf{x})$ *a.e.*,

$$(5.4)\qquad \boldsymbol{\varepsilon}'_{trip}(\mathbf{x}, t) = \frac{3}{2}\boldsymbol{\sigma}^*(\mathbf{x}, t) \sum_{i=1}^m \kappa_i \frac{\partial \Phi_i}{\partial p_i}(p_i(\mathbf{x}, t))\max\{p'_i(\mathbf{x}, t), 0\}$$

$$(5.5)\qquad \boldsymbol{\varepsilon}_{cp}(\mathbf{x}, t) = \boldsymbol{\varepsilon}^*(\mathcal{S}_h\mathbf{u}(\mathbf{x}, t)) - \boldsymbol{\varepsilon}_{trip}(\mathbf{x}, t) - \frac{1}{2\mu}\boldsymbol{\sigma}^*(\mathbf{x}, t)$$

f.a.a. $(\mathbf{x}, t) \in \Omega_T$ *and* $\boldsymbol{\varepsilon}_{trip}(\mathbf{x}, 0) = \mathbf{0}$ *a.e.*, $\boldsymbol{\varepsilon}_{cp}(\mathbf{x}, 0) = \mathbf{0}$ *a.e.*,

$$(5.6)\qquad (\boldsymbol{\sigma}^*)'(t) + \partial\chi_\mathrm{K}(\boldsymbol{\sigma}^*(t)) \ni 2\mu\left(\boldsymbol{\varepsilon}^*(\mathcal{S}_h\mathbf{u}'(t)) - \boldsymbol{\varepsilon}'_{trip}(t)\right)$$

f.a.a. $t \in S$, $\boldsymbol{\sigma}^*(0) = \boldsymbol{\sigma}_0^* \in \mathbf{K}$, $\mathbf{K} := \{\boldsymbol{\tau} \in \mathbf{H}_\sigma : \boldsymbol{\tau}(\mathbf{x}) \in \mathbf{K}_F$ *f.a.a.* $\mathbf{x} \in \Omega\}$,
$\mathbf{K}_F := \left\{\boldsymbol{\tau} \in \mathbb{R}^{3\times 3}_{sym}, \mathrm{tr}(\boldsymbol{\tau}) = 0 : F(\boldsymbol{\tau}) \le 0\right\}$, $F(\boldsymbol{\sigma}) := \sqrt{\tfrac{3}{2}\boldsymbol{\sigma}^* : \boldsymbol{\sigma}^*} - (R_0 + R)$.

5.2 Existence and Uniqueness Results

In this section, we summarise the existence and uniqueness results for two different settings of Problem ($\mathbf{P}_A$). First, the situation with thermal and without intrinsic dissipation is considered. Furthermore, we look at the situation with thermal and intrinsic dissipation.

5.2.1 Situation with Thermal and without Intrinsic Dissipation

5.2.1 Theorem (Existence and Uniqueness for Problem ($\mathbf{P}_A$))
Let Assumptions 4.3.1 be valid and assume in addition that the intrinsic dissipation vanishes, i.e. $\boldsymbol{\sigma} : (\boldsymbol{\varepsilon}'_{trip} + \boldsymbol{\varepsilon}'_{cp}) = 0$. *Then the Problem* ($\mathbf{P}_A$) *possesses a unique weak solution* $(\mathbf{u}, \theta, \mathbf{p}, \boldsymbol{\varepsilon}_{cp}, \boldsymbol{\varepsilon}_{trip}) \in \mathcal{V}_\mathrm{u} \cap \mathfrak{H}_\mathrm{u} \times \mathfrak{H}_\theta \times \mathfrak{H}_\mathrm{p} \times \mathfrak{H}_\sigma \times \mathfrak{H}_\sigma$.

5.2.2 Remark
We conclude this section with some remarks:

(1) *We only consider the case of constant* $\mathbf{K}$ *in the proof of Theorem 5.2.1, but the proof in the situation of time- or parameter-dependent* $\mathbf{K}$ *works similarly and in the case of a parameter-dependent* $\mathbf{K}$*, e.g.* $\mathbf{K}$ *depending on* θ *or* $\boldsymbol{\varepsilon}_{cp}$*, one can integrate this parameter in the fixed-point argumentation.*

(2) *We also obtain (without going into details at this point) the continuous dependence of the solution on the parameters using suitable assumptions, i.e. it holds*

$$\begin{aligned}\|\theta_1-\theta_2\|_{\mathcal{V}_\theta^\infty\cap\mathfrak{H}_\theta}+\|\mathbf{u}_1-\mathbf{u}_2\|_{\mathcal{V}_\mathrm{u}^\infty\cap\mathfrak{H}_\mathrm{u}^\infty}+\|\mathbf{p}_1-\mathbf{p}_2\|_{\mathfrak{H}_\mathrm{p}}+\|\boldsymbol{\sigma}_1-\boldsymbol{\sigma}_2\|_{\mathcal{H}_\sigma}+\|\boldsymbol{\varepsilon}_{trip,1}-\boldsymbol{\varepsilon}_{trip,2}\|_{\mathfrak{H}_\sigma}\\ +\|\boldsymbol{\varepsilon}_{cp,1}-\boldsymbol{\varepsilon}_{cp,2}\|_{\mathfrak{H}_\sigma}\le c\Big\{\|\mathbf{u}_{0,1}-\mathbf{u}_{0,2}\|_{\mathrm{V_u}}+\|\mathbf{u}_{1,1}-\mathbf{u}_{1,2}\|_{\mathrm{H_u}}+\|\mathbf{f}_1-\mathbf{f}_2\|_{\mathcal{H}_\mathrm{u}}\\ +\|\theta_{0,1}-\theta_{0,2}\|_{H_\theta}+\|\theta_{\Gamma,1}-\theta_{\Gamma,2}\|_{V_\theta}+\|r_1-r_2\|_{H_\theta}+\|\mathbf{p}_{0,1}-\mathbf{p}_{0,2}\|_{\mathrm{H_p}}\Big\}.\end{aligned}$$

(3) *Note that the embedding* $\mathcal{V}_\theta\cap\mathfrak{H}_\theta \hookrightarrow \mathbf{X}$ *is not compact. Therefore we have chosen a technique of proving the existence that does not need any compactness arguments. The proof of the Problem* ($\mathbf{P}_{FP3}$) *does not work with the Schauder fixed-point argument in this setting, because of this lack of compactness.*

(4) *The application of the existence and uniqueness theorems in [Akagi and Ôtani 2004, 2005] for the differential inclusion that determines* $\boldsymbol{\sigma}^*$ *(or rather* $\boldsymbol{\varepsilon}_{cp}$*) are not leading*

to the desired results. Let $\bar{\mathbf{u}} \in \mathcal{V}_{\mathrm{u}} \cap \mathfrak{H}_{\mathrm{u}}$. *This leads to* $\boldsymbol{\sigma} \in \mathcal{V}_\sigma \cap W^{1,2}(S;\mathbf{V}_\sigma^*) \cap \mathcal{X}_\sigma^\infty$. *Applying the results in [Akagi and Ôtani 2004, 2005] gives* $\boldsymbol{\varepsilon}_{cp} \in \mathcal{H}_\sigma \cap W^{1,2}(S;\mathbf{V}_\sigma^*))$ *and the following estimate holds* f.a.a. $t \in S$*:*

$$\|\mathbf{u}'(t)\|_{\mathrm{H_u}} + \|\mathbf{u}(t)\|_{\mathrm{V_u}} \leq c\Big\{ \int_0^t \big(\mathbf{f}(s), \mathbf{u}'(s)\big)_{\mathrm{H_u}} \mathrm{d}s + \int_0^t \big(\theta(s), \nabla \mathbf{u}'(s)\big)_{\mathrm{H_u}} \mathrm{d}s$$
$$+ \int_0^t \big(\mathbf{p}(s), \nabla \mathbf{u}'(s)\big)_{\mathrm{H_u}} \mathrm{d}s + \int_0^t \big(\boldsymbol{\varepsilon}_{cp}(s), \nabla \mathbf{u}'(s)\big)_{\mathrm{H}_\sigma} \mathrm{d}s + \int_0^t \big(\boldsymbol{\sigma}^*(s), \nabla \mathbf{u}'(s)\big)_{\mathrm{H}_\sigma} \mathrm{d}s \Big\}.$$

Because of the lack of regularity in the last two integral terms, it is not clear how to treat these terms in order to continue this approach. A possibility could be the regularisation of $\boldsymbol{\varepsilon}_{cp}$, *cf. Remark 5.6.12.*

(5) *A-posteriori estimates provide no further information since these are not independent of the regularisation parameter.*

5.2.2 Situation with Thermal and Intrinsic Dissipation

5.2.3 Theorem (Existence for Problem ($\mathbf{P}_A$))
Let Assumptions 4.3.1 be valid and assume in addition that the evolution equations for the phase transitions do not dependent on the time derivative of temperature, but on the temperature itself. Then the Problem ($\mathbf{P}_A$) *possesses at least one weak solution* $(\mathbf{u}, \theta, \mathbf{p}, \boldsymbol{\varepsilon}_{cp}, \boldsymbol{\varepsilon}_{trip}) \in \mathcal{V}_{\mathrm{u}} \cap \mathfrak{H}_{\mathrm{u}} \times \mathfrak{H}_\theta \times \mathfrak{H}_{\mathrm{p}} \times \mathfrak{H}_\sigma \times \mathfrak{H}_\sigma$.

5.2.4 Remark (Uniqueness for Problem ($\mathbf{P}_A$))
In addition to the assumptions of Theorem 5.2.3 we assume that either the intrinsic dissipation vanishes, i.e. $\boldsymbol{\sigma} : (\boldsymbol{\varepsilon}'_{trip} + \boldsymbol{\varepsilon}'_{cp}) = 0$ *(cf. the setting in Subsection 5.2.1) or that at least one solution possesses better regularity than obtained in the proof of Theorem 5.2.3, e.g.* $\boldsymbol{\sigma} \in \mathbf{X}_\sigma^\infty$ *and* $\boldsymbol{\varepsilon}_{trip}, \boldsymbol{\varepsilon}_{cp} \in \mathcal{X}_\sigma^\infty$. *Then the weak solution* $(\mathbf{u}, \theta, \mathbf{p}, \boldsymbol{\varepsilon}_{cp}, \boldsymbol{\varepsilon}_{trip}) \in \mathcal{V}_{\mathrm{u}} \cap \mathfrak{H}_{\mathrm{u}} \times \mathfrak{H}_\theta \times \mathfrak{H}_{\mathrm{p}} \times \mathfrak{H}_\sigma \times \mathfrak{H}_\sigma$ *of Problem* ($\mathbf{P}_A$) *corresponding to Theorem 5.2.3 is unique.*

5.3 Proof of Theorem 5.2.1

This section provides the proof of Theorem 5.2.1. Subsection 5.3.1 gives the outline of the proof, whereas the proof itself is divided into Subsections 5.3.2 – 5.3.9.

5.3.1 Outline of the Proof

The main idea of the proof is based on classical arguments of functional analysis concerning variational problems and fixed-point arguments, cf. Section 4.1. The proof will be done in several steps (cf. Figure 5.1 for a schematic representation of the proof). In order to prove the unique existence of a weak solution $(\mathbf{u}, \boldsymbol{\sigma}, \boldsymbol{\varepsilon}_{trip}, \boldsymbol{\varepsilon}_{cp}, \theta, \mathbf{p})$ of Problem $(\mathbf{P}_A)$, we apply the following strategy.

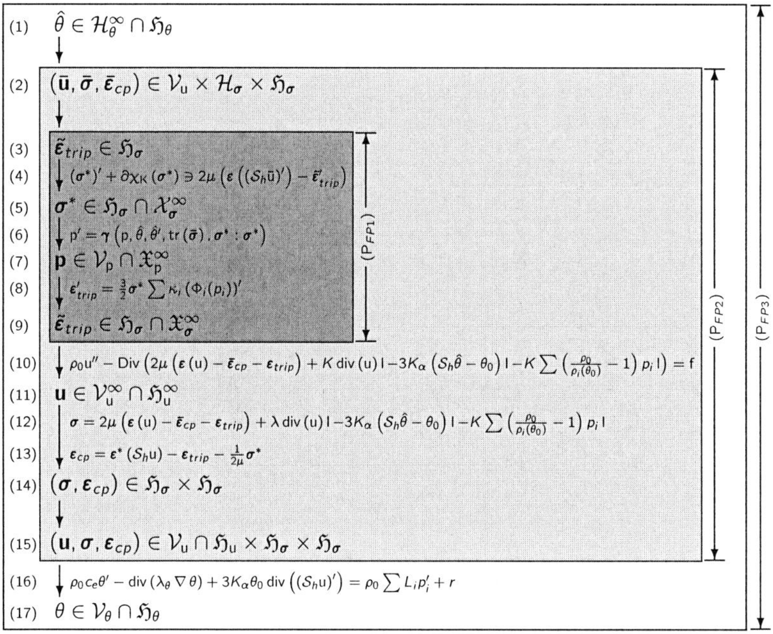

Figure 5.1: Scheme of the proof of existence and uniqueness for the Steklov regularised problem using the Banach fixed-point theorem for Problems $(\mathbf{P}_{FPi})$, $i = 1, 2, 3$.

We use the subproblems summarised in Tables 4.2 and 4.4. 'Subproblem' means that we look at each single equation (resp. inequality) separately and treat the dependent variables as 'data', i.e.

- For fixed $(\mathbf{u}, \boldsymbol{\varepsilon}_{trip}) \in \mathcal{V}_u \times \mathfrak{H}_\sigma$ we prove the unique existence of a solution $\boldsymbol{\varepsilon}_{cp} \in \mathfrak{H}_\sigma \cap \mathcal{X}_\sigma^\infty$ of the Problem $(\mathbf{P}_{A,\boldsymbol{\varepsilon}_{cp}})$. We apply the results for differential inclusions given in Appendix C. This subproblem is called 'Subproblem of Classical Plasticity' and discussed in Subsection 5.3.2.
- For fixed $\theta \in \mathcal{H}_\theta$ (and additional internal variables if necessary) we show the unique existence of a solution $\mathbf{p} \in \mathfrak{V}_p \cap \mathfrak{X}_p^\infty$ of the Problem $(\mathbf{P}_{A,p})$. The idea of the proof is based on Banach's fixed-point theorem and presented in detail in [Hüßler 2007]. This subproblem is called 'Subproblem of Phase Transitions' and discussed in Subsection 5.3.3.
- For fixed $(\mathbf{u}, \boldsymbol{\sigma}, \boldsymbol{\varepsilon}_{cp}) \in \mathcal{V}_u \times \mathcal{H}_\sigma \times \mathfrak{H}_\sigma$ we prove the unique existence of a solution $\boldsymbol{\varepsilon}_{trip} \in \mathfrak{H}_\sigma \cap \mathcal{X}_\sigma^\infty$ of the Problem $(\mathbf{P}_{A,\boldsymbol{\varepsilon}_{trip}})$. This subproblem is called 'Subproblem of TRIP' and discussed in Subsection 5.3.4.
- For fixed $(\theta, \mathbf{p}, \boldsymbol{\varepsilon}_{trip}, \boldsymbol{\varepsilon}_{cp}) \in \mathfrak{V}_\theta \times \mathfrak{X}_p^\infty \times \mathfrak{H}_\sigma \times \mathfrak{H}_\sigma$ we show the unique existence of a solution $\mathbf{u} \in \mathcal{V}_u^\infty \cap \mathfrak{H}_u$ of the Problem $(\mathbf{P}_{A,u})$ using the Galerkin method. This subproblem is called 'Subproblem of the Displacement Equation of Linearised Theory of Elasticity' and discussed in Subsection 5.3.6. Note that for given $(\theta, \mathbf{p}, \boldsymbol{\varepsilon}_{trip}, \boldsymbol{\varepsilon}_{cp}) \in \mathfrak{V}_\theta \times \mathfrak{X}_p^\infty \times \mathfrak{H}_\sigma \times \mathfrak{H}_\sigma$ and $\mathbf{u} \in \mathcal{V}_u^\infty \cap \mathfrak{H}_u$ the stress $\boldsymbol{\sigma} \in \mathcal{V}_\sigma$ is known. In order to prove the statement for the quasi-static problem and some regularity results we apply known results for elliptic systems, cited in Appendix B.
- For fixed $(\mathbf{u}, \boldsymbol{\sigma}, \boldsymbol{\varepsilon}_{trip}, \boldsymbol{\varepsilon}_{cp}, \mathbf{p}) \in \mathcal{V}_u \times \mathcal{V}_\sigma \times \mathfrak{H}_\sigma \cap \mathcal{X}_\sigma^\infty \times \mathfrak{H}_\sigma \cap \mathcal{X}_\sigma^\infty \times \mathfrak{X}_p^\infty$ we show the unique existence of a solution $\theta \in \mathcal{V}_\theta \cap \mathfrak{H}_\theta$ of the problem $(\mathbf{P}_{A,\theta})$ via the Galerkin method. This subproblem is called 'Subproblem of the Heat Equation' and discussed in Subsection 5.3.8.

The subproblems will be extended sequentially by the other variables until the fully coupled problem is considered. The further procedure is as follows:

- We begin with the subproblem $(\boldsymbol{\sigma}^*, \boldsymbol{\varepsilon}_{trip}, \mathbf{p})$ treating θ, $\mathbf{u}$, $\boldsymbol{\sigma}$ and $\boldsymbol{\varepsilon}_{cp}$ as data (cf. Problem $(\mathbf{P}_{A,FP1})$ in Subsection 5.3.5),
- then consider the subsystem $\mathbf{u}$, $\boldsymbol{\sigma}$ and $\boldsymbol{\varepsilon}_{cp}$ with data θ (cf. Problem $(\mathbf{P}_{A,FP2})$ in Subsection 5.3.7).
- finally obtain existence and uniqueness for the complete system based on the foregoing results (cf. Problem $(\mathbf{P}_{A,FP3})$ in Subsection 5.3.9).

In the sequel, we introduce the following three fixed-point steps (i.e. Problems $(\mathbf{P}_{A,FPi})$, $i = 1, 2, 3$) for Problem $(\mathbf{P}_A)$. The modifications in each different setting in Section 5.2 are highlighted in gray colour and explained in detail in Sections 5.3 and 5.4 respectively.

In Problem $(\mathbf{P}_{A,FP1})$ we

- start with the differential inclusion for $\boldsymbol{\sigma}^*$ treating $\mathbf{u}$ and $\boldsymbol{\varepsilon}_{trip}$ as given data (cf. Problem $(\mathbf{P}_{A,\boldsymbol{\varepsilon}_{cp}})$),
- deal then with the evolution equations for the phase fractions $\mathbf{p}$ with data θ, $\boldsymbol{\sigma}$ and $\boldsymbol{\sigma}^*$ (cf. Problem $(\mathbf{P}_{A,\mathrm{p}})$),
- finally obtain unique existence for the complete problem (cf. Problem $(\mathbf{P}_{A,\boldsymbol{\varepsilon}_{trip}})$).

Problem $(\mathbf{P}_{A,FP1})$
Let $(\hat{\theta}, \bar{\mathbf{u}}, \bar{\boldsymbol{\sigma}}, \bar{\boldsymbol{\varepsilon}}_{cp})$ *be given and find* $(\boldsymbol{\sigma}^*, \boldsymbol{\varepsilon}_{trip}, \mathbf{p})$ *s.t.*

$$\frac{\partial \boldsymbol{\sigma}^*}{\partial t} + \partial \chi_K(\boldsymbol{\sigma}^*) \ni 2\mu \left(\varepsilon(\mathcal{S}_h \bar{\mathbf{u}}') - \boldsymbol{\varepsilon}'_{trip}\right) \quad \text{in} \quad \Omega_T$$

$$\frac{\partial \mathbf{p}}{\partial t} = \boldsymbol{\gamma}(\mathbf{p}, \hat{\theta}, \hat{\theta}', \mathrm{tr}(\bar{\boldsymbol{\sigma}}), \boldsymbol{\sigma}^* : \boldsymbol{\sigma}^*) \quad \text{in} \quad \Omega_T$$

$$\boldsymbol{\varepsilon}'_{trip} = 3\mu \left(\boldsymbol{\varepsilon}^*(\bar{\mathbf{u}}) - \boldsymbol{\varepsilon}_{cp} - \boldsymbol{\varepsilon}_{trip}\right) \sum_{i=1}^{m} \kappa_i \frac{\partial \Phi_i}{\partial p_i}(p_i) \max\{p'_i, 0\} \quad \text{in} \quad \Omega_T$$

including conditions (2.4), (2.10), (2.11), (2.15) *and initial values* $(2.24)_1$, $(2.24)_2$, $(2.25)_1$, $(2.25)_2$, (2.26).

In Problem $(\mathbf{P}_{A,FP2})$

- we start with the Problem $(\mathbf{P}_{A,FP1})$,
- then consider the system of equations for $\mathbf{u}$ and $\boldsymbol{\sigma}$ with data θ, $\mathbf{p}$, $\boldsymbol{\varepsilon}_{trip}$, $\boldsymbol{\varepsilon}_{cp}$ and $\boldsymbol{\sigma}^*$ (cf. Problem $(\mathbf{P}_{A,\mathrm{u}})$),
- then consider the equation for $\boldsymbol{\varepsilon}_{cp}$ with data $\mathbf{u}$ and $\boldsymbol{\sigma}$
- and finally achieve unique existence for the complete problem.

Problem $(\mathbf{P}_{A,FP2})$
Let $\hat{\theta}$ *be given, let* $(\boldsymbol{\sigma}^*, \boldsymbol{\varepsilon}_{trip}, \mathbf{p})$ *be the solution of Problem* $(\mathbf{P}_{FP1})$ *and find* $(\mathbf{u}, \boldsymbol{\sigma}, \boldsymbol{\varepsilon}_{cp})$ *s.t.*

$$\rho_0 \frac{\partial^2 \mathbf{u}}{\partial t^2} - \mathrm{Div}(\boldsymbol{\sigma}) = \mathbf{f} \quad \text{in} \quad \Omega_T$$

$$\boldsymbol{\sigma} = 2\mu\boldsymbol{\varepsilon}_{te} + \lambda\,\mathrm{div}(\mathbf{u})\,\mathbf{I} - 3K_\alpha\left(\mathcal{S}_h\hat{\theta} - \theta_0\right)\mathbf{I} - K\sum_{i=1}^{m}\left(\frac{\rho_0}{\rho_i(\theta_0)} - 1\right)p_i\,\mathbf{I} \quad \textit{in} \quad \Omega_T$$

$$\boldsymbol{\varepsilon}_{cp} = \boldsymbol{\varepsilon}^*(\mathcal{S}_h\mathbf{u}) - \boldsymbol{\varepsilon}_{trip} - \frac{1}{2\mu}\boldsymbol{\sigma}^* \quad \textit{in} \quad \Omega_T$$

including conditions (2.4), (2.10), (2.11) *as well as initial and boundary values* $(2.24)_1$, $(2.24)_2$, $(2.25)_1$, $(2.25)_2$, (2.28).

The derivatives $(\boldsymbol{\sigma}', \boldsymbol{\varepsilon}'_{te}, \boldsymbol{\varepsilon}'_{cp})$ are given by

$$\boldsymbol{\sigma}' = 2\mu\boldsymbol{\varepsilon}'_{te} + \lambda\,\mathrm{div}(\mathbf{u}')\,\mathbf{I} - 3K_\alpha\mathcal{S}_h\hat{\theta}'\,\mathbf{I} - K\sum_{i=1}^{m}\left(\frac{\rho_0}{\rho_i(\theta_0)} - 1\right)p_i'\,\mathbf{I} \quad \text{in} \quad \Omega_T,$$

$$\boldsymbol{\varepsilon}'_{te} = \boldsymbol{\varepsilon}(\mathcal{S}_h\mathbf{u}') - \boldsymbol{\varepsilon}'_{trip} - \boldsymbol{\varepsilon}'_{cp} \quad \text{in} \quad \Omega_T,$$

$$\boldsymbol{\varepsilon}'_{cp} = \boldsymbol{\varepsilon}^*(\mathcal{S}_h\mathbf{u}') - \boldsymbol{\varepsilon}'_{trip} - \frac{1}{2\mu}(\boldsymbol{\sigma}^*)' \quad \text{in} \quad \Omega_T.$$

The fully coupled problem is solved in the following way:

- we start with the Problem ($\mathbf{P}_{A,FP1}$),
- then consider the Problem ($\mathbf{P}_{A,FP2}$),
- and finally achieve unique existence for the complete problem (cf. Problem ($\mathbf{P}_{A,\theta}$)).

Problem ($\mathbf{P}_{A,FP3}$)
Let $(\boldsymbol{\sigma}^*, \boldsymbol{\varepsilon}_{trip}, \mathbf{p})$ *be given by Problem* ($\mathbf{P}_{FP1}$), *let* $(\mathbf{u}, \boldsymbol{\sigma}, \boldsymbol{\varepsilon}_{cp})$ *given by Problem* ($\mathbf{P}_{FP2}$) *and find* θ *s.t.*

$$\rho_0 c_e\frac{\partial\theta}{\partial t} - \mathrm{div}(\lambda_\theta\nabla\theta) + 3K_\alpha\theta_0\,\mathrm{div}\left(\mathcal{S}_h\mathbf{u}'\right) = \boldsymbol{\sigma} : \left(\boldsymbol{\varepsilon}'_{trip} + \boldsymbol{\varepsilon}'_{cp}\right) + \rho_0\sum_{i=2}^{m}L_i p_i' + r \quad \textit{in} \quad \Omega_T$$

including initial and boundary values $(2.24)_3$, (2.26), (2.29).

Actually, the Problem ($\mathbf{P}_{A,FP3}$) coincides with the fully coupled problem, i.e. solving Problem ($\mathbf{P}_{A,FP3}$) completes the proof of the fully coupled problem.

In a second step we define the following three fixed-point operators and the corresponding solution operators:

$\mathbf{T}_{A,FP1} : \tilde{\boldsymbol{\varepsilon}}_{trip} \mapsto \boldsymbol{\sigma}^* \mapsto \mathbf{p} \mapsto \boldsymbol{\varepsilon}_{trip}$ for given $\hat{\theta}$, $\bar{\mathbf{u}}$, $\bar{\boldsymbol{\sigma}}$, $\bar{\boldsymbol{\varepsilon}}_{cp}$
$\mathbf{T}_{A,FP2} : (\bar{\mathbf{u}}, \bar{\boldsymbol{\sigma}}, \bar{\boldsymbol{\varepsilon}}_{cp}) \mapsto \mathbf{S}_{FP1}(\hat{\theta}, \bar{\mathbf{u}}, \bar{\boldsymbol{\sigma}}, \bar{\boldsymbol{\varepsilon}}_{cp}) \mapsto \mathbf{u} \mapsto (\boldsymbol{\sigma}, \boldsymbol{\varepsilon}_{cp}) \mapsto (\mathbf{u}, \boldsymbol{\sigma}, \boldsymbol{\varepsilon}_{cp})$ for given $\hat{\theta}$
$\mathbf{T}_{A,FP3} : \hat{\theta} \mapsto \mathbf{S}_{FP2}(\hat{\theta}) \mapsto \theta$ for given data

$\mathbf{S}_{A,FP1} : (\hat{\theta}, \bar{\mathbf{u}}, \bar{\boldsymbol{\sigma}}, \bar{\boldsymbol{\varepsilon}}_{cp}) \mapsto (\boldsymbol{\sigma}^*, \mathbf{p}, \boldsymbol{\varepsilon}_{trip})$
$\mathbf{S}_{A,FP2} : \hat{\theta} \mapsto (\boldsymbol{\sigma}^*, \mathbf{p}, \boldsymbol{\varepsilon}_{trip}, \mathbf{u}, \boldsymbol{\sigma}, \boldsymbol{\varepsilon}_{cp})$
$\mathbf{S}_{A,FP3} : \text{data} \mapsto (\boldsymbol{\sigma}^*, \mathbf{p}, \boldsymbol{\varepsilon}_{trip}, \mathbf{u}, \boldsymbol{\sigma}, \boldsymbol{\varepsilon}_{cp}, \theta)$

In order to prove the unique existence of a weak solution $(\mathbf{u}, \boldsymbol{\sigma}, \boldsymbol{\varepsilon}_{trip}, \boldsymbol{\varepsilon}_{cp}, \theta, \mathbf{p})$ of Problem $(\mathbf{P}_A)$, it is sufficient to show that each operator $\mathbf{T}_{A,FPi}, i = 1, 2, 3$ has a (unique) fixed-point. In order to show that the operators $\mathbf{T}_{A,FP1}$ and $\mathbf{T}_{A,FP2}$ have a fixed-point, we apply Banach's fixed-point theorem. In order to prove this property for the operator $\mathbf{T}_{A,FP3}$ we have the choice – either using Banach's fixed-point theorem or using Schauder's fixed point theorem. In case of Schauder's fixed-point theorem we have to prove additionally the uniqueness, because the theorem only gives a existence statement. Finally we get existence (and uniqueness) for the complete system based on the foregoing lemmas.

We remark, that various different fixed-point settings seem to be possible or reasonable, but we have not found any benefit in using them.

5.3.2 Step 1: Analysis of Problem $(\mathbf{P}_{A,\boldsymbol{\varepsilon}_{cp}})$ for constant K

We start with the solvability of Problem $(\mathbf{P}_{A,\boldsymbol{\varepsilon}_{cp}})$ in case of constant $\mathbf{K}$ for $(\mathbf{u}, \boldsymbol{\sigma}, \boldsymbol{\varepsilon}_{trip}) \in \mathcal{V}_u \times \mathcal{H}_\sigma \times \mathfrak{H}_\sigma$ considered as data. We define $F : \mathbb{R}^{3\times3} \to \mathbb{R}$ via

$$F(\boldsymbol{\sigma}) := \sqrt{\frac{3}{2}\boldsymbol{\sigma}^* : \boldsymbol{\sigma}^*} - (R_0 + R)$$

for given constant $R_0, R \in \mathbb{R}^+$. The set of all admissible $\boldsymbol{\sigma}$ is convex (cf. [Han and Reddy 1999b]). Moreover, we define

$$\mathbf{K}_F := \Big\{\boldsymbol{\tau} \in \mathbb{R}^{3\times 3}_{\text{sym}}, \text{tr}(\boldsymbol{\tau}) = 0 : F(\boldsymbol{\tau}) \leq 0\Big\}, \quad \mathbf{K} := \Big\{\boldsymbol{\tau} \in \mathbf{H}_{\boldsymbol{\sigma}} : \boldsymbol{\tau}(\mathbf{x}) \in \mathbf{K}_F \text{ f.a.a. } \mathbf{x} \in \Omega\Big\}.$$

Consider the following problem (cf. Subsection 3.2.3):

Problem ($\mathbf{P}_{A,\boldsymbol{\varepsilon}_{cp}}$)
Find the stress deviator $\boldsymbol{\sigma}^* : \overline{\Omega}_T \to \mathbb{R}^{3\times 3}_{sym}$, s.t.

$$\Big(\boldsymbol{\sigma}^*(t)\Big)' + \partial\chi_K(\boldsymbol{\sigma}^*(t)) \ni 2\mu\Big(\boldsymbol{\varepsilon}^*(\mathcal{S}_h\mathbf{u}'(t)) - \boldsymbol{\varepsilon}'_{trip}(t)\Big) \text{ f.a.a. } t \in S \tag{5.7}$$

$$\boldsymbol{\sigma}^*(0) = \boldsymbol{\sigma}^*_0 := 2\mu\boldsymbol{\varepsilon}^*(\mathbf{u}_0). \tag{5.8}$$

5.3.1 Lemma
Let Assumption 4.3.1$_{(A3)}$ be valid. Then **K** *is a nonempty closed convex subset of* $\mathbf{H}_{\boldsymbol{\sigma}}$.

Proof. The proof contains three steps:

(1) Since $R_0 > 0$, $R \geq 0$ one concludes that $\mathbf{0} \in \mathbf{K}_F$. Hence, $\mathbf{K}$ is nonempty.

(2) $\mathbf{K}$ is convex. Let $\boldsymbol{\sigma}, \boldsymbol{\tau} \in \mathbf{K}$ and $\lambda \in [0,1]$. Obviously, $\text{tr}(\lambda\boldsymbol{\sigma} + (1-\lambda)\boldsymbol{\tau}) = 0$ for $\text{tr}(\boldsymbol{\sigma}) = \text{tr}(\boldsymbol{\tau}) = 0$ and $\lambda\boldsymbol{\sigma} + (1-\lambda)\boldsymbol{\tau} = (\lambda\boldsymbol{\sigma} + (1-\lambda)\boldsymbol{\tau})^T$ for $\boldsymbol{\sigma} = \boldsymbol{\sigma}^T$ and $\boldsymbol{\tau} = \boldsymbol{\tau}^T$. Furthermore, with the help of the trace condition, one gets

$$\begin{aligned}
F(\lambda\boldsymbol{\sigma} + (1-\lambda)\boldsymbol{\tau}) &\leq \sqrt{\frac{3}{2}(\lambda\boldsymbol{\sigma} + (1-\lambda)\boldsymbol{\tau})^* : (\lambda\boldsymbol{\sigma} + (1-\lambda)\boldsymbol{\tau})^*} - (R_0 + R) \\
&\leq \sqrt{\frac{3}{2}(\lambda\boldsymbol{\sigma}^* + (1-\lambda)\boldsymbol{\tau}^*) : (\lambda\boldsymbol{\sigma}^* + (1-\lambda)\boldsymbol{\tau}^*)} - (R_0 + R) \\
&\leq \lambda\sqrt{\frac{3}{2}\boldsymbol{\sigma}^* : \boldsymbol{\sigma}^*} + (1-\lambda)\sqrt{\frac{3}{2}\boldsymbol{\tau}^* : \boldsymbol{\tau}^*} - (R_0 + R) \\
&\leq \lambda F(\boldsymbol{\sigma}) + (1-\lambda)F(\boldsymbol{\tau}) \leq 0.
\end{aligned}$$

(3) We show that $\mathbf{K}$ is closed in $\mathbf{H}_{\boldsymbol{\sigma}}$. Let $\boldsymbol{\sigma}_n \to \boldsymbol{\sigma}$ in $\mathbf{H}_{\boldsymbol{\sigma}}$. Then, $\boldsymbol{\sigma}^{ij}_n \to \boldsymbol{\sigma}^{ij}$ in $L^2(\Omega)$ f.a. $i, j = \{1,2,3\} =: M$. Define a bijective map $g : N \to M \times M$, $N := \{1,\ldots,9\}$ s.t. $\hat{\boldsymbol{\sigma}}^k := \boldsymbol{\sigma}^{g(k)}$. Hence, $\hat{\boldsymbol{\sigma}}^k_n \to \hat{\boldsymbol{\sigma}}^k$ in $L^2(\Omega)$ f.a. $k \in N$. The theorem of Fischer-Riesz (cf. [Werner 2005]) yields the existence of a subsequence $(n_l)_{l\in\mathbb{N}}$ s.t. $\hat{\boldsymbol{\sigma}}^1_{n_l}(\mathbf{x}) \to \hat{\boldsymbol{\sigma}}^1(\mathbf{x})$ and $\hat{\boldsymbol{\sigma}}^k_{n_l}(\mathbf{x}) \to \hat{\boldsymbol{\sigma}}^k(\mathbf{x})$ f.a.a. $\mathbf{x} \in \Omega$ and f.a. $k \in N \setminus \{1\}$. With the help of this

argumentation (denote all subsequences with $(n)_{n\in\mathbb{N}}$) it follows that $\hat{\boldsymbol{\sigma}}_n^k(\mathbf{x}) \to \hat{\boldsymbol{\sigma}}^k(\mathbf{x})$ f.a. $k \in N$ f.a.a. $\mathbf{x} \in \Omega$. Hence, $\boldsymbol{\sigma}_n^{(ij)}(\mathbf{x}) \to \boldsymbol{\sigma}^{(ij)}(\mathbf{x})$ f.a.a. $\mathbf{x} \in \Omega$ f.a. $i, j \in M$.
Let $\boldsymbol{\sigma}_n^{(ij)}(x) \in \mathbf{K}$. Obviously, $\boldsymbol{\sigma}^{(ij)}(\mathbf{x}) = \boldsymbol{\sigma}^{(ji)}(\mathbf{x})$ and $\mathrm{tr}(\boldsymbol{\sigma}(\mathbf{x})) = 0$ f.a.a. $\mathbf{x} \in \Omega$.
Moreover, $\boldsymbol{\sigma}_n^{*(ij)}(\mathbf{x}) \to \boldsymbol{\sigma}^{*(ij)}(\mathbf{x})$. Because of the continuity of F

$$|F(\boldsymbol{\sigma}_n) - F(\boldsymbol{\sigma})| \leq \sqrt{\frac{2}{3}}\left|\sqrt{\boldsymbol{\sigma}_n^* : \boldsymbol{\sigma}_n^*} - \sqrt{\boldsymbol{\sigma}^* : \boldsymbol{\sigma}^*}\right| \leq \frac{1}{2(R_0 + R)}|\boldsymbol{\sigma}_n^* : \boldsymbol{\sigma}_n^* - \boldsymbol{\sigma}^* : \boldsymbol{\sigma}^*|$$

it follows $F(\boldsymbol{\sigma}) \leq 0$ and therefore $\boldsymbol{\sigma} \in \mathbf{K}$.

□

5.3.2 Lemma

Let Assumption 4.3.1$_{(A3)}$ be valid. Then

(1) *the indicator function χ_K is proper, convex and lower semi-continuous.*

(2) *$\partial\chi_K : \mathbf{H}_\sigma \to \mathcal{P}(\mathbf{H}_\sigma)$ is maximal monotone.*

(3) *$D(\chi_K) = D(\partial\chi_K) = \mathbf{K}$.*

Proof. A detailed proof can be found in [Růžička 2004], we sketch the ideas:

(1) Obviously, $\chi_K \not\equiv \infty$ because $\mathbf{K}$ is nonempty. Since $\mathbf{K}$ is convex, it holds f.a. $\mathbf{u}, \mathbf{v} \in \mathbf{K}$, $\lambda \in [0, 1]$:

$$\chi_K((1-\lambda)\mathbf{u} + \lambda\mathbf{v}) = 0 = (1-\lambda)\chi_K(\mathbf{u}) + \lambda\chi_K(\mathbf{v}).$$

If $\mathbf{u} \neq \mathbf{K}$ and/or $\mathbf{v} \in \mathbf{K}$, then

$$\chi_K((1-\lambda)\mathbf{u} + \lambda\mathbf{v}) \leq \infty = (1-\lambda)\chi_K(\mathbf{u}) + \lambda\chi_K(\mathbf{v}).$$

χ_K is lower semi-continuous, because the preimage $\chi_K^{-1}(]-\infty, r[)$ is closed f.a. $r \in \mathbb{R}$:

$$\chi_K^{-1}(]-\infty, r[) = \{u | \chi_K \leq r\} = \begin{cases} \emptyset, & \text{if } r > 0 \\ \mathbf{K}, & \text{if } r \geq 0 \end{cases}$$

(2) It is comparatively easy to show that $\partial\chi_K : \mathbf{K} \to \mathcal{P}(\mathbf{H}_\sigma)$ is maximal monotone (cf. [Růžička 2004]). For $(\mathbf{u}, \mathbf{u}^*), (\mathbf{v}, \mathbf{v}^*) \in G(\partial\chi_K)$, it holds f.a. $\mathbf{w} \in \mathbf{K}$:

$$\langle \mathbf{u}^*, \mathbf{u} - \mathbf{w}\rangle \geq 0 \qquad \text{and} \qquad \langle \mathbf{v}^*, \mathbf{v} - \mathbf{w}\rangle \geq 0.$$

Hence, $\partial\chi_K$ is monotone. Let $(\mathbf{u}, \mathbf{u}^*) \in \mathbf{K} \times \mathbf{H}_\sigma$ and $\langle \mathbf{u}^* - \mathbf{v}^*, \mathbf{u} - \mathbf{v}\rangle \geq 0$ f.a. $(\mathbf{v}, \mathbf{v}^*) \in G(\partial\chi_K)$. Then it follows $\langle \mathbf{u}^*, \mathbf{u} - \mathbf{v}\rangle \geq 0$, i.e. $\mathbf{u}^* \in \partial\chi_K(\mathbf{u})$ because of $0 \in \partial\chi_K(\mathbf{v})$.
In order to prove that $\partial\chi_K : \mathbf{H}_\sigma \to \mathcal{P}(\mathbf{H}_\sigma)$ is maximal monotone, we apply the theorem of Rockafellar, cf. [Růžička 2004; Zeidler 1990b].

(3) The definition of the effective domain yields $D(\chi_K) = D(\partial\chi_K) = \mathbf{K}$.

□

5.3.3 Lemma (Existence and Uniqueness)
Let Assumption 4.3.1$_{(A3)}$ be valid. Assume $(\mathbf{u}, \boldsymbol{\varepsilon}_{trip}) \in \mathcal{V}_u \times \mathfrak{H}_\sigma$. Then there exists a unique solution $\boldsymbol{\sigma}^* \in C(\bar{S}; \mathbf{H}_\sigma) \cap \mathfrak{H}_\sigma$ of Problem $(\mathbf{P}_{A,\boldsymbol{\varepsilon}_{cp}})$ satisfying $\boldsymbol{\sigma}^*(t) \in \mathbf{K}$ f.a.a. $t \in S$ and

$$\|\boldsymbol{\sigma}^*(t) - \boldsymbol{\sigma}^*(0)\|_{\mathrm{H}_\sigma} \le c(h^{-1}) \left\{ \int_0^t \|\mathbf{u}(s)\|_{\mathrm{V}_u} \,\mathrm{d}s + \int_0^t \|\boldsymbol{\varepsilon}'_{trip}(s)\|_{\mathrm{H}_\sigma} \,\mathrm{d}s \right\} \text{ f.a.a. } t \in S,$$
$$\|(\boldsymbol{\sigma}^*)'(t)\|_{\mathrm{H}_\sigma} \le c(h^{-1}) \left\{ \|\mathbf{u}(t)\|_{\mathrm{V}_u} + \|\boldsymbol{\varepsilon}'_{trip}(t)\|_{\mathrm{H}_\sigma} \right\} \text{ f.a.a. } t \in S.$$

Proof. Using Lemma 5.3.1, Lemma 5.3.2 and applying Lemma C.2.1 proves the result.

□

5.3.4 Remark (Boundedness of the Solution)
Let Assumption 4.3.1$_{(A3)}$ be valid and assume in addition $\boldsymbol{\sigma}^*(t) \in \mathbf{K}$ f.a.a. $t \in S$. Then $F(\boldsymbol{\sigma}) \le 0$, which means

$$|\boldsymbol{\sigma}^*(t, \mathbf{x})|^2 \le \frac{3}{2}\Big(R_0 + R\Big)^2 < \infty$$

f.a.a. $t \in S$ and f.a. $\mathbf{x} \in \Omega$. Therefore, $\boldsymbol{\sigma}^* \in \mathcal{X}_\sigma^\infty$.

5.3.5 Remark
For $\mathbf{u}' \in \mathcal{H}_u$ and $\boldsymbol{\varepsilon}'_{trip} \in \mathcal{V}_\sigma^*$ there exists a solution $\boldsymbol{\sigma} \in \mathcal{H}_\sigma$ (cf. Remark C.2.3).

5.3.6 Lemma (Continuous Dependence on Parameters $\mathbf{u}$ and $\boldsymbol{\varepsilon}_{trip}$)
Let Assumption 4.3.1$_{(A3)}$ be valid. Then it follows for two different solutions $\boldsymbol{\sigma}_i^*$ of Problem $(\mathbf{P}_{A,\boldsymbol{\varepsilon}_{cp}})$ in the sense of Lemma 5.3.3 corresponding to the data $(\mathbf{u}_i, \boldsymbol{\varepsilon}_{trip,i})$, $i = 1, 2$

$$\|\boldsymbol{\sigma}_1^*(t) - \boldsymbol{\sigma}_2^*(t)\|_{\mathrm{H}_\sigma} \le \|\boldsymbol{\sigma}_1^*(0) - \boldsymbol{\sigma}_2^*(0)\|_{\mathrm{H}_\sigma}$$
$$+ c(h^{-1}) \Big\{ \int_0^t \|\mathbf{u}_1(s) - \mathbf{u}_2(s)\|_{\mathrm{V}_u} \,\mathrm{d}s + \int_0^t \|\boldsymbol{\varepsilon}'_{trip,1}(s) - \boldsymbol{\varepsilon}'_{trip,2}(s)\|_{\mathrm{H}_\sigma} \,\mathrm{d}s \Big\} \text{ f.a.a. } t \in S.$$

Proof. Applying Remark C.2.4 gives the result. □

5.3.3 Step 2: Analysis of Problem $(\mathbf{P}_{A,\mathbf{p}})$

We continue with the solvability of Problem $(\mathbf{P}_{A,\mathbf{p}})$ for given $\theta \in \mathcal{H}_\theta$ (as a fixed parameter or data). Consider the following problem:

Problem ($\mathbf{P}_{A,\mathrm{p}}$)
Find the phase fractions $\mathbf{p} : \overline{\Omega}_T \to \mathbb{R}^m$ s.t.

$$\frac{\partial p_i}{\partial t}(\mathbf{x}, t) = \gamma_i(\mathbf{p}(\mathbf{x}, t), \theta(\mathbf{x}, t)), \qquad (\mathbf{x}, t) \in \Omega_T \tag{5.9}$$

$$p_i(\mathbf{x}, 0) = p_{0i}(\mathbf{x}), \qquad \mathbf{x} \in \Omega \tag{5.10}$$

for given $i = 1, \ldots, m$.

5.3.7 Remark (Additional Variables)
The consideration of the following differential equation

$$\frac{\partial p_i}{\partial t}(\mathbf{x}, t) = \gamma_i(\mathbf{p}(\mathbf{x}, t), \theta(\mathbf{x}, t), \boldsymbol{\xi}(\mathbf{x}, t), \mathbf{x}, t), \qquad (\mathbf{x}, t) \in \Omega_T$$

with additional variables $\boldsymbol{\xi} \in L^2(S; [L^2(\Omega)]^M)$ and explicit dependence on $(\mathbf{x}, t) \in \Omega_T$ would be possible as well (cf. [Hüßler 2007]). For example, let $\boldsymbol{\xi} = (\theta', \boldsymbol{\sigma}_{vM}, \boldsymbol{\sigma}_m)$ (cf. Remark 2.1.2) with

$$\boldsymbol{\sigma}_{vM} := \sqrt{\frac{3}{2}|\boldsymbol{\sigma}^*|^2} \leq R_0 + R \quad \text{and}$$

$$\boldsymbol{\sigma}_m := \frac{1}{3}\operatorname{tr}(\boldsymbol{\sigma}) = \frac{1}{3}\boldsymbol{\sigma} : \mathbf{I}, \quad \|\boldsymbol{\sigma}_m\|_{\mathcal{H}_\sigma} \leq c\left\{\|\mathbf{u}\|_{\mathcal{V}_u} + \|\theta\|_{\mathcal{H}_\theta} + \|\mathbf{p}\|_{\mathcal{H}_p}\right\}.$$

We note that in the ODEs (5.9), *the spatial variable* $\mathbf{x}$ *is only a parameter.*

5.3.8 Lemma (Existence and Uniqueness)
Let Assumption 4.3.1$_{(A5),(A6)}$ be valid. Assume $\theta \in \mathcal{H}_\theta$. Then the initial value problem (5.9), (5.10) *has a unique solution $\mathbf{p} \in \mathfrak{H}_p \cap \mathcal{X}_p^\infty$ satisfying $\frac{\partial \mathbf{p}}{\partial t} \in \mathcal{X}_p^\infty$.*

We only give a short sketch of the proof. Details can be found in [Hüßler 2007].

Proof. Define the operator $\mathbf{A} : \mathcal{X}_p^\infty \to \mathcal{X}_p^\infty$ via

$$\mathbf{p}(\mathbf{x}, t) \mapsto (\mathbf{A}\mathbf{p})(\mathbf{x}, t) := \mathbf{p}_0(\mathbf{x}) + \int_0^t \boldsymbol{\gamma}(\theta(\mathbf{x}, s), \mathbf{p}(\mathbf{x}, s))\, ds$$

f.a.a. $(\mathbf{x}, t) \in \Omega_T$, for $\mathbf{p} \in \mathcal{X}_p^\infty$ and for fixed $\theta \in \mathcal{H}_\theta$. Applying the Banach fixed-point theorem gives the unique existence of a solution

$$\mathbf{p}(\mathbf{x}, t) := \mathbf{p}_0(\mathbf{x}) + \int_0^t \boldsymbol{\gamma}(\theta(\mathbf{x}, s), \mathbf{p}(\mathbf{x}, s))\, ds$$

f.a.a. $(\mathbf{x}, t) \in \Omega_T$, for $\mathbf{p} \in \mathcal{X}_{\mathrm{p}}^{\infty}$ and for fixed $\theta \in \mathcal{H}_\theta$. The majorant (boundedness) and the measurability condition complete the proof. □

5.3.9 Remark (Alternative Proof)
Existence and uniqueness of a solution of the considered problem can also be derived from the theorem of Carathéodory (cf. [Fasano and Primicerio 1996; Fasano et al. 2007] and [Panizzi 2010] for details). One rewrites the problem (5.9), (5.10) as

$$\mathbf{p}' = \tilde{\boldsymbol{\gamma}}(\mathbf{p}, t), \qquad t \in S, \qquad \mathbf{p}(0) = 0$$

with $\tilde{\boldsymbol{\gamma}}(\mathbf{p}, \cdot) = \boldsymbol{\gamma}(\mathbf{p}, \theta(\cdot))$. The hypothesis of the existence theorem of Carathéodory (cf. [Roubíček 2005; Zeidler 1990b]) follow from the definition of $\tilde{\boldsymbol{\gamma}}$ as a consequence of the measurability of θ and the Lipschitz continuity of $\boldsymbol{\gamma}$.

5.3.10 Lemma (Continuous Dependence on the Parameter θ)
Let Assumption 4.3.1$_{(A5),(A6)}$ be valid and let $\mathbf{p}_i$ be solutions from Lemma 5.3.8 corresponding to θ_i, $i = 1, 2$, then

$$\sup_{t \in \bar{S}} \|\mathbf{p}_1(t) - \mathbf{p}_2(t)\|^2_{\mathrm{H}_p} \le c\, \|\theta_1 - \theta_2\|^2_{\mathcal{H}_\theta} \qquad \text{and} \qquad \|\mathbf{p}_1' - \mathbf{p}_2'\|^2_{\mathcal{H}_p} \le c\, \|\theta_1 - \theta_2\|^2_{\mathcal{H}_\theta},$$

where the positive constant does not depend on the data.

Proof. Consider the difference

$$\mathbf{p}_1' - \mathbf{p}_2' = \boldsymbol{\gamma}(\mathbf{p}_1, \theta_1) - \boldsymbol{\gamma}(\mathbf{p}_2, \theta_2), \qquad \mathbf{p}_1(0) - \mathbf{p}_2(0) = 0.$$

Integrating, squaring and using the Lipschitz continuity of $\boldsymbol{\gamma}$ gives

$$\|\mathbf{p}_1(t) - \mathbf{p}_2(t)\|^2_{\mathrm{H}_p} \le c_1\, \|\theta_1 - \theta_2\|^2_{\mathcal{H}_\theta} + c_2\, \|\mathbf{p}_1 - \mathbf{p}_2\|^2_{\mathcal{H}_p}.$$

The Gronwall inequality completes the proof (cf. [Hüßler 2007]). □

5.3.11 Remark (Continuous Dependence on Additional Parameters)
In the situation of Remark 5.3.7 we get analogously

$$\sup_{t \in \bar{S}} \|\mathbf{p}_1(t) - \mathbf{p}_2(t)\|^2_{\mathrm{H}_p} \le c \left(\|\theta_1 - \theta_2\|^2_{\mathcal{H}_\theta} + \|\boldsymbol{\xi}_1 - \boldsymbol{\xi}_2\|^2_{L^2(S;[L^2(\Omega)]^M)} \right),$$

$$\|\mathbf{p}_1' - \mathbf{p}_2'\|^2_{\mathcal{H}_p} \le c \left(\|\theta_1 - \theta_2\|^2_{\mathcal{H}_\theta} + \|\boldsymbol{\xi}_1 - \boldsymbol{\xi}_2\|^2_{L^2(S;[L^2(\Omega)]^M)} \right).$$

5.3.12 Remark (Assumptions)
For weaker assumptions and for the case $\theta \in L^\infty(\Omega_T)$ we refer to the literature (cf. [Hüßler 2007]).

5.3.4 Step 3: Analysis of Problem $(\mathbf{P}_{A,\boldsymbol{\varepsilon}_{trip}})$

We continue with the solvability of Problem $(\mathbf{P}_{A,\boldsymbol{\varepsilon}_{trip}})$ for either given $\boldsymbol{\sigma}^* \in \mathcal{V}_\sigma$ or (when using the identity (3.7), cf. Subsection 3.2.2) $(\mathbf{u}, \boldsymbol{\varepsilon}_{cp}) \in \mathcal{V}_\mathrm{u} \times \mathcal{H}_\sigma$ and $\mathbf{p} \in \mathcal{X}_\mathrm{p}^\infty$ with $\mathbf{p}' \in \mathcal{X}_\mathrm{p}^\infty$. Consider the following problem:

Problem $(\mathbf{P}_{A,\boldsymbol{\varepsilon}_{trip}})$
Find the strain $\boldsymbol{\varepsilon}_{trip} : \overline{\Omega}_T \to \mathbb{R}^{3\times 3}_{sym}$ *s.t.*

$$\text{(5.11)} \quad \boldsymbol{\varepsilon}'_{trip}(\mathbf{x}, t) = \frac{3}{2}\boldsymbol{\sigma}^*(\mathbf{x}, t) \sum_{i=1}^{m} \kappa_i \frac{\partial \Phi_i}{\partial p_i}(p_i(\mathbf{x}, t)) \max\{p_i'(\mathbf{x}, t), 0\}, \quad (\mathbf{x}, t) \in \Omega_T$$

$$\text{(5.12)} \quad \boldsymbol{\varepsilon}_{trip}(\mathbf{x}, 0) = \mathbf{0}, \qquad \mathbf{x} \in \Omega$$

Applying the Carathéodory theory, we get

5.3.13 Lemma (Existence and Uniqueness)
Let Assumption 4.3.1$_{(A4)}$ *be valid. Assume either* $\boldsymbol{\sigma}^* \in \mathcal{H}_\sigma$ *or (v.s.)* $(\mathbf{u}, \boldsymbol{\varepsilon}_{cp}) \in \mathcal{V}_\mathrm{u} \times \mathcal{H}_\sigma$. *Then the initial value problem* (3.8), (3.9) *has a unique solution* $\boldsymbol{\varepsilon}_{trip} \in \mathfrak{H}_\sigma$ *satisfying*

$$\text{(5.13)} \quad \|\boldsymbol{\varepsilon}_{trip}\|_{\mathcal{H}_\sigma} \le c\, \|\boldsymbol{\sigma}^*\|_{\mathcal{H}_\sigma} \quad \text{or rather} \quad \|\boldsymbol{\varepsilon}_{trip}\|_{\mathcal{H}_\sigma} \le c \left\{ \|\mathbf{u}\|_{\mathcal{V}_\mathrm{u}} + \|\boldsymbol{\varepsilon}_{cp}\|_{\mathcal{H}_\sigma} \right\},$$

$$\text{(5.14)} \quad \|\boldsymbol{\varepsilon}'_{trip}\|_{\mathcal{H}_\sigma} \le c\, \|\boldsymbol{\sigma}^*\|_{\mathcal{H}_\sigma} \quad \text{or rather} \quad \|\boldsymbol{\varepsilon}'_{trip}\|_{\mathcal{H}_\sigma} \le c \left\{ \|\mathbf{u}\|_{\mathcal{V}_\mathrm{u}} + \|\boldsymbol{\varepsilon}_{cp}\|_{\mathcal{H}_\sigma} \right\}.$$

If in addition $\boldsymbol{\sigma}^* \in \mathcal{X}_\sigma^\infty$, *then* $\boldsymbol{\varepsilon}_{trip} \in \mathcal{X}_\sigma^\infty$ *with* $\boldsymbol{\varepsilon}'_{trip} \in \mathcal{X}_\sigma^\infty$ *satisfying*

$$|\boldsymbol{\varepsilon}_{trip}(\mathbf{x}, t)| \le c, \qquad |\boldsymbol{\varepsilon}'_{trip}(\mathbf{x}, t)| \le c$$

for almost all $\mathbf{x} \in \Omega$ *and f.a.* $t > 0$.

and the solution reads as (cf. [Amann 1995])

$$\text{(5.15)} \quad \boldsymbol{\varepsilon}_{trip}(t) = \int_0^t b(t) \exp\Big(-\int_s^t b(\tau)\,\mathrm{d}\tau\Big) \Big(\boldsymbol{\varepsilon}^*(\mathbf{u}(s)) - \boldsymbol{\varepsilon}_{cp}(s)\Big)\,\mathrm{d}s, \qquad t \in S.$$

Proof. Using the boundedness of $\frac{\partial \Phi_i}{\partial p_i}$, $i = 1, \dots, m$ and $\mathbf{p}$ or rather $\mathbf{p}'$ gives

$$\int_0^t \int_\Omega |\boldsymbol{\varepsilon}'_{trip}(\mathbf{x}, s)|\,\mathrm{d}\mathbf{x}\,\mathrm{d}s \le \frac{3}{2} \int_0^t \int_\Omega |\boldsymbol{\sigma}^* \sum_{i=1}^{m} \kappa_i \frac{\partial \Phi_i}{\partial p_i}(p_i) \max\{p_i', 0\}|\,\mathrm{d}\mathbf{x}\,\mathrm{d}s$$

$$\leq c_1 \int_0^t \int_\Omega |\sigma^*| \underbrace{|\frac{\partial \Phi}{\partial p_i}(p_i)|_\infty}_{\leq M_\Phi} \underbrace{|p'|_\infty}_{\leq c_2} \,\mathrm{d}\mathbf{x}\,\mathrm{d}s$$

and therefore $(5.14)_1$ holds. Using the relation (3.7) we get the estimate $(5.14)_2$. Integration (or using the solution (5.15)) yields the estimates (5.13). □

5.3.14 Remark (Additional Estimates of the TRIP strain)
Some (different) estimates of the TRIP strain are derived in [Wolff and Böhm 2006a], e.g.

$$\|\boldsymbol{\varepsilon}_{trip}\|_{\mathcal{V}_\sigma} \leq c\Big\{\|\mathbf{u}_0\|_{\mathrm{V_u}} + \|\mathbf{u}\|_{\mathfrak{V}_\mathrm{u}} + \|\boldsymbol{\varepsilon}_{cp}\|_{\mathfrak{H}_\sigma}\Big\}.$$

5.3.15 Lemma (Continuous Dependence on Parameters $\mathbf{p}$ and $\boldsymbol{\sigma}^*$)
Let Assumption 4.3.1$_{(A4)}$ be valid. Assume in addition $(\boldsymbol{\sigma}^*, \mathbf{p}) \in \mathcal{X}_\sigma^\infty \times \mathcal{X}_\mathrm{p}^\infty$. Then it follows for two different solutions $\boldsymbol{\varepsilon}_{trip,i}$ from Lemma 5.3.13 corresponding to data $(\boldsymbol{\sigma}_i^*, \mathbf{p}_i)$, $i = 1, 2$

$$\|\boldsymbol{\varepsilon}_{trip,1} - \boldsymbol{\varepsilon}_{trip,2}\|_{\mathcal{H}_\sigma^\infty} \leq c\Big\{\|\mathbf{p}_1 - \mathbf{p}_2\|_{\mathfrak{H}_\mathrm{p}} + \|\boldsymbol{\sigma}_1^* - \boldsymbol{\sigma}_2^*\|_{\mathcal{H}_\sigma}\Big\},$$
$$\|\boldsymbol{\varepsilon}'_{trip,1} - \boldsymbol{\varepsilon}'_{trip,2}\|_{\mathcal{H}_\sigma} \leq c\Big\{\|\mathbf{p}_1 - \mathbf{p}_2\|_{\mathfrak{H}_\mathrm{p}} + \|\boldsymbol{\sigma}_1^* - \boldsymbol{\sigma}_2^*\|_{\mathcal{H}_\sigma}\Big\},$$

where the positive constant does not depend on the data.

Proof. Consider the difference

$$\begin{aligned}
&\int_0^t \int_\Omega |\boldsymbol{\varepsilon}'_{trip,1}(\mathbf{x}, s) - \boldsymbol{\varepsilon}'_{trip,2}(\mathbf{x}, s)|^2 \,\mathrm{d}\mathbf{x}\,\mathrm{d}s \\
&\leq c_1 \int_0^t \int_\Omega \Big|\frac{\mathrm{d}}{\mathrm{d}s}\Phi(\mathbf{p}_1(\mathbf{x}, s))\boldsymbol{\sigma}_1^*(\mathbf{x}, s) - \frac{\mathrm{d}}{\mathrm{d}s}\Phi(\mathbf{p}_2(\mathbf{x}, s))\boldsymbol{\sigma}_2^*(\mathbf{x}, s)\Big|_\infty^2 \,\mathrm{d}\mathbf{x}\,\mathrm{d}s \\
&\leq c_1 \Big\{ \int_0^t \int_\Omega \Big|\frac{\mathrm{d}}{\mathrm{d}s}\Phi(\mathbf{p}_1(\mathbf{x}, s)) - \frac{\mathrm{d}}{\mathrm{d}s}\Phi(\mathbf{p}_2(\mathbf{x}, s))\Big|_\infty^2 \underbrace{|\boldsymbol{\sigma}_1^*(\mathbf{x}, s)|_\infty^2}_{\leq c_2} \,\mathrm{d}\mathbf{x}\,\mathrm{d}s + \\
&\quad + \int_0^t \int_\Omega \underbrace{\Big|\frac{\mathrm{d}}{\mathrm{d}s}\Phi(\mathbf{p}_2(\mathbf{x}, s))\Big|_\infty^2}_{=\left|\frac{\partial \Phi}{\partial \mathrm{p}_2}(\mathrm{p}_2(\mathrm{x}, s))\right|_\infty^2 \left|\frac{\partial \mathrm{p}_2}{\partial t}(\mathrm{x}, s)\right|_\infty^2 \leq c_3} \Big|\boldsymbol{\sigma}_1^*(\mathbf{x}, s) - \boldsymbol{\sigma}_2^*(\mathbf{x}, s)\Big|_\infty^2 \,\mathrm{d}\mathbf{x}\,\mathrm{d}s \Big\} \\
&\leq c \Big\{ \int_0^t \|\mathbf{p}_1(s) - \mathbf{p}_2(s)\|_{\mathrm{H_p}}^2 \,\mathrm{d}s + \int_0^t \|\boldsymbol{\sigma}_1^*(s) - \boldsymbol{\sigma}_2^*(s)\|_{\mathrm{H}_\sigma}^2 \,\mathrm{d}s \Big\}.
\end{aligned}$$

Integration yields the first estimate in Lemma 5.3.15. □

5.3.5 Step 4: Analysis of Problem $(\mathbf{P}_{A,FP1})$

In the next step, we investigate the solvability of Problem $(\mathbf{P}_{A,FP1})$ for $\hat{\theta}$, $\bar{\mathbf{u}}$, $\bar{\boldsymbol{\sigma}}$ and $\bar{\boldsymbol{\varepsilon}}_{cp}$ considered as data.

5.3.16 Lemma (Existence and Uniqueness)
Let Assumption 4.3.1 be valid. Assume $(\hat{\theta}, \bar{\mathbf{u}}, \bar{\boldsymbol{\sigma}}, \bar{\boldsymbol{\varepsilon}}_{cp}) \in \mathfrak{H}_\theta \times \mathcal{V}_{\mathrm{u}} \times \mathcal{H}_\sigma \times \mathcal{H}_\sigma$. *Then there exists a weak solution* $(\boldsymbol{\varepsilon}_{trip}, \boldsymbol{\sigma}^*, \mathbf{p}) \in \mathfrak{H}_\sigma \times \mathfrak{H}_\sigma \times \mathfrak{X}_{\mathrm{p}}^\infty$ *of Problem* $(\mathbf{P}_{A,FP1})$.
Moreover, let $(\boldsymbol{\varepsilon}_{trip,i}, \boldsymbol{\sigma}_i^*, \mathbf{p}_i)$ *be solutions w.r.t. data* $(\hat{\theta}_i, \bar{\mathbf{u}}_i, \bar{\boldsymbol{\sigma}}_i, \bar{\boldsymbol{\varepsilon}}_{cp,i})$, $i = 1, 2$. *Then the following estimate holds:*

$$
\begin{aligned}
(5.16)\quad & \|\boldsymbol{\varepsilon}_{trip,1}(t) - \boldsymbol{\varepsilon}_{trip,2}(t)\|_{\mathrm{H}_\sigma} + \|\boldsymbol{\varepsilon}'_{trip,1}(t) - \boldsymbol{\varepsilon}'_{trip,2}(t)\|_{\mathrm{H}_\sigma} + \|\boldsymbol{\sigma}_1^*(t) - \boldsymbol{\sigma}_2^*(t)\|_{\mathrm{H}_\sigma} \\
& + \|\mathbf{p}_1(t) - \mathbf{p}_2(t)\|_{\mathrm{H}_{\mathrm{p}}} + \|\mathbf{p}'_1(t) - \mathbf{p}'_2(t)\|_{\mathrm{H}_{\mathrm{p}}} \le c(h^{-1}) \Big\{ \|\bar{\boldsymbol{\sigma}}_1(t) - \bar{\boldsymbol{\sigma}}_2(t)\|_{\mathrm{H}_\sigma} \\
& + \int_0^t \|\bar{\boldsymbol{\sigma}}_1(s) - \bar{\boldsymbol{\sigma}}_2(s)\|_{\mathrm{H}_\sigma}\, ds + \|\hat{\theta}_1(t) - \hat{\theta}_2(t)\|_{\mathrm{H}_\sigma} + \int_0^t \|\hat{\theta}_1(s) - \hat{\theta}_2(s)\|_{\mathrm{H}_\sigma}\, ds \\
& + \|\hat{\theta}'_1(t) - \hat{\theta}'_2(t)\|_{\mathrm{H}_\sigma} + \int_0^t \|\hat{\theta}'_1(s) - \hat{\theta}'_2(s)\|_{\mathrm{H}_\sigma}\, ds + \int_0^t \|\bar{\mathbf{u}}_1(s) - \bar{\mathbf{u}}_2(s)\|_{\mathrm{V}_{\mathrm{u}}}\, ds \Big\},
\end{aligned}
$$

f.a.a. $t \in S$, *where the positive constant* c *depends on the regularisation parameter* h, *but not on the data.*

Proof. We use a standard fixed-point argument (cf. e.g. [Amann and Escher 2006]). Before we start with the proof let us give the following results for the subproblems.
Let $(\mathbf{u}, \boldsymbol{\varepsilon}_{trip}) \in \mathcal{V}_{\mathrm{u}} \times \mathfrak{H}_\sigma$. The problem (5.6) possesses a unique solution $\boldsymbol{\sigma}^* \in \mathfrak{H}_\sigma \cap \mathcal{X}_\sigma^\infty$ s.t.

$$
\|\boldsymbol{\sigma}_1^*(t) - \boldsymbol{\sigma}_2^*(t)\|_{\mathrm{H}_\sigma} \le c(h^{-1}) \Big\{ \|\boldsymbol{\varepsilon}'_{trip,1} - \boldsymbol{\varepsilon}'_{trip,2}\|_{\mathcal{H}_\sigma} + \|\mathbf{u}_1 - \mathbf{u}_2\|_{\mathcal{V}_{\mathrm{u}}} \Big\}
$$

f.a.a. $t \in S$, where the positive constant c depends on the regularisation parameter h, but not on the data.
Let $(\theta, \boldsymbol{\sigma}) \in \mathfrak{H}_\theta \times \mathcal{H}_\sigma$. The problem (5.3) possesses a unique solution $\mathbf{p} \in \mathfrak{H}_{\mathrm{p}} \cap \mathcal{X}_{\mathrm{p}}^\infty$ s.t.

$$
\begin{aligned}
\|\mathbf{p}_1(t) - \mathbf{p}_2(t)\|_{\mathrm{H}_{\mathrm{p}}} &\le c_1 \Big\{ \|\hat{\theta}_1 - \hat{\theta}_2\|_{\mathfrak{H}_\theta} + \|\hat{\theta}'_1 - \hat{\theta}'_2\|_{\mathfrak{H}_\theta} + \|\bar{\boldsymbol{\sigma}}_1 - \bar{\boldsymbol{\sigma}}_2\|_{\mathcal{H}_\sigma} + \|\boldsymbol{\sigma}_1^* + \boldsymbol{\sigma}_2^*\|_{\mathcal{H}_\sigma} \Big\} \\
\|\mathbf{p}'_1 - \mathbf{p}'_2\|_{\mathcal{H}_{\mathrm{p}}} &\le c_2 \Big\{ \|\hat{\theta}_1 - \hat{\theta}_2\|_{\mathfrak{H}_\theta} + \|\hat{\theta}'_1 - \hat{\theta}'_2\|_{\mathfrak{H}_\theta} + \|\bar{\boldsymbol{\sigma}}_1 - \bar{\boldsymbol{\sigma}}_2\|_{\mathcal{H}_\sigma} + \|\boldsymbol{\sigma}_1^* + \boldsymbol{\sigma}_2^*\|_{\mathcal{H}_\sigma} \Big\}
\end{aligned}
$$

f.a.a. $t \in S$, where c_1, c_2 are positive constants.
Let $(\mathbf{p}, \boldsymbol{\sigma}) \in \mathfrak{H}_{\mathrm{p}} \times \mathcal{H}_\sigma$. The problem (5.4) possesses a unique solution $\boldsymbol{\varepsilon}_{trip} \in \mathfrak{H}_\sigma \cap \mathcal{X}_\sigma^\infty$ s.t.

$$
\|\boldsymbol{\varepsilon}_{trip,1}(t) - \boldsymbol{\varepsilon}_{trip,2}(t)\|_{\mathrm{H}_\sigma} \le c_1 \Big\{ \|\boldsymbol{\sigma}_1^* - \boldsymbol{\sigma}_2^*\|_{\mathcal{H}_\sigma} + \|\mathbf{p}_1 - \mathbf{p}_2\|_{\mathfrak{H}_{\mathrm{p}}} \Big\},
$$

$$\|\boldsymbol{\varepsilon}_{trip,1} - \boldsymbol{\varepsilon}_{trip,2}\|_{\mathfrak{H}_\sigma} \le c_2 \Big\{ \|\boldsymbol{\sigma}_1^* - \boldsymbol{\sigma}_2^*\|_{\mathcal{H}_\sigma} + \|\mathbf{p}_1 - \mathbf{p}_2\|_{\mathfrak{H}_p} \Big\},$$

where the c_1, c_2 are positive constants.

We use Banach's fixed-point theorem with the (complete) weighted-norm space

$$\mathbf{X} := \mathfrak{H}_\sigma, \quad \|\mathbf{x}\|_X := \|\mathbf{x}\|_{H_\sigma,\lambda} + \|\mathbf{x}'\|_{H_\sigma,\lambda}, \quad \|\mathbf{x}\|^2_{H_\sigma,\lambda} := \sup_{t\in\bar{S}} \Big\{ \exp(-\lambda t) \int_0^t \|\mathbf{x}(s)\|^2 \, ds \Big\}$$

with $\lambda > 0$. For $\tilde{\varepsilon}_{trip} \in \mathbf{X}$ given, we consider the linearised problem (5.6), (5.3), (5.4) for $\boldsymbol{\varepsilon}_{trip}$ in which the parameters $\hat{\theta}$, $\bar{\mathbf{u}}$, $\bar{\boldsymbol{\sigma}}$ and $\bar{\boldsymbol{\varepsilon}}_{cp}$ are fixed. By Lemma 5.3.13 there exists a unique solution $\tilde{\varepsilon}_{trip} \in \mathfrak{H}_\sigma \hookrightarrow \mathbf{X}$. We define a fixed-point operator as

$$\mathbf{T}_{A,FP1} : \mathbf{X} \to \mathbf{X}, \qquad \mathbf{T}_{A,FP1}(\tilde{\varepsilon}_{trip}) = \boldsymbol{\varepsilon}_{trip}$$

and consider solutions $\boldsymbol{\varepsilon}_{trip,1}$, $\boldsymbol{\varepsilon}_{trip,2}$ corresponding to different data $\tilde{\boldsymbol{\varepsilon}}_{trip,1}$, $\tilde{\boldsymbol{\varepsilon}}_{trip,2}$. Using the arguments from above we obtain f.a.a. $t \in S$ the following estimate

$$\begin{aligned}
&\|\boldsymbol{\varepsilon}_{trip,1}(t) - \boldsymbol{\varepsilon}_{trip,2}(t)\|_{H_\sigma} + \|\boldsymbol{\varepsilon}'_{trip,1}(t) - \boldsymbol{\varepsilon}'_{trip,2}(t)\|_{H_\sigma} \\
&\quad \le c \Big\{ \|\boldsymbol{\sigma}_1^*(t) - \boldsymbol{\sigma}_2^*(t)\|_{H_\sigma} + \|\mathbf{p}_1(t) - \mathbf{p}_2(t)\|_{H_p} + \|\mathbf{p}'_1(t) - \mathbf{p}'_2(t)\|_{H_p} \\
&\qquad + \int_0^t \big(\|\boldsymbol{\sigma}_1^*(s) - \boldsymbol{\sigma}_2^*(s)\|_{H_\sigma} + \|\mathbf{p}_1(s) - \mathbf{p}_2(s)\|_{H_p} + \|\mathbf{p}'_1(s) - \mathbf{p}'_2(s)\|_{H_p} \big) \, ds \Big\} \\
&\quad \le c \Big\{ \|\boldsymbol{\sigma}_1^*(t) - \boldsymbol{\sigma}_2^*(t)\|_{H_\sigma} + \int_0^t \|\boldsymbol{\sigma}_1^*(s) - \boldsymbol{\sigma}_2^*(s)\|_{H_\sigma} \, ds \Big\} \\
&\quad \le c \int_0^t \|\tilde{\boldsymbol{\varepsilon}}'_{trip,1}(s) - \tilde{\boldsymbol{\varepsilon}}'_{trip,2}(s)\|_{H_\sigma} \, ds \\
&\quad \le c \Big\{ \int_0^t \big(\|\tilde{\boldsymbol{\varepsilon}}_{trip,1}(s) - \tilde{\boldsymbol{\varepsilon}}_{trip,2}(s)\|_{H_\sigma} + \|\tilde{\boldsymbol{\varepsilon}}'_{trip,1}(s) - \tilde{\boldsymbol{\varepsilon}}'_{trip,2}(s)\|_{H_\sigma} \big) \, ds \Big\} \\
&\quad \le c \int_0^t \exp(\lambda t) \, ds \sup_{s\in\bar{S}} \Big\{ \exp(-\lambda t) \big(\|\tilde{\boldsymbol{\varepsilon}}_{trip,1}(s) - \tilde{\boldsymbol{\varepsilon}}_{trip,2}(s)\|^2_{H_\sigma} \\
&\qquad + \|\tilde{\boldsymbol{\varepsilon}}'_{trip,1}(s) - \tilde{\boldsymbol{\varepsilon}}'_{trip,2}(s)\|_{H_\sigma} \big) \Big\} \\
&\quad \le c \frac{\exp(\lambda t)}{\lambda} \|\tilde{\boldsymbol{\varepsilon}}_{trip,1} - \tilde{\boldsymbol{\varepsilon}}_{trip,2}\|_X.
\end{aligned}$$

It follows

$$\|\boldsymbol{\varepsilon}_{trip,1} - \boldsymbol{\varepsilon}_{trip,2}\|_X \le \sqrt{\frac{c}{\lambda}} \|\tilde{\boldsymbol{\varepsilon}}_{trip,1} - \tilde{\boldsymbol{\varepsilon}}_{trip,2}\|_X$$

and hence, for λ large enough, $\mathbf{T}_{A,FP1}$ is strictly contractive. The conditions for Banach's fixed-point theorem are fulfilled and Problem ($\mathbf{P}_{A,FP1}$) possesses a unique solution. Moreover, the estimate (5.16) can be shown in the same manner. □

5.3.6 Step 5: Analysis of Problem $(\mathbf{P}_{A,\mathrm{u}})$

We continue with the solvability of Problem $(\mathbf{P}_{A,\mathrm{u}})$. Similarly to thermo-elasticity, the stress $\boldsymbol{\sigma}$ is eliminated in Equation (2.1) via (2.12). We obtain a hyperbolic system for the displacement $\mathbf{u}$, similar to the Lamé equations, but with additional terms on the right-hand side. In [Anzellotti and Luckhaus 1987; Duvaut and Lions 1976] the representation for the stress tensor is not used. Due to some basic calculation rules one gets similar estimates for the displacement $\mathbf{u}$ and the stress deviator $\boldsymbol{\sigma}$.

In contrast to the approach in thermo-elasticity (cf. [Boettcher 2007; Gawinecki 1986]) there will be no investigation of the coupled thermo-mechanical system via a simultaneous Galerkin method in this work. The coupling terms in the heat equation and the stress tensor do not vanish. There is a lack of regularity for $\mathbf{u}'$ (in the differential inclusion for $\boldsymbol{\varepsilon}_{cp}$ and in the heat equation) and therefore we discuss the following three different approaches to handle this problem.

We briefly recall the definition and some properties of the stress tensor:

5.3.17 Remark (Stress Tensor and Deviator)
The definition of the stress tensor $\boldsymbol{\sigma}$ in Equation (2.12) is equivalent to

$$\boldsymbol{\sigma} = 2\mu\left(\boldsymbol{\varepsilon}(\mathbf{u}) - \boldsymbol{\varepsilon}_{cp} - \boldsymbol{\varepsilon}_{trip}\right) + \lambda\,\mathrm{tr}(\boldsymbol{\varepsilon}(\mathbf{u}))\,\mathbf{I} - 3K_\alpha\,(\theta-\theta_0)\,\mathbf{I} - K\sum_{i=1}^{m}\left(\frac{\rho_0}{\rho_i(\theta_0)} - 1\right)p_i\,\mathbf{I}$$

and one can calculate

$$\mathrm{tr}(\boldsymbol{\sigma}) = (2\mu+3\lambda)\,\mathrm{div}(\mathbf{u}) - 9K_\alpha\,(\theta-\theta_0) - 3K\sum_{i=1}^{m}\left(\frac{\rho_0}{\rho_i(\theta_0)} - 1\right)p_i,$$

$$\boldsymbol{\sigma}^* = 2\mu(\boldsymbol{\varepsilon}^*(\mathbf{u}) - \boldsymbol{\varepsilon}_{cp} - \boldsymbol{\varepsilon}_{trip}) \quad \text{or} \quad \boldsymbol{\sigma}^* = 2\mu\left(\boldsymbol{\varepsilon}^*(\mathbf{u}) - \boldsymbol{\varepsilon}_{cp}\right) - 3\mu\int_0^t\sum_{i=1}^{m}\kappa_i\frac{\mathrm{d}\Phi_i}{\mathrm{d}t}\boldsymbol{\sigma}^*\,\mathrm{d}s.$$

Therefore, we get (with the help of Gronwall's inequality in the second case)

$$|\boldsymbol{\sigma}^*|^2 = \left(\boldsymbol{\sigma} - \frac{1}{3}\mathrm{tr}(\boldsymbol{\sigma})\,\mathbf{I}\right) : \left(\boldsymbol{\sigma} - \frac{1}{3}\mathrm{tr}(\boldsymbol{\sigma})\,\mathbf{I}\right) = |\boldsymbol{\sigma}|^2 - \frac{1}{3}(\mathrm{tr}(\boldsymbol{\sigma}))^2 \leq |\boldsymbol{\sigma}|^2,$$
$$\|\boldsymbol{\sigma}^*(t)\|^2_{H_\sigma} \leq 2\mu\exp\left(3\mu M_\Phi T\|\kappa\|_\infty\|\mathbf{p}\|_{\mathfrak{X}_p^\infty}\right)\left(\|\mathbf{u}(t)\|^2_{\mathrm{V_u}} + \|\boldsymbol{\varepsilon}_{cp}(t)\|^2_{\mathrm{H}_\sigma}\right) \quad \text{f.a.a. } t\in S.$$

One can calculate the stress deviator if one knows the stress tensor via $\boldsymbol{\sigma}^* = \boldsymbol{\sigma} - \frac{1}{3}\mathrm{tr}(\boldsymbol{\sigma})\,\mathbf{I}$. Unfortunately one cannot compute the stress tensor if one only knows the stress deviator, because $\boldsymbol{\sigma}^*_{ij} = \boldsymbol{\sigma}_{ij}$, $i \neq j$ and $(\boldsymbol{\sigma}^*_{ii})_i = \frac{1}{3}\mathbf{S}\,(\boldsymbol{\sigma}_{ii})_i$, $\mathbf{S} = \begin{bmatrix} 2 & -1 & -1 \\ -1 & 2 & -1 \\ -1 & -1 & 2\end{bmatrix}$ for $i,j = 1,2,3$ with $\mathrm{rank}(A) = 2$ *and hence* $\det(A) = 0$.

Assume $(\theta, \mathbf{p}, \boldsymbol{\varepsilon}_{trip}, \boldsymbol{\varepsilon}_{cp}) \in \mathcal{H}_\theta \times \mathfrak{X}_{\mathrm{p}}^\infty \times \mathfrak{H}_\sigma \times \mathfrak{H}_\sigma$ are given data. Consider the following problem:

Problem ($\mathbf{P}_{A,\mathrm{u}}$)
Find the displacement field $\mathbf{u} : \overline{\Omega}_T \to \mathbb{R}^3$ *and the stress field* $\boldsymbol{\sigma} : \overline{\Omega}_T \to \mathbb{R}^{3\times 3}_{sym}$ *s.t.*

$$
\begin{aligned}
&\rho_0 \frac{\partial^2 \mathbf{u}}{\partial t^2} - 2\,\mathrm{Div}(\mu \boldsymbol{\varepsilon}(\mathbf{u})) - \mathrm{grad}(\lambda\, \mathrm{div}(\mathbf{u})) + 3\,\mathrm{grad}(K_\alpha (\mathcal{S}_h\theta - \theta_0)) \\
&\quad + \mathrm{grad}\left(K \sum_{i=1}^{m} \left(\frac{\rho_0}{\rho_i(\theta_0)} - 1\right) p_i\right) + 2\,\mathrm{Div}(\mu \boldsymbol{\varepsilon}_{trip}) + 2\,\mathrm{Div}(\mu \boldsymbol{\varepsilon}_{cp}) = \mathbf{f} \text{ in } \Omega_T
\end{aligned} \tag{5.17}
$$

including conditions (2.10), (2.11) *as well as initial conditions* $(2.24)_1$, $(2.24)_2$ *and boundary values*

$$
\mathbf{u} = 0 \quad \text{on} \quad \Gamma_1, \qquad\qquad \boldsymbol{\sigma} \cdot \nu = \mathbf{0} \quad \text{on} \quad \Gamma_2.
$$

The aim of this subsection is to introduce an abstract formulation of the corresponding initial and boundary value problems that arise in order to treat Problem ($\mathbf{P}_{A,\mathrm{u}}$) and state related existence and uniqueness results. We use the Galerkin method in the sequel. For other possibilities like semi-group methods, the Rothe method etc. we refer to the literature (especially hyperbolic equations are the topic of [Baiocchi 1967; D'Ancona 1995; Lasiecka and Triggiani 1983]; coupled hyperbolic-parabolic equations are treated in [Hsiao and Jiang 2004]) and to the overview in Section 4.5.

Preliminaries

First of all, we state some calculation rules which will be needed in this subsection:

$$
\begin{aligned}
\mathrm{div}(\mathbf{u}) &= \mathrm{tr}(\boldsymbol{\varepsilon}(\mathbf{u})) = \mathrm{tr}(\nabla \mathbf{u}), && \forall \mathbf{u} \in \mathbf{V}_{\mathrm{u}}, \\
\boldsymbol{\sigma} : \boldsymbol{\varepsilon}(\mathbf{u}) &= \frac{1}{3}\mathrm{tr}(\boldsymbol{\sigma})\,\mathrm{div}(\mathbf{u}) + \boldsymbol{\sigma}^* : \boldsymbol{\varepsilon}^*(\mathbf{u}), && \forall \mathbf{u} \in \mathbf{V}_{\mathrm{u}},\ \boldsymbol{\sigma} \in \mathbf{H}_\sigma.
\end{aligned}
$$

If $\boldsymbol{\sigma} \in \mathbf{H}_\sigma$, there exists an element $\boldsymbol{\sigma} \cdot \nu \in [W^{-\frac{1}{2},2}(\partial\Omega)]^3$ s.t. the following Green formula holds

$$
\int_\Omega \boldsymbol{\sigma} : \boldsymbol{\varepsilon}(\mathbf{u})\,\mathrm{d}\mathbf{x} + \int_\Omega \mathbf{u}\,\mathrm{Div}(\boldsymbol{\sigma})\,\mathrm{d}\mathbf{x} = \int_{\partial\Omega} \mathrm{tr}(\mathbf{u})\,\boldsymbol{\sigma} \cdot \boldsymbol{\nu}\,\mathrm{d}\sigma_x, \qquad \forall \mathbf{u} \in \mathbf{V}_{\mathrm{u}}.
$$

Moreover,

$$\frac{1}{3}\int_\Omega \mathrm{tr}(\boldsymbol{\sigma})\,\mathrm{tr}(\boldsymbol{\varepsilon}(\mathbf{u}))\,\mathrm{d}\mathbf{x} = -\int_\Omega \boldsymbol{\sigma}^* : \boldsymbol{\varepsilon}^*(\mathbf{u})\,\mathrm{d}\mathbf{x} - \int_\Omega \mathbf{u}\,\mathrm{Div}(\boldsymbol{\sigma})\,\mathrm{d}\mathbf{x}, \qquad \forall\,\mathbf{u} \in \mathbf{V}_{\mathrm{u}}.$$

Let us introduce two bilinear continuous symmetric forms $a : \mathbf{V}_{\mathrm{u}} \times \mathbf{V}_{\mathrm{u}} \to \mathbb{R}$ defined by

$$a(\mathbf{u},\mathbf{v}) := \lambda \int_\Omega \mathrm{div}(\mathbf{u})\,\mathrm{div}(\mathbf{v})\,\mathrm{d}\mathbf{x} + 2\mu \int_\Omega \boldsymbol{\varepsilon}(\mathbf{u}) : \boldsymbol{\varepsilon}(\mathbf{v})\,\mathrm{d}\mathbf{x} \qquad \forall\,\mathbf{v},\mathbf{w} \in \mathbf{V}_{\mathrm{u}}$$

and $b : \mathbf{V}_{\mathrm{u}} \times \mathbf{V}_{\mathrm{u}} \to \mathbb{R}$ defined by

$$b(\mathbf{u},\mathbf{v}) := \int_\Omega \boldsymbol{\varepsilon}(\mathbf{u}) : \boldsymbol{\varepsilon}(\mathbf{v})\,\mathrm{d}\mathbf{x} \qquad \forall\,\mathbf{v},\mathbf{w} \in \mathbf{V}_{\mathrm{u}}.$$

Since $\mathrm{meas}(\Gamma_1) > 0$, Korn's inequality (cf. [Zeidler 1988] for an overview and [Ciarlet and Ciarlet Jr. 2004; Duvaut and Lions 1976; Fuchs 2010; Neff 2002; Nitsche 1981] for details) holds and thus, there exists a positive constant c depending only on Ω, Γ s.t.

$$\|\boldsymbol{\varepsilon}(\mathbf{u})\|_{\mathrm{H}_\sigma} \geq c\|\mathbf{u}\|_{\mathrm{V_u}} \qquad \forall\,\mathbf{u} \in \mathbf{V}_{\mathrm{u}} \qquad (\text{equivalent norm in } \mathbf{V}_{\mathrm{u}}).$$

Hence, we infer from Korn's inequality that there exists a positive constant c, depending on λ, μ and Ω, s.t.

$$a(\mathbf{u},\mathbf{u}) \geq c\|\mathbf{u}\|^2_{\mathrm{V_u}} \qquad \forall\,\mathbf{u} \in \mathbf{V}_{\mathrm{u}}.$$

Also note that

$$b(\mathbf{u},\mathbf{u}) \geq c\|\mathbf{u}\|^2_{\mathrm{V_u}} \qquad \forall\,\mathbf{u} \in \mathbf{V}_{\mathrm{u}}.$$

Next, to set the problem in the abstract framework of the dual space $\mathbf{V}_{\mathrm{u}}^*$, we introduce the operators

$$\begin{aligned} &\mathbf{A}_{\mathrm{u}} : \mathbf{V}_{\mathrm{u}} \to \mathbf{V}_{\mathrm{u}}^*, && \langle \mathbf{A}_{\mathrm{u}}\mathbf{u},\mathbf{v}\rangle_{\mathrm{V_u^*V_u}} = a(\mathbf{u},\mathbf{v}), && \mathbf{u},\mathbf{v} \in \mathbf{V}_{\mathrm{u}},\\ &\mathbf{B}_{\mathrm{u}} : \mathbf{V}_{\mathrm{u}} \to \mathbf{V}_{\mathrm{u}}^*, && \langle \mathbf{B}_{\mathrm{u}}\mathbf{u},\mathbf{v}\rangle_{\mathrm{V_u^*V_u}} = b(\mathbf{u},\mathbf{v}), && \mathbf{u},\mathbf{v} \in \mathbf{V}_{\mathrm{u}}. \end{aligned}$$

The operator $\mathbf{A}_{\mathrm{u}} : \mathbf{V}_{\mathrm{u}} \to \mathbf{V}_{\mathrm{u}}^*$ is linear, symmetric, continuous (i.e. there exists a $c > 0$ s.t. for all $\mathbf{u},\mathbf{v} \in \mathbf{V}_{\mathrm{u}} : |\langle \mathbf{A}_{\mathrm{u}}\mathbf{u},\mathbf{v}\rangle_{\mathrm{V_u^*V_u}}| \leq c\|\mathbf{u}\|_{\mathrm{V_u}}\|\mathbf{v}\|_{\mathrm{V_u}}$) and strongly positive (i.e. there exists a $c > 0$ s.t. f.a. $\mathbf{v} \in \mathbf{V}_{\mathrm{u}} : \langle \mathbf{A}_{\mathrm{u}}\mathbf{v},\mathbf{v}\rangle_{\mathrm{V_u^*V_u}} \geq c\|\mathbf{v}\|^2_{\mathrm{V_u}}$). Linearity and symmetry are obvious, continuity is standard and coercivity follows from Korn's inequality. The same properties hold for the operator $\mathbf{B}_{\mathrm{u}} : \mathbf{V}_{\mathrm{u}} \to \mathbf{V}_{\mathrm{u}}^*$.

5.3.18 Remark (Equivalent Representation)
The following representations are equivalent:

$$\langle \mathbf{A}_{\mathrm{u}}\mathbf{v},\mathbf{w}\rangle_{\mathrm{V_u^*V_u}} = \int_\Omega \boldsymbol{\sigma}(\mathbf{v}) : \boldsymbol{\varepsilon}(\mathbf{w})\,\mathrm{d}x = \int_\Omega \boldsymbol{\sigma}(\mathbf{v}) : \nabla\mathbf{w}\,\mathrm{d}x \quad \forall\,\mathbf{v},\mathbf{w} \in \mathbf{V}_{\mathrm{u}}.$$

Existence and Uniqueness

5.3.19 Definition (Weak Formulation)
Under the Assumption 4.3.1$_{(A1)}$ the function $\mathbf{u} \in \mathcal{U}_u$ is called a weak solution of Problem ($\mathbf{P}_{A,u}$), if

$$(5.18)\quad \left\langle \rho_0 \frac{\partial^2 \mathbf{u}(t)}{\partial t^2}, \mathbf{v} \right\rangle_{V_u^* V_u} + 2\int_\Omega \mu\, \boldsymbol{\varepsilon}(\mathbf{u}(t)) : \boldsymbol{\varepsilon}(\mathbf{v})\, \mathrm{d}\mathbf{x} + \int_\Omega \lambda\, \operatorname{div}(\mathbf{u}(t))\, \operatorname{div}(\mathbf{v})\, \mathrm{d}\mathbf{x}$$
$$= 3\int_\Omega K_\alpha\, (\mathcal{S}_h\theta(t) - \theta_0)\, \operatorname{div}(\mathbf{v})\, \mathrm{d}\mathbf{x} + \int_\Omega K \sum_{i=1}^m \left(\frac{\rho_0}{\rho_i(\theta_0)} - 1\right) p_i(t)\, \operatorname{div}(\mathbf{v})\, \mathrm{d}\mathbf{x}$$
$$+ 2\int_\Omega \mu\, \boldsymbol{\varepsilon}_{trip}(t) : \nabla \mathbf{v}\, \mathrm{d}\mathbf{x} + 2\int_\Omega \mu\, \boldsymbol{\varepsilon}_{cp}(t) : \nabla \mathbf{v}\, \mathrm{d}\mathbf{x} + \int_\Omega \mathbf{f}(t)\, \mathbf{v}\, \mathrm{d}\mathbf{x}$$

f.a. $\mathbf{v} \in \mathbf{V}_u$, f.a.a. $t \in S$ and $\mathbf{u}(\mathbf{x}, 0) = \mathbf{u}_0(\mathbf{x})$ a.e., $\mathbf{u}'(\mathbf{x}, 0) = \mathbf{u}_1(\mathbf{x})$ a.e.

5.3.20 Remark (Equivalent Weak Formulation)
The equation (5.18) *can be rewritten as*

$$\frac{\mathrm{d}^2}{\mathrm{d}t^2}\langle \mathbf{u}(t), \mathbf{v}\rangle_{V_u^* V_u} = \langle \underbrace{\mathbf{f}(t) + 2\mu\, \operatorname{Div}\left(\boldsymbol{\varepsilon}_{trip}(t) + \boldsymbol{\varepsilon}_{cp}(t)\right)}_{\tilde{\mathbf{f}}_1(t)}$$
$$\underbrace{+ \operatorname{grad}\left(K \sum_{i=1}^m \left(\frac{\rho_0}{\rho_i(\theta_0)} - 1\right) p_i(t)\right) - 3\operatorname{grad}\left(K_\alpha\, (\theta(t) - \theta_0)\right)}_{=:\tilde{\mathbf{f}}_2(t)}, \mathbf{v}\rangle_{V_u^* V_u} - \langle \mathbf{A}_u \mathbf{u}(t), \mathbf{v}\rangle_{V_u^* V_u}$$

f.a.a. $t \in S$ *and* f.a. $\mathbf{v} \in \mathbf{V}_u$. *Since $\mathbf{u} \in \mathcal{V}_u$ and $A_u : \mathbf{V}_u \to \mathbf{V}_u^*$ linear and continuous it holds $A_u \mathbf{u}(t) \in \mathbf{V}_u^*$. Hence $\tilde{\mathbf{f}}_1(t) + \tilde{\mathbf{f}}_2(t) - A_u \mathbf{u}(t) \in \mathbf{V}_u^*$ and therefore $\mathbf{u}'' \in \mathcal{V}_u^*$ and $\|\mathbf{u}''\|_{\mathcal{V}_u^*} \leq c$.*

We prove the a-priori estimate for the original Problem ($\mathbf{P}_u$) (without the Steklov regularisation for θ) and make some remarks for the considered situation of Problem ($\mathbf{P}_{A,u}$) afterwards.

5.3.21 Proposition (A-priori Estimates)
Let Assumption 4.3.1$_{(A1)}$ be valid, assume in addition $\theta \in \mathfrak{H}_\theta$ and let $\mathbf{u} \in \mathcal{H}_u^\infty \cap \mathcal{V}_u^\infty$ be a weak solution of Problem ($\mathbf{P}_u$). Then there exist $c_1, c_2 > 0$ s.t.

$$\|\mathbf{u}'\|_{\mathcal{H}_u^\infty} + \|\mathbf{u}\|_{\mathcal{V}_u^\infty} \leq c_1 + c_2 \Big\{ \|\mathbf{f}\|_{\mathcal{H}_u} + \|\theta\|_{\mathcal{H}_\theta^\infty} + \|\theta'\|_{\mathcal{H}_\theta} + \|\mathbf{p}\|_{\mathcal{H}_p^\infty}$$
$$+ \|\mathbf{p}'\|_{\mathcal{H}_p} + \|\boldsymbol{\varepsilon}_{trip}\|_{\mathcal{H}_\sigma^\infty} + \|\boldsymbol{\varepsilon}'_{trip}\|_{\mathcal{H}_\sigma} + \|\boldsymbol{\varepsilon}_{cp}\|_{\mathcal{H}_\sigma^\infty} + \|\boldsymbol{\varepsilon}'_{cp}\|_{\mathcal{H}_\sigma} \Big\}.$$

Proof. The formal derivation of the estimate is achieved by testing Equation (5.18) with $\mathbf{v} = \mathbf{u}'$. Since this choice is usually not allowed, (5.18) can be approximated by a Galerkin scheme, cf. Lemma 5.3.25. Multiplying the Galerkin equations by $g'_{ni}(t)$ and adding these equations for $i = 1, \dots, n$ leads to

$$\left\langle \rho_0 \, \mathbf{u}''(t), \mathbf{u}'(t) \right\rangle_{\mathrm{V}_u^*\mathrm{V}_u} + \left\langle \mathbf{A}_u \mathbf{u}(t), \mathbf{u}'(t) \right\rangle_{\mathrm{V}_u^*\mathrm{V}_u} = 3 \int_\Omega K_\alpha \left(\theta(t) - \theta_0 \right) \operatorname{div}(\mathbf{u}'(t)) \, \mathrm{d}\mathbf{x}$$
$$+ \int_\Omega K \sum_{i=1}^{m} \left(\frac{\rho_0}{\rho_i(\theta_0)} - 1 \right) p_i(t) \operatorname{div}(\mathbf{u}'(t)) \, \mathrm{d}\mathbf{x} + 2 \int_\Omega \mu \, \boldsymbol{\varepsilon}_{trip}(t) : \nabla \mathbf{u}'(t) \, \mathrm{d}\mathbf{x}$$
$$+ 2 \int_\Omega \mu \, \boldsymbol{\varepsilon}_{cp}(t) : \nabla \mathbf{u}'(t) \, \mathrm{d}\mathbf{x} + \left(\mathbf{f}(t), \mathbf{u}'(t) \right)_{\mathrm{H}_u}.$$

Because of $\langle \mathbf{u}'(t), \mathbf{u}(t) \rangle = \frac{1}{2} \frac{\mathrm{d}}{\mathrm{d}t} |\mathbf{u}(t)|^2$ and integrating over S one gets

$$(5.19) \quad \frac{\rho_0}{2} \|\mathbf{u}'(t)\|^2_{\mathrm{H}_u} + \frac{c}{2} \|\mathbf{u}(t)\|_{\mathrm{V}_u} = \underbrace{3 \int_0^t \int_\Omega K_\alpha \left(\theta(s) - \theta_0 \right) \operatorname{div}(\mathbf{u}'(s)) \, \mathrm{d}\mathbf{x} \, \mathrm{d}s}_{=:\mathbb{I}_1}$$
$$+ \underbrace{\int_0^t \int_\Omega K \sum_{i=1}^{m} \left(\frac{\rho_0}{\rho_i(\theta_0)} - 1 \right) p_i(s) \operatorname{div}(\mathbf{u}'(s)) \, \mathrm{d}\mathbf{x} \, \mathrm{d}s}_{=:\mathbb{I}_2} + \underbrace{2 \int_0^t \int_\Omega \mu \, \boldsymbol{\varepsilon}_{trip}(s) : \nabla \mathbf{u}'(s) \, \mathrm{d}\mathbf{x} \, \mathrm{d}s}_{=:\mathbb{I}_3}$$
$$+ \underbrace{2 \int_0^t \int_\Omega \mu \, \boldsymbol{\varepsilon}_{cp}(s) : \nabla \mathbf{u}'(s) \, \mathrm{d}\mathbf{x} \, \mathrm{d}s}_{=:\mathbb{I}_4} + \underbrace{\int_0^t \left(\mathbf{f}(s), \mathbf{u}'(s) \right)_{\mathrm{H}_u} \mathrm{d}s}_{=:\mathbb{I}_5} + \frac{\rho_0}{2} \|\mathbf{u}_1\|^2_{\mathrm{H}_u} + \frac{c}{2} \|\mathbf{u}_0\|_{\mathrm{V}_u}.$$

The right-hand side of (5.19) is majorised by (using integration by parts w.r.t. time and standard inequalities):

$$|\mathbb{I}_1| = \left| 3 \int_\Omega K_\alpha \left(\theta(t) - \theta_0 \right) \operatorname{div}(\mathbf{u}(t)) \, \mathrm{d}\mathbf{x} - 3 \int_0^t \int_\Omega K_\alpha \, \theta'(s) \operatorname{div}(\mathbf{u}'(s)) \, \mathrm{d}\mathbf{x} \, \mathrm{d}s \right|$$
$$\leq \frac{9K_\alpha^2}{4\varepsilon_1} \|\theta(t)\|^2_{H_\theta} + \varepsilon_1 \|\mathbf{u}(t)\|^2_{\mathrm{V}_u} + \frac{9k_\alpha^2 \theta_0^2}{4\varepsilon_2} \operatorname{meas}(\Omega) + \varepsilon_2 \|\mathbf{u}(t)\|^2_{\mathrm{V}_u}$$
$$+ \frac{3K_\alpha}{2} \int_0^t \|\theta'(s)\|^2_{H_\theta} \, \mathrm{d}s + \frac{3K_\alpha}{2} \int_0^t \|\mathbf{u}(s)\|^2_{\mathrm{V}_u} \, \mathrm{d}s,$$
$$|\mathbb{I}_2| = \left| \int_\Omega K \sum_{i=1}^{m} \left(\frac{\rho_0}{\rho_i(\theta_0)} - 1 \right) p_i(t) \operatorname{div}(\mathbf{u}(t)) \, \mathrm{d}\mathbf{x} \right.$$
$$- \int_\Omega K \sum_{i=1}^{m} \left(\frac{\rho_0}{\rho_i(\theta_0)} - 1 \right) p_i(0) \operatorname{div}(\mathbf{u}(0)) \, \mathrm{d}\mathbf{x}$$
$$\left. - \int_0^t \int_\Omega K \sum_{i=1}^{m} \left(\frac{\rho_0}{\rho_i(\theta_0)} - 1 \right) p_i'(s) \operatorname{div}(\mathbf{u}(s)) \, \mathrm{d}\mathbf{x} \, \mathrm{d}s \right|$$

$$\leq \frac{K^2\left\|\frac{\rho_0}{\rho(\theta_0)}\right\|_\infty^2}{4\varepsilon_3}\|\mathbf{p}(t)\|_{\mathrm{H_p}}^2 + \varepsilon_3\|\mathbf{u}(t)\|_{\mathrm{V_u}}^2 + \frac{K\left\|\frac{\rho_0}{\rho(\theta_0)}\right\|_\infty}{2}\|\mathbf{p}_0\|_{\mathrm{H_p}}^2 + \frac{K\left\|\frac{\rho_0}{\rho(\theta_0)}\right\|_\infty}{2}\|\mathbf{u}_0\|_{\mathrm{V_u}}^2$$
$$+ \frac{K\left\|\frac{\rho_0}{\rho(\theta_0)}\right\|_\infty}{2}\int_0^t \|\mathbf{p}'(s)\|_{\mathrm{H_p}}^2\,\mathrm{d}s + \frac{K\left\|\frac{\rho_0}{\rho(\theta_0)}\right\|_\infty}{2}\int_0^t \|\mathbf{u}(s)\|_{\mathrm{V_u}}^2\,\mathrm{d}s,$$
$$|\mathbb{I}_3| = \left|2\int_\Omega \mu\,\boldsymbol{\varepsilon}_{trip}(t) : \nabla\,\mathbf{u}(t)\,\mathrm{d}\mathbf{x} - 2\int_0^t\int_\Omega \mu\,\boldsymbol{\varepsilon}'_{trip}(s) : \nabla\,\mathbf{u}(s)\,\mathrm{d}\mathbf{x}\,\mathrm{d}s\right|$$
$$\leq \frac{\mu^2}{\varepsilon_5}\|\boldsymbol{\varepsilon}_{trip}\|_{\mathrm{H}_\sigma}^2 + \varepsilon_5\|\mathbf{u}(t)\|_{\mathrm{V_u}}^2 + \mu\int_0^t \|\boldsymbol{\varepsilon}'_{trip}(s)\|_{\mathrm{H}_\sigma}^2\,\mathrm{d}s + \mu\int_0^t \|\mathbf{u}(s)\|_{\mathrm{V_u}}^2\,\mathrm{d}s,$$
$$|\mathbb{I}_4| = \left|2\int_\Omega \mu\,\boldsymbol{\varepsilon}_{cp}(t) : \nabla\,\mathbf{u}(t)\,\mathrm{d}\mathbf{x} - 2\int_0^t\int_\Omega \mu\,\boldsymbol{\varepsilon}'_{cp}(s) : \nabla\,\mathbf{u}(s)\,\mathrm{d}\mathbf{x}\,\mathrm{d}s\right|$$
$$\leq \frac{\mu^2}{\varepsilon_4}\|\boldsymbol{\varepsilon}_{cp}\|_{\mathrm{H}_\sigma}^2 + \varepsilon_4\|\mathbf{u}(t)\|_{\mathrm{V_u}}^2 + \mu\int_0^t \|\boldsymbol{\varepsilon}'_{cp}(s)\|_{\mathrm{H}_\sigma}^2\,\mathrm{d}s + \mu\int_0^t \|\mathbf{u}(s)\|_{\mathrm{V_u}}^2\,\mathrm{d}s,$$
$$|\mathbb{I}_5| \leq \frac{1}{2}\int_0^t \|\mathbf{f}(s)\|_{\mathrm{H_u}}^2\,\mathrm{d}s + \frac{1}{2}\int_0^t \|\mathbf{u}'(s)\|_{\mathrm{H_u}}^2\,\mathrm{d}s$$

for arbitrary $\varepsilon_i > 0$, $i = 1, \ldots, 5$. Therefore, using the Gronwall Lemma, it follows

$$\|\mathbf{u}'(t)\|_{\mathrm{H_u}} + \|\mathbf{u}(t)\|_{\mathrm{V_u}} \leq c\Big\{\|\mathbf{u}_0\| + \|\mathbf{u}_1\| + \|\mathbf{f}\|_{\mathcal{H}_\mathrm{u}} + \|\theta\|_{\mathcal{H}_\theta} + \|\theta'\|_{\mathcal{H}_\theta} + \|\mathbf{p}\|_{\mathcal{H}_\mathrm{p}^\infty}$$
$$+ \|\mathbf{p}'\|_{\mathcal{H}_\mathrm{p}} + \|\boldsymbol{\varepsilon}'_{trip}\|_{\mathcal{H}_\sigma} + \|\boldsymbol{\varepsilon}_{cp}\|_{\mathcal{H}_\sigma^\infty} + \|\boldsymbol{\varepsilon}'_{cp}\|_{\mathcal{H}_\sigma}\Big\}.$$

Hence,

$$\sup_{t\in\bar{S}}\|\mathbf{u}'(t)\|_{\mathrm{H_u}} \leq c, \qquad \sup_{t\in\bar{S}}\|\mathbf{u}(t)\|_{\mathrm{V_u}} \leq c,$$

where c is a generic constant. □

5.3.22 Remark
The operator $\mathbf{A}_\mathrm{u}$ is not coercive in case of $\mathrm{meas}(\Gamma_1) = 0$, *but then there exists $c > 0$ s.t. $\langle \mathbf{A}_\mathrm{u}\mathbf{u}, \mathbf{u}\rangle_{\mathrm{V_u^*V_u}} + \|\mathbf{u}\|_{\mathrm{H_u}}^2 \geq c\,\|\mathbf{u}\|_{\mathrm{V_u}}^2$ for all $\mathbf{u} \in \mathbf{V}_\mathrm{u}$ (cf. e.g. [Duvaut and Lions 1976; Nitsche 1981]). Using this inequality and the estimate $\|\mathbf{u}(t)\|_{\mathrm{H_u}}^2 \leq 2\|\mathbf{u}_0\|_{\mathrm{H_u}}^2 + 2t\int_0^t \|\mathbf{u}'(s)\|_{\mathrm{H_u}}^2\,\mathrm{d}s$, $t \in S$ gives the similar estimate in the preceding proposition.*

5.3.23 Remark (Estimate for the Stress Tensor)
Let Assumption 4.3.1$_{(A1)}$ be valid, assume in addition $\theta \in \mathfrak{H}_\theta$ and let $\mathbf{u} \in \mathcal{H}_\mathrm{u}^\infty \cap \mathcal{V}_\mathrm{u}^\infty$ be a weak solution of Problem ($\mathbf{P}_\mathrm{u}$). Then, by using the definition of the stress tensor and calculation of the norm, there exists $c_1, c_2 > 0$ s.t.

$$\|\boldsymbol{\sigma}\|_{\mathcal{H}_\sigma} \leq c_1 + c_2\left\{\|\mathbf{u}\|_{\mathcal{V}_\mathrm{u}} + \|\theta\|_{\mathcal{H}_\theta} + \|\mathbf{p}\|_{\mathcal{H}_\mathrm{p}} + \|\boldsymbol{\varepsilon}_{trip}\|_{\mathcal{H}_\sigma} + \|\boldsymbol{\varepsilon}_{cp}\|_{\mathcal{H}_\sigma}\right\},$$
$$\|\boldsymbol{\sigma}'\|_{\mathcal{H}_\sigma} \leq c_1 + c_2\left\{\|\mathbf{u}'\|_{\mathcal{V}_\mathrm{u}} + \|\theta'\|_{\mathcal{H}_\theta} + \|\mathbf{p}'\|_{\mathcal{H}_\mathrm{p}} + \|\boldsymbol{\varepsilon}'_{trip}\|_{\mathcal{H}_\sigma} + \|\boldsymbol{\varepsilon}'_{cp}\|_{\mathcal{H}_\sigma}\right\}.$$

5.3.24 Remark (A-priori Estimate for Problem ($\mathbf{P}_{A,\mathrm{u}}$))
Let Assumption 4.3.1$_{(A1)}$ be valid, assume in addition $\theta \in \mathcal{H}_\theta$ and let $\mathbf{u} \in \mathcal{H}_\mathrm{u}^\infty \cap \mathcal{V}_\mathrm{u}^\infty$ be a weak solution of Problem ($\mathbf{P}_{A,\mathrm{u}}$). Then, there exists $c_1, c_2 > 0$ s.t.

$$\|\mathbf{u}'\|_{\mathcal{H}_\mathrm{u}^\infty} + \|\mathbf{u}\|_{\mathcal{V}_\mathrm{u}^\infty} \leq c_1(h^{-1}) + c_2(h^{-1})\Big\{\|\mathbf{f}\|_{\mathcal{H}_\mathrm{u}} + \|\theta\|_{\mathcal{H}_\theta^\infty} + \|\mathbf{p}\|_{\mathcal{H}_\mathrm{p}^\infty} + \|\mathbf{p}'\|_{\mathcal{H}_\mathrm{p}} + \|\boldsymbol{\varepsilon}'_{trip}\|_{\mathcal{H}_\sigma} + \|\boldsymbol{\varepsilon}_{cp}\|_{\mathcal{H}_\sigma^\infty} + \|\boldsymbol{\varepsilon}'_{cp}\|_{\mathcal{H}_\sigma}\Big\}.$$

The proof is similar to the proof of Proposition 5.3.21. Using the Steklov average of the temperature changes only the following estimate

$$|\mathbb{I}_1^h| \leq \frac{9K_\alpha^2}{4\varepsilon_1 h}\|\theta(t)\|_{H_\theta}^2 + \varepsilon_1\|\mathbf{u}(t)\|_{\mathrm{V_u}}^2 + \frac{9k_\alpha^2\theta_0^2}{4\varepsilon_2}\,\mathrm{meas}(\Omega) + \varepsilon_2\|\mathbf{u}(t)\|_{\mathrm{V_u}}^2 + \frac{3K_\alpha}{2h}\int_0^t \|\theta(s)\|_{H_\theta}^2\,\mathrm{d}s + \frac{3K_\alpha}{2}\int_0^t \|\mathbf{u}(s)\|_{\mathrm{V_u}}^2\,\mathrm{d}s.$$

An alternative would by to consider 'smallness' conditions for the parameter K_α. In this case the a-priori estimate is independent of the regularisation parameter and passing to the limit would be possible:

$$|\tilde{\mathbb{I}}_1^h| \leq \frac{3K_\alpha}{2}\|\theta(t)\|_{H_\theta}^2 + \frac{3K_\alpha}{2}\|\mathbf{u}(t)\|_{\mathrm{V_u}}^2 + \frac{3K_\alpha\theta_0}{2}\,\mathrm{meas}(\Omega) + \frac{3K_\alpha\theta_0}{2}\|\mathbf{u}(t)\|_{\mathrm{V_u}}^2 + \frac{3K_\alpha}{2}\int_0^t \|\theta(s)\|_{H_\theta}^2\,\mathrm{d}s + \frac{3K_\alpha}{2}\int_0^t \|\mathbf{u}(s)\|_{\mathrm{V_u}}^2\,\mathrm{d}s.$$

5.3.25 Lemma (Existence and Uniqueness)
Let Assumption 4.3.1$_{(A1)}$ be valid and assume in addition $\theta \in \mathcal{H}_\theta$. Then there exists a unique solution $\mathbf{u} \in \mathcal{V}_\mathrm{u}$ satisfying (5.18) Moreover,

$$\mathbf{u} \in \mathcal{V}_\mathrm{u}^\infty, \qquad \mathbf{u}' \in \mathcal{H}_\mathrm{u}^\infty, \qquad \mathbf{u}'' \in \mathcal{V}_\mathrm{u}^*.$$

Proof. The proof is based on the (Faedo-)Galerkin method.

(1) *Existence of Galerkin solutions:*
Let $(\mathbf{v}_j)_{j\in\mathbb{N}}$ be a Galerkin basis in the separable Banach space $\mathbf{V}_\mathrm{u}$ with $(\mathbf{V}_{\mathrm{u},n})_{n\in\mathbb{N}}$, $\mathbf{V}_{\mathrm{u},n} := \mathrm{span}\{\mathbf{v}_1, \ldots, \mathbf{v}_n\}$, $\mathbf{V}_\mathrm{u} = \overline{\bigcup_{n\in\mathbb{N}} \mathbf{V}_{\mathrm{u},n}}$ the corresponding Galerkin scheme. Furthermore, let $(\mathbf{u}_{0,n})_{n\in\mathbb{N}}$ and $(\mathbf{u}_{1,n})_{n\in\mathbb{N}}$ sequences in $\mathbf{V}_\mathrm{u}$ with

(5.20) $\quad \mathbf{u}_{0,n} \in \mathbf{V}_{\mathrm{u},n} \quad \forall n \in \mathbb{N}, \quad \mathbf{u}_{0,n} \to \mathbf{u}_0$ in $\mathbf{V}_\mathrm{u}$ for $\quad n \to \infty$,

(5.21) $\quad \mathbf{u}_{1,n} \in \mathbf{V}_{\mathrm{u},n} \quad \forall n \in \mathbb{N}, \quad \mathbf{u}_{1,n} \to \mathbf{u}_1$ in $\mathbf{H}_\mathrm{u}$ for $\quad n \to \infty$,

We are seeking Galerkin solutions $\mathbf{u}_n : \bar{S} \to \mathbf{V}_{\mathrm{u},n}$ looking like

(5.22) $$\mathbf{u}_n(t) = \sum_{i=1}^n g_{ni}(t)\,\mathbf{v}_i, \qquad g_{nj} \in W^{2,2}(S), \qquad \forall\, t \in S, \qquad \forall\, 1 \leq i \leq n$$

which satisfy

$$(5.23)\quad \Big(\rho_0\,\mathbf{u}_n''(t),\mathbf{v}_j\Big)_{\mathrm{H}_{un}} + \Big\langle \mathbf{A}_u\mathbf{u}_n(t),\mathbf{v}_j\Big\rangle_{\mathrm{V}_{un}^*\mathrm{V}_{um}} = 3\int_\Omega K_\alpha\Big(\mathcal{S}_h\theta(t)-\theta_0\Big)\,\mathrm{div}(\mathbf{v}_j)\,\mathrm{d}\mathbf{x}$$
$$+\int_\Omega K\sum_{i=1}^N\left(\frac{\rho_0}{\rho_i(\theta_0)}-1\right)p_i(t)\,\mathrm{div}(\mathbf{v}_j)\,\mathrm{d}\mathbf{x} + 2\int_\Omega \mu\,\boldsymbol{\varepsilon}_{trip} : \nabla\mathbf{v}_j\,\mathrm{d}\mathbf{x}$$
$$+2\int_\Omega \mu\,\boldsymbol{\varepsilon}_{cp} : \nabla\mathbf{v}_j\,\mathrm{d}\mathbf{x} + \Big(\mathbf{f}(t),\mathbf{v}_j\Big)_{\mathrm{H}_{um}},\ j=1,\dots,n,$$

and the initial conditions

$$(5.24)\qquad \mathbf{u}_n(0)=\mathbf{u}_{0,n}, \qquad\qquad \mathbf{u}_n'(0)=\mathbf{u}_{1,n},$$

where $\mathbf{u}_{0,n}$ is, e.g., the orthogonal projection in $\mathbf{H}_u$ of $\mathbf{u}_0$ on the space $\mathbf{V}_{u,n}$. In general, $\mathbf{u}_{0,n}$ can be any element of the space $\mathbf{V}_{u,n}$ s.t. $\mathbf{u}_{0,n}\to\mathbf{u}_0$ in the norm of $\mathbf{H}_u$, as $n\to\infty$.
It is well-known, that the matrix $\mathbf{G}_{un} := ((\mathbf{v}_i,\mathbf{v}_j)_{\mathrm{H}_{un}})_{i,j=1,\dots,m}$ is non-singular; hence by inverting this matrix we reduce (5.23),(5.24) to a linear system with constant coefficients

$$(5.25)\qquad g_{nj}''(t)+\sum_{i=1}^n \alpha_{ij}g_{nj}(t) = \sum_{i=1}^n \beta_{ij}\Big(\mathbf{F}(t),\mathbf{v}_j\Big)_{\mathrm{H}_{un}},$$

where $\alpha_{ij},\beta_{ij}\in\mathbb{R}$ and $\mathbf{F} := \tilde{\mathbf{f}}_1+\tilde{\mathbf{f}}_2$. The initial condition is equivalent to $\zeta_{unj}(0)$, which is the jth component of $\mathbf{u}_{0,n}$. The linear second-order ODE system (5.25) together with the initial conditions (5.24) defines uniquely the g_{nj} on the whole interval S. Since the scalar functions $t\mapsto\langle\mathbf{f}(t),\mathbf{w}_j\rangle$ are square integrable, so are the functions g_{nj} and therefore, for each n, $\mathbf{u}_n\in\mathcal{V}_u$, $\mathbf{u}_n'\in\mathcal{H}_u$. Due to the given assumptions the theorem of Carathéodory (cf. [Roubíček 2005] for a global result and [Coddington and Levinson 1955; Fillipov 1988; Walter 2000; Emmrich 2004; Zeidler 1990b] for alternatives) is applicable to this initial value problem (5.25),(5.24) and there exists exactly one absolutely continuous solution $\mathbf{u}_n\in AC^1(\bar{S};\mathbf{V}_{u,n})$.
The Peano existence theorem and the Picard-Lindelöf theorem are not applicable in this case, because the right-hand side $\mathbf{F}$ is not (Lipschitz-) continuous in t.

(2) *A-priori Estimates:*
By standard techniques, we get the analogous a-priori estimates as in Proposition 5.3.21, namely

$$\sup_{t\in\bar{S}}\|\mathbf{u}_n'(t)\|_{\mathrm{H}_u}\le c, \qquad\qquad \sup_{t\in\bar{S}}\|\mathbf{u}_n(t)\|_{\mathrm{V}_u}\le c,$$

where the positive constant c does not depend on m.

(3) *Passage to the Limit $n \to \infty$:*
The a-priori estimates show the existence of an element $\mathbf{u} \in \mathcal{V}_\mathrm{u}$ and a subsequence (cf. [Werner 2005] for the argumentation via Eberlein-Schmulyan theorem or rather Banach-Alaoglu theorem) of $(\mathbf{u}_n)_{n\in\mathbb{N}}$ – still denoted by $(\mathbf{u}_n)_{n\in\mathbb{N}}$ – s.t.

$$\mathbf{u}_n \rightharpoonup \mathbf{u} \quad \text{in} \quad \mathcal{V}_u \quad \text{and} \quad \mathbf{u}_n \overset{*}{\rightharpoonup} \mathbf{u} \quad \text{in} \quad \mathcal{V}_u^\infty,$$
$$\mathbf{u}'_n \rightharpoonup \mathbf{w}' \quad \text{in} \quad \mathcal{H}_u \quad \text{and} \quad \mathbf{u}'_n \overset{*}{\rightharpoonup} \mathbf{v}' \quad \text{in} \quad \mathcal{H}_u^\infty.$$

(Only $\mathbf{u}'$ can be the weak*-limit of $(\mathbf{u}'_m)_{m\in\mathbb{N}}$, cf. [Zeidler 1990a].) Furthermore, it follows from the weak lower semi-continuity of norms that

$$\|\mathbf{u}'\|_{\mathcal{H}_u^\infty} + \|\mathbf{u}\|_{\mathcal{V}_u^\infty} < \infty.$$

It remains to prove that $\mathbf{u}$ is a solution of the original problem.
In order to pass to the limit in the Galerkin equations, let $\varphi_\mathrm{u} \in C^1(\bar{S})$ an arbitrary scalar function with $\varphi_u(T) = 0$. Multiplying Equation (5.18) by φ_u and integration by parts over S yields for $k \in \mathbb{N}$:

$$\begin{aligned}
&\int_0^T \Big(\rho_0\, \mathbf{u}''_n(t), \varphi(t)\mathbf{v}_k\Big)_{\mathrm{H_u}} \mathrm{d}t + \int_0^T \Big\langle \mathbf{A}_\mathrm{u}\mathbf{u}_n(t), \varphi(t)\mathbf{v}_k\Big\rangle_{\mathrm{V_u^*V_u}} \mathrm{d}t \\
&\quad = \int_0^T \Big(\mathbf{f}(t), \varphi(t)\mathbf{v}_k\Big)_{\mathrm{H_u}} \mathrm{d}t + 3\int_0^T\int_\Omega K_\alpha \Big(\mathcal{S}_h\theta(t) - \theta_0\Big)\, \mathrm{div}(\varphi(t)\mathbf{v}_k)\, \mathrm{d}\mathbf{x}\, \mathrm{d}t \\
&\qquad + \int_0^T\int_\Omega K \sum_{i=1}^m \left(\frac{\rho_0}{\rho_i(\theta_0)} - 1\right) p_i(t)\, \mathrm{div}(\varphi(t)\mathbf{v}_k)\, \mathrm{d}\mathbf{x}\, \mathrm{d}t \\
&\qquad + 2\int_0^T\int_\Omega \mu\, \boldsymbol{\varepsilon}_{trip} : \nabla\, \varphi(t)\mathbf{v}_k\, \mathrm{d}\mathbf{x}\, \mathrm{d}t + 2\int_0^T\int_\Omega \mu\, \boldsymbol{\varepsilon}_{cp} : \nabla\, \varphi(t)\mathbf{v}_k\, \mathrm{d}\mathbf{x}\, \mathrm{d}t.
\end{aligned}$$

Integration by parts yields

$$\begin{aligned}
\int_0^T \Big(\rho_0\, \mathbf{u}''_n(t), \varphi(t)\mathbf{v}_k\Big)_{\mathrm{H_u}} \mathrm{d}t = -\int_0^T \Big(\rho_0\mathbf{u}'_n(t), \varphi'(t)\mathbf{v}_k\Big)_{\mathrm{H_u}} \mathrm{d}t + \\
- \Big(\rho_0\, \mathbf{u}'_n(0), \varphi(0)\mathbf{v}_k\Big)_{\mathrm{H_u}}.
\end{aligned}$$

The passage to the limit for $n \to \infty$ in the integrals on the left-hand side is easy, using the weak star convergence as well as the linearity and the continuity of the operator $\mathbf{A}_\mathrm{u}$. Observe $\mathbf{u}_{0,n} \to \mathbf{u}_0$ in $\mathbf{H}_\mathrm{u}$. Hence

$$\begin{aligned}
&-\int_0^T \Big(\rho_0\mathbf{u}'(t), \varphi'(t)\mathbf{v}_k\Big)_{\mathrm{H_u}} \mathrm{d}t - \Big(\rho_0\, \mathbf{u}'(0), \varphi(0)\mathbf{v}_k\Big)_{\mathrm{H_u}} + \int_0^T \Big\langle \mathbf{A}_\mathrm{u}\mathbf{u}(t), \varphi(t)\mathbf{v}_k\Big\rangle_{\mathrm{V_u^*V_u}} \mathrm{d}t \\
&\quad = \int_0^T \Big(\mathbf{f}(t), \varphi(t)\mathbf{v}_k\Big)_{\mathrm{H_u}} \mathrm{d}t + 3\int_0^T\int_\Omega K_\alpha \Big(\mathcal{S}_h\theta(t) - \theta_0\Big)\, \mathrm{div}(\varphi(t)\mathbf{v}_k)\, \mathrm{d}\mathbf{x}\, \mathrm{d}t
\end{aligned}$$

$$+\int_0^T \int_\Omega K \sum_{i=1}^m \left(\frac{\rho_0}{\rho_i(\theta_0)} - 1 \right) p_i(t) \operatorname{div}(\varphi(t)\mathbf{v}_k) \, \mathrm{d}\mathbf{x} \, \mathrm{d}t$$

$$+2\int_0^T \int_\Omega \mu \, \boldsymbol{\varepsilon}_{trip} : \nabla \varphi(t)\mathbf{v}_k \, \mathrm{d}\mathbf{x} \, \mathrm{d}t + 2\int_0^T \int_\Omega \mu \, \boldsymbol{\varepsilon}_{cp} : \nabla \varphi(t)\mathbf{v}_k \, \mathrm{d}\mathbf{x} \, \mathrm{d}t,$$

The linear hull of the functions $\mathbf{v}_k$ is dense in $\mathbf{V}_\mathrm{u}$. Therefore, the functions $\varphi(t)\mathbf{v}_k$ are dense in $\mathcal{V}_u$ and admissionable test functions (cf. [Naumann 2005a]).
Finally, it remains to check that $\mathbf{u}(0) = \mathbf{u}_0$ and $\mathbf{u}'(0) = \mathbf{u}_1$. Integration by parts yields

$$\int_0^T (\mathbf{u}_n'(t), \varphi(t)\mathbf{v}_k)_{\mathrm{H_u}} \, \mathrm{d}t = -\int_0^T (\mathbf{u}_n(t), \varphi'(t)\mathbf{v}_k)_{\mathrm{H_u}} \, \mathrm{d}t + (\mathbf{u}_{0,n}, \varphi(0)\mathbf{v}_k)_{\mathrm{H_u}}$$

for $k = 1, \ldots, n$ and

$$\int_0^T (\mathbf{u}'(t), \varphi(t)\mathbf{v}_k)_{\mathrm{H_u}} \, \mathrm{d}t = -\int_0^T (\mathbf{u}(t), \varphi'(t)\mathbf{v}_k)_{\mathrm{H_u}} \, \mathrm{d}t + (\mathbf{u}(0), \varphi(0)\mathbf{v}_k)_{\mathrm{H_u}}$$

for $k \in \mathbb{N}$. The passage to the limit for $m \to \infty$ gives

$$(\mathbf{u}_{0,n}, \varphi(0)\mathbf{v}_k)_{\mathrm{H_u}} \overset{n\to\infty}{\longrightarrow} (\mathbf{u}(0), \varphi(0)\mathbf{v}_k)_{\mathrm{H_u}} \quad \forall \; k \in \mathbb{N}.$$

A density argument yields $\mathbf{u}_{0,n} \rightharpoonup \mathbf{u}(0)$ for $n \to \infty$ in $\mathbf{H}_\mathrm{u}$. Due to $\mathbf{u}_{0,n} \to \mathbf{u}_0$ for $n \to \infty$ in $\mathbf{V}_\mathrm{u}$ it follows $\mathbf{u}(0) = \mathbf{u}_0$.
Analogously, integration by parts yields:

$$\int_0^T (\mathbf{u}_n''(t), \varphi(t)\mathbf{v}_k)_{\mathrm{H_u}} \, \mathrm{d}t = -\int_0^T (\mathbf{u}_n'(t), \varphi'(t)\mathbf{v}_k)_{\mathrm{H_u}} \, \mathrm{d}t + (\mathbf{u}_{1,n}, \varphi(0)\mathbf{v}_k)_{\mathrm{H_u}}$$

for $k = 1, \ldots, n$ and

$$\int_0^T (\mathbf{u}''(t), \varphi(t)\mathbf{v}_k)_{\mathrm{H_u}} \, \mathrm{d}t = -\int_0^T (\mathbf{u}'(t), \varphi'(t)\mathbf{v}_k)_{\mathrm{H_u}} \, \mathrm{d}t + (\mathbf{u}'(0), \varphi(0)\mathbf{v}_k)_{\mathrm{H_u}}$$

for $k \in \mathbb{N}$. The passage to the limit for $m \to \infty$

$$(\mathbf{u}_{1,n}, \varphi(0)\mathbf{v}_k)_{\mathrm{H_u}} \overset{n\to\infty}{\longrightarrow} (\mathbf{u}'(0), \varphi(0)\mathbf{v}_k)_{\mathrm{H_u}} \quad \forall \; k \in \mathbb{N}.$$

The density argument yields $\mathbf{u}_{1,n} \rightharpoonup \mathbf{u}'(0)$ for $n \to \infty$ in $\mathbf{H}_\mathrm{u}$. Due to $\mathbf{u}_{1,n} \to \mathbf{u}_1$ for $n \to \infty$ in $\mathbf{V}_\mathrm{u}$ it follows $\mathbf{u}'(0) = \mathbf{u}_1$.
Thus, $\mathbf{u}$ is a weak solution of the original problem with the specified properties. Because of the weak lower semi-continuity of norms, the solution is also bounded in the appropriate spaces.

(4) *Uniqueness:*
Let $\mathbf{u}_1$ and $\mathbf{u}_2$ be two solutions of Problem $(\mathbf{P}_{A,\mathrm{u}})$ and let $\mathbf{u} := \mathbf{u}_1 - \mathbf{u}_2$. Then $\mathbf{u}$ belongs to the same spaces as $\mathbf{u}_1$ and $\mathbf{u}_2$ and fulfills (5.18). Linearity gives $\mathbf{u}(0) = 0$, $\mathbf{u}'(0) = 0$ and

$$\int_0^T \left\langle \rho_0 \mathbf{u}''(t), \mathbf{v}(t) \right\rangle_{\mathrm{V}_\mathrm{u}^*\mathrm{V}_\mathrm{u}} \mathrm{d}t + \int_0^T \left\langle \mathbf{A}_\mathrm{u}\mathbf{u}(t), \mathbf{v}(t) \right\rangle_{\mathrm{V}_\mathrm{u}^*\mathrm{V}_\mathrm{u}} \mathrm{d}t = 0 \tag{5.26}$$

f.a. $\mathbf{v} \in \mathcal{V}_\mathrm{u}$. Testing the equation (5.26) with $\mathbf{v}(t) = \chi_{[0,\vartheta]}\, \mathbf{u}'(t)$ f.a.a. $t \in \bar{S}$ gives:

$$\rho_0 \int_0^\vartheta \left\langle \mathbf{u}''(t), \mathbf{u}'(t) \right\rangle_{\mathrm{V}_\mathrm{u}^*\mathrm{V}_\mathrm{u}} \mathrm{d}t + \int_0^\vartheta \left\langle \mathbf{A}_\mathrm{u}\mathbf{u}(t), \mathbf{u}'(t) \right\rangle_{\mathrm{V}_\mathrm{u}^*\mathrm{V}_\mathrm{u}} \mathrm{d}t = 0.$$

The a-priori estimates provide overall:

$$\frac{\rho_0}{2}\|\mathbf{u}'(\vartheta)\|^2_{\mathrm{H}_\mathrm{u}} + \frac{c}{2}\|\mathbf{u}(\vartheta)\|^2_{\mathrm{V}_\mathrm{u}} \leq 0.$$

Hence, $\|\mathbf{u}(\vartheta)\|_{\mathrm{V}_\mathrm{u}} = 0$ f.a.a. $t \in S$. Therefore, $\mathbf{u} = 0$ a.e. on S, i.e. $\mathbf{u}_1 = \mathbf{u}_2$ a.e. on S.

□

5.3.26 Lemma (Continuous Dependence on Parameters $\mathbf{u}_0$, $\mathbf{u}_1$, $\mathbf{f}$, θ, $\mathbf{p}$, $\boldsymbol{\varepsilon}_{trip}$ and $\boldsymbol{\varepsilon}_{cp}$)
Let Assumption 4.3.1$_{(A1)}$ be valid and let $\mathbf{u}_i$ be two different solutions of Problem $(\mathbf{P}_{A,\mathrm{u}})$ from Lemma 5.3.25 corresponding to initial data $\mathbf{u}_{0,i}$, $\mathbf{u}_{1,i}$ and parameters $\mathbf{f}_i$, θ_i, $\mathbf{p}_i$, $\boldsymbol{\varepsilon}_{trip,i}$, $\boldsymbol{\varepsilon}_{cp,i}$, $i = 1, 2$. Then there exists a constant $c > 0$ s.t.

$$\begin{aligned}\|\mathbf{u}_1' - \mathbf{u}_2'\|_{\mathcal{H}_\mathrm{u}^\infty} + \|\mathbf{u}_1 - \mathbf{u}_2\|_{\mathcal{V}_\mathrm{u}^\infty} \leq c(h^{-1}) \Big\{ &\|\mathbf{u}_{1,1} - \mathbf{u}_{1,2}\|_{\mathrm{H}_\mathrm{u}} + \|\mathbf{u}_{0,1} - \mathbf{u}_{0,2}\|_{\mathrm{V}_\mathrm{u}} + \|\mathbf{f}_1 - \mathbf{f}_2\|_{\mathcal{H}_\mathrm{u}} \\ &+ \|\theta_1 - \theta_2\|_{\mathcal{H}_\theta} + \|\mathbf{p}_1 - \mathbf{p}_2\|_{\mathcal{H}_\mathrm{p}^\infty} + \|\mathbf{p}_1' - \mathbf{p}_2'\|_{\mathcal{H}_\mathrm{p}} + \|\boldsymbol{\varepsilon}_{trip,1} - \boldsymbol{\varepsilon}_{trip,2}\|_{\mathcal{H}_\sigma^\infty} \\ &+ \|\boldsymbol{\varepsilon}_{trip,1}' - \boldsymbol{\varepsilon}_{trip,2}'\|_{\mathcal{H}_\sigma} + \|\boldsymbol{\varepsilon}_{cp,1} - \boldsymbol{\varepsilon}_{cp_2}\|_{\mathcal{H}_\sigma^\infty} + \|\boldsymbol{\varepsilon}_{cp,1}' - \boldsymbol{\varepsilon}_{cp,2}'\|_{\mathcal{H}_\sigma} \Big\}.\end{aligned}$$

Proof. The estimates are obtained analogously to Proposition 5.3.21 and 5.3.24. □

5.3.7 Step 6: Analysis of Problem $(\mathbf{P}_{A,FP2})$

We continue with studying the solvability of Problem $(\mathbf{P}_{A,FP2})$ for fixed data $\hat{\theta}$.

5.3.27 Lemma (Existence and Uniqueness)
Let Assumption 4.3.1 be valid. Assume $\hat{\theta} \in \mathcal{H}_\theta^\infty \cap \mathcal{V}_\theta$. Then there exists a weak solution $(\mathbf{u}, \boldsymbol{\sigma}, \boldsymbol{\varepsilon}_{cp}) \in \mathcal{V}_\mathrm{u} \cap \mathfrak{H}_\mathrm{u} \times \mathcal{H}_\sigma \times \mathfrak{H}_\sigma$ of Problem $(\mathbf{P}_{A,FP2})$.

Moreover, let $(\mathbf{u}_i, \boldsymbol{\sigma}_i, \boldsymbol{\varepsilon}_{cp,i})$ *be solutions w.r.t. the data* $\hat{\theta}_i$, $i = 1, 2$. *Then the following estimate holds:*

$$(5.27)\quad \|\mathbf{u}_1 - \mathbf{u}_2\|_{\mathcal{V}_u} + \|\mathbf{u}_1' - \mathbf{u}_2'\|_{\mathcal{H}_u} + \|\boldsymbol{\sigma}_1 - \boldsymbol{\sigma}_2\|_{\mathcal{H}_\sigma} + \|\boldsymbol{\varepsilon}_{cp,1} - \boldsymbol{\varepsilon}_{cp,2}\|_{\mathfrak{H}_\sigma} \le c\left\{\|\hat{\theta}_1 - \hat{\theta}_2\|_{\mathcal{H}_\theta^\infty \cap \mathcal{V}_\theta} + \|\hat{\theta}_1' - \hat{\theta}_2'\|_{\mathcal{H}_\theta}\right\},$$

where the positive constant does not depend on the data.

Proof. Again, we use a standard fixed-point argument. Before we start with the proof let us cite the following results for the subproblems.
Let $(\theta, \mathbf{p}, \boldsymbol{\varepsilon}_{cp}, \boldsymbol{\varepsilon}_{trip}) \in \mathbf{H}_\theta^\infty \cap \mathfrak{H}_\theta \times \mathbf{H}_p^\infty \cap \mathfrak{H}_p \times \mathbf{H}_\sigma^\infty \cap \mathfrak{H}_\sigma \times \mathbf{H}_\sigma^\infty \cap \mathfrak{H}_\sigma$. The problem (5.1) possesses a unique solution $(\mathbf{u}, \boldsymbol{\sigma}) \in \mathfrak{H}_u^\infty \cap \mathcal{V}_u^\infty \times \mathcal{H}_\sigma^\infty$ s.t.

$$\begin{aligned}\|\mathbf{u}_1 - \mathbf{u}_2\|_{\mathcal{V}_u^\infty} + \|\mathbf{u}_1' - \mathbf{u}_2'\|_{\mathcal{H}_u^\infty} &\le c_1\left\{\|\hat{\theta}_1 - \hat{\theta}_2\|_{\mathcal{H}_\theta^\infty \cap \mathfrak{H}_\theta} + \|\mathbf{p}_1 - \mathbf{p}_2\|_{\mathcal{H}_p^\infty \cap \mathfrak{H}_p}\right.\\ &\quad \left.+ \|\bar{\boldsymbol{\varepsilon}}_{cp,1} - \bar{\boldsymbol{\varepsilon}}_{cp,2}\|_{\mathcal{H}_\sigma^\infty \cap \mathfrak{H}_\sigma} + \|\boldsymbol{\varepsilon}_{trip,1} - \boldsymbol{\varepsilon}_{trip,2}\|_{\mathcal{H}_\sigma^\infty \cap \mathfrak{H}_\sigma}\right\}\\ \text{and } \|\boldsymbol{\sigma}_1 - \boldsymbol{\sigma}_2\|_{\mathcal{H}_\sigma} &\le c_2\left\{\|\mathbf{u}_1 - \mathbf{u}_2\|_{\mathcal{V}_u} + \|\hat{\theta}_1 - \hat{\theta}_2\|_{\mathcal{H}_\theta} + \|\mathbf{p}_1 - \mathbf{p}_2\|_{\mathcal{H}_p}\right.\\ &\quad \left.+ \|\bar{\boldsymbol{\varepsilon}}_{cp,1} - \bar{\boldsymbol{\varepsilon}}_{cp,2}\|_{\mathcal{H}_\sigma} + \|\boldsymbol{\varepsilon}_{trip,1} - \boldsymbol{\varepsilon}_{trip,2}\|_{\mathcal{H}_\sigma}\right\},\end{aligned}$$

where c_1 and c_2 are positive constants.
Using the definition of the stress deviator and the preceding estimate one get

$$\begin{aligned}\|\boldsymbol{\sigma}_1^* - \boldsymbol{\sigma}_2^*\|_{\mathcal{H}_\sigma} &\le c_1\left\{\|\mathbf{u}_1 - \mathbf{u}_2\|_{\mathcal{V}_u} + \|\theta_1 - \theta_2\|_{\mathcal{H}_\theta} + \|\mathbf{p}_1 - \mathbf{p}_2\|_{\mathcal{H}_p}\right.\\ &\quad \left.+ \|\boldsymbol{\varepsilon}_{trip,1} - \boldsymbol{\varepsilon}_{trip,2}\|_{\mathcal{H}_\sigma} + \|\boldsymbol{\varepsilon}_{cp,1} - \boldsymbol{\varepsilon}_{cp,2}\|_{\mathcal{H}_\sigma}\right\},\\ \|(\boldsymbol{\sigma}^*)_1' - (\boldsymbol{\sigma}^*)_2'\|_{\mathcal{H}_\sigma} &\le c_2(h^{-1})\left\{\|\mathbf{u}_1 - \mathbf{u}_2\|_{\mathcal{V}_u} + \|\theta_1' - \theta_2'\|_{\mathcal{H}_\theta} + \|\mathbf{p}_1' - \mathbf{p}_2'\|_{\mathcal{H}_p}\right.\\ &\quad \left.+ \|\boldsymbol{\varepsilon}_{trip,1}' - \boldsymbol{\varepsilon}_{trip,2}'\|_{\mathcal{H}_\sigma} + \|\boldsymbol{\varepsilon}_{cp,1}' - \boldsymbol{\varepsilon}_{cp,2}'\|_{\mathcal{H}_\sigma}\right\}.\end{aligned}$$

Let $(\mathbf{u}, \boldsymbol{\sigma}, \boldsymbol{\varepsilon}_{trip}) \in \mathcal{V}_u \times \mathcal{H}_\sigma \times \mathfrak{H}_\sigma$. The problem (5.6) possesses a unique solution $\boldsymbol{\varepsilon}_{cp} \in \mathfrak{X}_\sigma^\infty \cap \mathcal{H}_\sigma$ s.t.

$$\|\boldsymbol{\varepsilon}_{cp,1} - \boldsymbol{\varepsilon}_{cp,2}\|_{\mathfrak{H}_\sigma} \le c(h^{-1})\left\{\|\mathbf{u}_1 - \mathbf{u}_2\|_{\mathcal{V}_u} + \|\boldsymbol{\varepsilon}_{trip,1} - \boldsymbol{\varepsilon}_{trip,2}\|_{\mathfrak{H}_\sigma} + \|\boldsymbol{\sigma}_1^* - \boldsymbol{\sigma}_2^*\|_{\mathcal{H}_\sigma}\right\},$$

where c is a positive constant depending on the regularisation parameter h.

According to Lemma 5.3.16 we use Banach's fixed-point theorem with a weighted-norm space. Define $\mathbf{X} := \mathcal{V}_u \times \mathfrak{H}_\sigma \times \mathfrak{H}_\sigma$. For $(\bar{\mathbf{u}}, \bar{\boldsymbol{\sigma}}, \bar{\boldsymbol{\varepsilon}}_{cp}) \in \mathbf{X}$ given, we consider the linearised problem (5.1) for $(\mathbf{u}, \boldsymbol{\sigma}, \boldsymbol{\varepsilon}_{cp}) \in \mathbf{X}$ in which the parameter $\hat{\theta}$ is fixed and $\boldsymbol{\varepsilon}_{trip}$ is the solution of Problem $(\mathbf{P}_{A,FP1})$ according to Lemma 5.3.16. Then there exists a unique solution $(\bar{\mathbf{u}}, \bar{\boldsymbol{\sigma}}, \bar{\boldsymbol{\varepsilon}}_{cp}) \in \mathcal{V}_{\mathrm{u}}^{\infty} \cap \mathfrak{H}_{\mathrm{u}}^{\infty} \times \mathfrak{H}_{\sigma} \times \mathfrak{H}_{\sigma} \hookrightarrow \mathbf{X}$. We define a fixed-point operator as

$$(5.28) \qquad \mathbf{T}_{A,FP2} : \mathbf{X} \to \mathbf{X}, \qquad \mathbf{T}_{A,FP2}(\bar{\mathbf{u}}, \bar{\boldsymbol{\sigma}}, \bar{\boldsymbol{\varepsilon}}_{cp}) = (\mathbf{u}, \boldsymbol{\sigma}, \boldsymbol{\varepsilon}_{cp})$$

and consider two different solutions $(\mathbf{u}_1, \boldsymbol{\sigma}_1, \boldsymbol{\varepsilon}_{cp,1})$, $(\mathbf{u}_2, \boldsymbol{\sigma}_2, \boldsymbol{\varepsilon}_{cp,2})$ corresponding to different data $(\bar{\mathbf{u}}_1, \bar{\boldsymbol{\sigma}}_1, \bar{\boldsymbol{\varepsilon}}_{cp,1})$, $(\bar{\mathbf{u}}_2, \bar{\boldsymbol{\sigma}}_2, \bar{\boldsymbol{\varepsilon}}_{cp,2})$. By similar arguments as for Lemma 5.3.16 we obtain f.a.a. $t \in S$ the following estimate

$$\begin{aligned}
&\|\mathbf{u}_1(t) - \mathbf{u}_2(t)\|_{\mathrm{V_u}} + \|\mathbf{u}_1'(t) - \mathbf{u}_2'(t)\|_{\mathrm{H_u}} + \|\boldsymbol{\sigma}_1(t) - \boldsymbol{\sigma}_2(t)\|_{\mathrm{H}_\sigma} \\
&+ \|\boldsymbol{\varepsilon}_{cp,1}(t) - \boldsymbol{\varepsilon}_{cp,2}(t)\|_{\mathfrak{H}_\sigma} + \|\boldsymbol{\varepsilon}_{cp,1}'(t) - \boldsymbol{\varepsilon}_{cp,2}'(t)\|_{\mathfrak{H}_\sigma} \\
&\quad \le c \Big\{ \|\mathbf{p}_1(t) - \mathbf{p}_2(t)\|_{\mathrm{H_p}} + \|\mathbf{p}_1'(t) - \mathbf{p}_2'(t)\|_{\mathrm{H_p}} + \int_0^t \|\mathbf{p}_1'(s) - \mathbf{p}_2'(s)\|_{\mathrm{H_p}}\, ds \\
&\qquad + \|\bar{\boldsymbol{\varepsilon}}_{cp,1}(t) - \bar{\boldsymbol{\varepsilon}}_{cp,2}(t)\|_{\mathrm{H}_\sigma} + \int_0^t \|\bar{\boldsymbol{\varepsilon}}_{cp,1}'(s) - \bar{\boldsymbol{\varepsilon}}_{cp,2}'(s)\|_{\mathrm{H}_\sigma}\, ds \\
&\qquad + \|\boldsymbol{\sigma}_1^*(t) - \boldsymbol{\sigma}_2^*(t)\|_{\mathrm{H}_\sigma} + \|(\boldsymbol{\sigma}_1^*)'(t) - (\boldsymbol{\sigma}_2^*)'(t)\|_{\mathrm{H}_\sigma} \Big\} \\
&\quad \le c(h^{-1}) \Big\{ \|\bar{\boldsymbol{\varepsilon}}_{cp,1}(t) - \bar{\boldsymbol{\varepsilon}}_{cp,2}(t)\|_{\mathrm{H}_\sigma} + \int_0^t \|\bar{\boldsymbol{\varepsilon}}_{cp,1}'(s) - \bar{\boldsymbol{\varepsilon}}_{cp,2}'(s)\|_{\mathrm{H}_\sigma}\, ds \\
&\qquad + \|\bar{\boldsymbol{\sigma}}_1(t) - \bar{\boldsymbol{\sigma}}_2(t)\|_{\mathrm{H}_\sigma} + \int_0^t \|\bar{\boldsymbol{\sigma}}_1(s) - \bar{\boldsymbol{\sigma}}_2(s)\|_{\mathrm{H}_\sigma}\, ds + \int_0^t \|\bar{\mathbf{u}}_1(s) - \bar{\mathbf{u}}_2(s)\|_{\mathrm{V_u}}\, ds \Big\} \\
&\quad \le c(h^{-1}) \Big\{ \int_0^t \|\bar{\boldsymbol{\varepsilon}}_{cp,1}'(s) - \bar{\boldsymbol{\varepsilon}}_{cp,2}'(s)\|_{\mathrm{H}_\sigma}\, ds + \int_0^t \|\bar{\boldsymbol{\sigma}}_1(s) - \bar{\boldsymbol{\sigma}}_2(s)\|_{\mathrm{H}_\sigma}\, ds \\
&\qquad + \int_0^t \|\bar{\boldsymbol{\sigma}}_1'(s) - \bar{\boldsymbol{\sigma}}_2'(s)\|_{\mathrm{H}_\sigma}\, ds + \int_0^t \|\bar{\mathbf{u}}_1(s) - \bar{\mathbf{u}}_2(s)\|_{\mathrm{V_u}}\, ds \Big\} \\
&\quad \le c(h^{-1}) \Big\{ \|\bar{\mathbf{u}}_1 - \bar{\mathbf{u}}_2\|_{\mathcal{V}_\mathrm{u}} + \|\bar{\boldsymbol{\sigma}}_1 - \bar{\boldsymbol{\sigma}}_2\|_{\mathcal{H}_\sigma} + \|\bar{\boldsymbol{\varepsilon}}_{cp,1} - \bar{\boldsymbol{\varepsilon}}_{cp,2}\|_{\mathfrak{H}_\sigma} \Big\}.
\end{aligned}$$

It follows (analogously to the proof of Lemma 5.3.16)

$$\|(\mathbf{u}_1, \boldsymbol{\sigma}_1, \boldsymbol{\varepsilon}_{cp,1}) - (\mathbf{u}_2, \boldsymbol{\sigma}_2, \boldsymbol{\varepsilon}_{cp,2})\|_{\mathrm{X}} \le \sqrt{\frac{c(h^{-1})}{\lambda}}\, \|(\bar{\mathbf{u}}_1, \bar{\boldsymbol{\sigma}}_1, \bar{\boldsymbol{\varepsilon}}_{cp,1}) - (\bar{\mathbf{u}}_2, \bar{\boldsymbol{\sigma}}_2, \bar{\boldsymbol{\varepsilon}}_{cp,2})\|_{\mathrm{X}}$$

and hence, for λ large enough, $\mathbf{T}_{A,FP2}$ is strictly contractive. The conditions for Banach's fixed-point theorem are fulfilled and Problem $(\mathbf{P}_{A,FP2})$ possesses a unique solution. Moreover, the estimate (5.27) holds. □

5.3.8 Step 7: Analysis of Problem $(\mathbf{P}_{A,\theta})$

We continue with Problem $(\mathbf{P}_{A,\theta})$ for given $(\mathbf{u}, \mathbf{p}) \in \mathfrak{H}_\mathrm{u} \times \mathfrak{X}_\mathrm{p}^{\infty}$. Consider the (linearised) parabolic problem:

Problem ($\mathbf{P}_{A,\theta}$)
Find the temperature $\theta : \overline{\Omega}_T \to \mathbb{R}$ *s.t.*

$$\rho_0 c_e \frac{\partial \theta}{\partial t} - \operatorname{div}(\lambda_\theta \nabla \theta) + 3K_\alpha \theta_0 \operatorname{div}\left(\frac{\partial \mathcal{S}_h \mathbf{u}}{\partial t}\right) = +\rho_0 \sum_{i=2}^{m} L_i p_i' + r \quad \text{in} \quad \Omega_T \tag{5.29}$$

$$\theta(0) = \theta_0 \quad \text{in} \quad \Omega \qquad \text{and} \qquad -\lambda_\theta \nabla \theta \nu = \delta(\theta - \theta_\Gamma) \quad \text{on} \quad \partial\Omega \tag{5.30}$$

5.3.28 Remark (Mixed Boundary Conditions)
Mixed boundary conditions are also possible (and mathematically feasible), e.g.

$$-\lambda_\theta \nabla \theta \nu = \delta(\theta - \theta_{\Gamma_1}) \quad \text{on} \quad \Gamma_1 \qquad \text{and} \qquad \theta = \theta_{\Gamma_2} \quad \text{on} \quad \Gamma_2,$$

where Γ_1 *is a closed subset of the boundary* $\partial\Omega$ *with positive surface measure and* $\Gamma_2 := \partial\Omega \setminus \Gamma_1$.

5.3.29 Definition (Weak Formulation)
Under the Assumption 4.3.1$_{(A2)}$ *the function* $\theta \in \mathcal{U}_\theta$ *is called a weak solution of the Problem* ($\mathbf{P}_{A,\theta}$), *if* $\theta(\mathbf{x}, 0) = \theta_0(\mathbf{x})$ *a.e. and*

$$\begin{aligned} \left\langle \rho_0 \, c_e \frac{\partial}{\partial t} \theta(t), \vartheta \right\rangle_{V_\theta^* V_\theta} &+ \int_\Omega \lambda_\theta \, \nabla \theta(t) \, \nabla \vartheta \, \mathrm{d}\mathbf{x} + 3 \int_\Omega K_\alpha \, \theta_0 \operatorname{div}\left(\frac{\partial \mathcal{S}_h \mathbf{u}(t)}{\partial t}\right) \vartheta \, \mathrm{d}\mathbf{x} \\ &= \int_\Omega \rho_0 \sum_{i=2}^{m} L_i \, p_i'(t) \, \vartheta \, \mathrm{d}\mathbf{x} + \int_\Omega r(t) \, \vartheta \, \mathrm{d}\mathbf{x} + \int_{\partial\Omega} \delta \, (\theta_\Gamma - \theta(t)) \, \vartheta \, \mathrm{d}\sigma_x \end{aligned} \tag{5.31}$$

f.a. $\vartheta \in V_\theta$, *f.a.a.* $t \in S$.

We prove the (unique) existence of a weak solution of Problem ($\mathbf{P}_{A,\theta}$) with the help of the Galerkin method. Therefore, we introduce the operator

$$A_\theta : V_\theta \to V_\theta^*, \qquad \langle A_\theta \theta, \vartheta \rangle_{V_\theta^* V_\theta} = \lambda_\theta \int_\Omega \nabla \theta : \nabla \vartheta \, \mathrm{d}x + \delta \int_{\partial\Omega} \theta \vartheta \, \mathrm{d}\sigma_x, \qquad \theta, \vartheta \in V_\theta.$$

The operator $A_\theta : V_\theta \to V_\theta^*$ is linear, symmetric, continuous and strongly positive. Linearity and symmetry are obvious, continuity is standard and positivity is a consequence of the equivalence of the corresponding norms.

5.3.30 Remark (Thermal Dissipation)
Sometimes the thermal dissipation (3.4) *is considered instead of the simplification* (3.1). *Therefore, it might be useful to define the operator*

$$B_\theta : H_\theta \to \mathbf{V}_u^* \qquad \langle B_\theta \theta, \mathbf{u} \rangle_{\mathbf{V}_u^* \mathbf{V}_u} = \int_\Omega \theta \operatorname{div}(\mathbf{u}) \, \mathrm{d}\mathbf{x}, \qquad \theta \in H_\theta, \; \mathbf{u} \in \mathbf{V}_u.$$

In particular, B_θ is a continuous and linear operator $\mathbf{V}_u \subset \mathbf{H}_u \to \mathbf{H}_u^* \subset \mathbf{V}_u^*$, i.e. there exists a positive constant c s.t. $\|B_\theta \theta\|_{\mathbf{V}_u^*} \leq c\|\theta\|_{H_\theta}$ f.a. $\theta \in V_\theta$, cf. [Rincon et al. 2005]. Another possibility to treat the term of thermal dissipation (3.4) would be to integrate it into the fixed-point argumentation, cf. Remark 3.2.3.

Moreover, it is convenient to introduce the function $\tilde{r} \in \mathcal{V}_\theta^*$ via

$$\text{(5.32)} \qquad \langle \tilde{r}(t), v\rangle_{V_\theta^* V_\theta} := \int_\Omega r(t) v \, \mathrm{d}\mathbf{x} + \delta \int_{p\Omega} \theta_\Gamma(t) v \, \mathrm{d}\sigma_x, \qquad v \in V_\theta, \qquad \text{f.a.a. } t \in S.$$

5.3.31 Remark (Equivalent Weak Formulation)
Equation (5.31) can be rewritten as

$$\text{(5.33)} \quad \frac{\mathrm{d}}{\mathrm{d}t}\Big\langle \rho_0 c_e \theta(t), \vartheta \Big\rangle_{V_\theta^* V_\theta} = \Big\langle \underbrace{\tilde{r}(t) + \rho_0 \sum_{i=2}^{m} L_i p_i'(t) - A_\theta \theta(t) - 3K_\alpha \theta_0 \operatorname{div}(\mathbf{u}'(t))}_{=:\bar{r}(t)}, \vartheta \Big\rangle_{V_\theta^* V_\theta}$$

f.a. $\vartheta \in V_\theta$. Since $\theta \in \mathcal{V}_\theta$ and $A_\theta : V_\theta \to V_\theta^*$ is linear and continuous it holds $A_\theta \theta \in \mathcal{V}_\theta^*$. Hence, $\bar{\mathbf{r}} \in \mathcal{V}_\theta^*$ and therefore $\theta' \in \mathcal{V}_\theta^*$ and θ is a.e. equal to an absolutely continuous function $S \to V_\theta^*$, cf. [Zeidler 1990a] for details. Any function satisfying $\theta \in \mathcal{V}_\theta$ and equation (5.31) is (after modification on a set of measure zero) a continuous function $S \to V_\theta^*$, therefore $\theta(0) = \theta_0$ makes sense. For given $\bar{\mathbf{r}} \in \mathcal{V}_\theta^*$ and $\theta \in \mathcal{V}_\theta$ satisfying Equation (5.33) it follows $\theta' \in \mathcal{V}_\theta^*$.

We prove the a-priori estimate for the original Problem ($\mathbf{P}_\theta$) (without the Steklov regularisation for $\mathbf{u}'$) and make some remarks for the considered situation of Problem ($\mathbf{P}_{A,\theta}$) afterwards.

5.3.32 Proposition (A-priori Estimate)
Let Assumption 4.3.1$_{(A2)}$ be valid and $\theta \in \mathcal{H}_\theta^\infty \cap \mathcal{V}_\theta$ be a weak solution of the original Problem ($\mathbf{P}_\theta$). Then there exists $c > 0$ s.t.

$$\|\theta\|_{\mathcal{H}_\theta^\infty} + \|\theta\|_{\mathcal{V}_\theta} \leq c\Big\{\|\mathbf{u}'\|_{\mathcal{V}_u} + \|\mathbf{p}'\|_{\mathcal{H}_p} + \|\tilde{r}\|_{\mathcal{V}_\theta^*}\Big\}.$$

Proof. Taking $v = \theta(t)$ in the equation (5.31) , we obtain

$$\Big\langle \rho_0 \, c_e \, \theta'(t), \theta(t)\Big\rangle_{V_\theta^* V_\theta} + \Big\langle A_\theta \theta(t), \theta(t)\Big\rangle_{V_\theta^* V_\theta} + \underbrace{3\int_\Omega K_\alpha \, \theta_0 \, \operatorname{div}(\mathbf{u}'(t))\, \theta(t) \, \mathrm{d}\mathbf{x}}_{=:\mathbb{I}_1}$$

$$= \underbrace{\int_\Omega \rho_0 \sum_{i=1}^m L_i\, p_i'(t)\,\theta(t)\,\mathrm{d}\mathbf{x}}_{=:\mathbb{I}_2} + \underbrace{\left(\tilde{r}(t),\theta(t)\right)_{H_\theta}}_{=:\mathbb{I}_3}.$$

The following estimates hold for arbitrary $\varepsilon > 0$:

$$|\mathbb{I}_1| \leq \frac{9K_\alpha^2\theta_0^2}{2}\|\mathbf{u}'(t)\|^2_{\mathrm{V_u}} + \frac{1}{2}\|\theta\|^2_{H_\theta},$$
$$|\mathbb{I}_2| \leq \frac{\rho_0^2\|L\|^2_\infty}{2}\|\mathbf{p}'(t)\|^2_{\mathrm{H_p}} + \frac{1}{2}\|\theta(t)\|^2_{H_\theta},$$
$$|\mathbb{I}_3| \leq \frac{1}{2}\|r(t)\|^2_{H_\theta} + \frac{1}{2}\|\theta(t)\|^2_{H_\theta} + \frac{1}{4\varepsilon}\|\theta_\Gamma(t)\|^2_{L^2(\Gamma_T)} + \varepsilon\|\theta(t)\|^2_{\mathrm{V}_\theta}.$$

Integrating equation over S and using the estimates $\mathbb{I}_i$, $i = 1, 2, 3$ gives for $\varepsilon < c$

$$\frac{\rho_0\, c_e}{2}\|\theta(t)\|^2_{H_\theta} + \left(c-\varepsilon\right)\int_0^t \|\theta(s)\|^2_{\mathrm{V}_\theta}\,\mathrm{d}s \leq \frac{9K_\alpha^2\theta_0^2}{2}\int_0^t \|\mathbf{u}'(s)\|^2_{\mathrm{V_u}}\,\mathrm{d}s + \frac{1}{2}\int_0^t \|r(s)\|^2_{H_\theta}\,\mathrm{d}s$$
$$+ \frac{\rho_0^2\|L\|^2_\infty}{2}\int_0^t \|\mathbf{p}'(s)\|^2_{\mathrm{H_p}}\,\mathrm{d}s + \frac{1}{4\varepsilon}\int_0^t \|\theta_\Gamma(s)\|^2_{L^2(\Gamma)}\,\mathrm{d}s + \frac{3}{2}\int_0^t \|\theta(s)\|^2_{H_\theta}\,\mathrm{d}s.$$

Application of the Gronwall Lemma leads to:

$$\|\theta(t)\|^2_{H_\theta} + \int_0^t \|\theta(s)\|^2_{\mathrm{V}_\theta}\,\mathrm{d}s \leq c\left\{\int_0^t \|\mathbf{u}'(s)\|^2_{\mathrm{V_u}}\,\mathrm{d}s + \int_0^t \|\mathbf{p}'(s)\|^2_{\mathrm{H_p}}\,\mathrm{d}s \right.$$
$$\left. + \int_0^t \|r(s)\|^2_{H_\theta}\,\mathrm{d}s + \int_0^t \|\theta_\Gamma(s)\|^2_{L^2(\Gamma)}\,\mathrm{d}s\right\}.$$

Hence,

$$\sup_{t\in\bar{S}} \|\theta(t)\|_{H_\theta} \leq c, \qquad \int_0^T \|\theta(t)\|^2_{\mathrm{V}_\theta}\,\mathrm{d}t \leq c,$$

where c is a generic constant. □

5.3.33 Remark (A-priori Estimate for Problem ($\mathbf{P}_{A,\theta}$))
The a-priori estimate can also be proven for $\mathbf{u}' \in \mathcal{H}_\mathrm{u}$ if $\mathbf{u}'|_{\partial\Omega} = 0$ or $\nabla\theta\nu|_{\partial\Omega} = 0$.
For the Problem ($\mathbf{P}_{A,\theta}$) we use the approximation $\mathbf{u}' \approx \mathcal{D}_h\mathbf{u}$ with $\|\mathcal{D}_h\mathbf{u}\|_{\mathcal{V}_\mathrm{u}} \leq \frac{2}{h}\|\mathbf{u}\|_{\mathcal{V}_\mathrm{u}}$ and therefore we obtain for $c_1, c_2 > 0$:

$$\|\theta\|_{\mathcal{H}_\theta^\infty} + \|\theta\|_{\mathcal{V}_\theta} \leq c_1(h^{-1}) + c_2(h^{-1})\left\{\|\mathbf{u}\|_{\mathcal{V}_\mathrm{u}} + \|\mathbf{p}'\|_{\mathcal{H}_\mathrm{p}} + \|\tilde{r}\|_{\mathcal{V}_\theta^*}\right\}.$$

5.3.34 Lemma (Existence and Uniqueness)
Under Assumption 4.3.1$_{(A2)}$ the Problem ($\mathbf{P}_{A,\theta}$) admits a unique solution $\theta : \overline{\Omega}_T \to \mathbb{R}$, satisfying the following conditions:

$$\theta \in \mathcal{H}_\theta^\infty \cap \mathcal{V}_\theta, \qquad \theta' \in \mathcal{V}_\theta^*, \qquad \|\theta\|_{\mathcal{H}_\theta^\infty\cap\mathcal{V}_\theta} \leq c.$$

Proof. We use a Galerkin approximation of the Problem $(\mathbf{P}_{A,\theta})$.

(1) *Existence of Galerkin solutions:*
Let $\{v_k\}_{k\in\mathbb{N}}$ be a Galerkin basis of V_θ and $\{V_{\theta,i}\}_{i\in\mathbb{N}}$ with $V_{\theta,n} := \operatorname{span}\{v_1, \dots, v_n\}$ be the corresponding Galerkin scheme of dimension $n \in \mathbb{N}$. Moreover, let $\theta_{0,n}$ be a sequence with

$$\theta_{0,n} \in V_{\theta,n} \qquad \forall\, n \in \mathbb{N}, \qquad \theta_{0,n} \to \theta_0 \quad \text{in} \quad V_\theta \quad \text{for} \quad n \to \infty.$$

The Galerkin approximation consists of finding functions

$$\theta_n \in L^2(S; V_{\theta,n}), \qquad \theta_n = \theta_n(t) := \sum_{i=1}^{n} g_{ni}(t) v_i(\mathbf{x}), \qquad n \in \mathbb{N} \tag{5.34}$$

where g_{ni} are absolutely continuous functions, s.t.

$$\begin{aligned}(5.35)\quad & \left(\rho_0\, c_e\, \theta_n'(t), v_i\right)_{H_{\theta,n}} + \left\langle A_\theta \theta_n(t), v_i\right\rangle_{V_{\theta,n}^* V_{\theta,n}} \\ & - 3\int_\Omega K_\alpha\, \theta_0 \operatorname{div}((\mathcal{S}_h \mathbf{u})'(t))\, v_i\, \mathrm{d}\mathbf{x} = \int_\Omega \rho_0 \sum_{i=1}^{m} L_i\, p_i'(t)\, v_i\, \mathrm{d}\mathbf{x} + \left(\tilde{r}(t)\, v_i\right)_{H_{\theta,n}},\end{aligned}$$

$i = 1, \dots, n$ are satisfied f.a.a. $t \in S$ and f.a. $\phi \in V_{\theta,n}$. The approximate initial condition are

$$\theta_n(0) = \theta_{0,n}. \tag{5.36}$$

Since the $\{v_k\}_{k\in\mathbb{N}}$ are linearly independent and $\langle \rho_0 c_e \theta_n, v_n\rangle := (\rho c_e \theta_n, v_n)$ with $\theta_n, v_n \in V_{\theta n}$ is a scalar product in V_θ, it follows that the Gramian matrix

$$\mathbf{G}_{\theta n} := ((v_i, v_j)_{H_{\theta n}})_{i,j=1,\dots,n}$$

is invertible f.a. $t \in S$. Therefore, the system (5.35), (5.36) can be formulated (using the representation (5.34)) as a linear first order ODE system together with the initial condition g_{ni} (the ith component of θ_{0n}) for its coefficients $g_n : \bar{S} \to \mathbb{R}$ f.a. $i = 1, \dots, n$, $n \in \mathbb{N}$. Due to the given assumptions it follows by the Carathéodory theorem for ODEs (cf. e.g. [Coddington and Levinson 1955; Fillipov 1988; Walter 2000; Emmrich 2004]) that there exists a unique solution $\theta_n \in AC(\bar{S}; V_{\theta n})$ on the whole interval $\bar{S}$. Since the scalar functions $t \mapsto \langle \tilde{r}(t), w_j\rangle$ are square integrable, so are the functions g_{ni} and therefore, $\theta_n \in V_\theta$, $\theta_n' \in V_\theta^*$ f.a. $n \in \mathbb{N}$.

(2) *A-priori Estimates:*
By standard techniques, i.e. testing the Galerkin equations with θ_n, we obtain the analogous a-priori estimates as in Proposition 5.3.32, namely

$$\sup_{t\in\bar{S}} \|\theta_n'(t)\|_{H_\theta} \le c, \qquad \int_0^T \|\theta_n'(t)\|_{V_\theta}^2\, \mathrm{d}t \le c, \tag{5.37}$$

where the constant c does not depend on n.

(3) *Passage to the Limit $n \to \infty$:*
The a-priori estimate shows the existence of an element θ in $\mathcal{H}_\theta^\infty$ and a subsequence of $(\theta_n)_{n\in\mathbb{N}}$ – still denoted $(\theta_n)_{n\in\mathbb{N}}$ – s.t.

$$\theta_n \overset{*}{\rightharpoonup} \theta \quad \text{in} \quad \mathcal{H}_\theta^\infty,$$

i.e. f.a. $\vartheta \in L^1(S; H_\theta)$ it holds

$$\int_0^T (\theta_n(t) - \theta(t), \vartheta(t))\,\mathrm{d}t \to 0, \qquad n \to \infty. \tag{5.38}$$

By (5.37) the subsequence θ_n belongs to a bounded set of $\mathcal{V}_\theta$; therefore another passage to a subsequence shows the existence of some $\theta^* \in \mathcal{V}_\theta$ and some subsequence (still denoted θ_n) s.t.

$$\theta_n \rightharpoonup \theta \quad \text{in} \quad \mathcal{V}_\theta,$$

i.e.

$$\int_0^T \langle \theta_n(t) - \theta^*(t), \vartheta(t)\rangle\,\mathrm{d}t \to 0 \qquad \forall\, \vartheta \in \mathcal{V}_\theta^*.$$

In particular,

$$\int_0^T \langle \theta_n(t), \vartheta(t)\rangle\,\mathrm{d}t \to \int_0^T \langle \theta^*(t), \vartheta(t)\rangle\,\mathrm{d}t \qquad \forall\, \vartheta \in \mathcal{V}_\theta^*.$$

Comparing with (5.38) it holds:

$$\int_0^T (\theta_n(t) - \theta^*(t), \vartheta(t))\,\mathrm{d}t \to 0 \qquad \forall\, \vartheta \in \mathcal{V}_\theta^*.$$

Hence, $\theta = \theta^* \in \mathcal{V}_\theta \cap \mathcal{H}_\theta^\infty$. It remains to show that θ is a solution of the original problem.
Let $\varphi \in C^1(\bar{S})$ an arbitrary function with $\varphi(T) = 0$. Multiplication of (5.35) with φ and integration over S yields for $j \in \mathbb{N}$:

$$\begin{aligned}\int_0^T \big(\rho_0\, c_e\, \theta_n'(t), \varphi(t) v_j\big)_{H_\theta}\,\mathrm{d}t &+ \int_0^T \big\langle A_\theta \theta_n(t), \varphi(t) v_j\big\rangle_{V_\theta^* V_\theta}\,\mathrm{d}t \\ &= 3\int_0^T \int_\Omega K_\alpha\, \theta_0\, \mathrm{div}((\mathcal{S}_h \mathbf{u})'(t))\varphi(t) v_j\,\mathrm{d}\mathbf{x}\,\mathrm{d}t \\ &\quad + \int_0^T (\tilde{r}(t), \varphi(t) v_j)_{H_\theta}\,\mathrm{d}t + \int_0^T \int_\Omega \rho_0 \sum_{i=2}^m L_i\, p_i'(t)\, \varphi(t) v_j\,\mathrm{d}x\,\mathrm{d}t.\end{aligned}$$

Integration by parts yields

$$\int_0^T \left(\rho_0\, c_e\, \theta_n'(t), \varphi(t) v_j\right)_{H_\theta} \mathrm{d}t = -\int_0^T \left(\rho_0\, c_e\, \theta_n(t), \varphi'(t) v_j\right)_{H_\theta} \mathrm{d}t + \\ - \left(\rho_0\, c_e\, \theta_n(0), \varphi(0) v_j\right)_{H_\theta}.$$

Because of the weak-* convergence, the passage to the limit for $n \to \infty$ gives

$$-\int_0^T \left(\rho_0\, c_e\, \theta_n(t), \varphi'(t) v_j\right)_{H_\theta} \mathrm{d}t + \int_0^T \left\langle A_\theta \theta_n(t), \varphi(t) v_j\right\rangle_{V_\theta^* V_\theta} \mathrm{d}t \\ = \left(\rho_0\, c_e\, \theta_n(0), \varphi(0) v_j\right)_{H_\theta} + 3\int_0^T \int_\Omega K_\alpha\, \theta_0\, \operatorname{div}((\mathcal{S}_h \mathbf{u})'(t)) \varphi(t) v_j \,\mathrm{d}\mathbf{x}\,\mathrm{d}t \\ + \int_0^T \left(\tilde{r}(t), \varphi(t) v_j\right)_{H_\theta} \mathrm{d}t + \int_0^T \int_\Omega \rho_0 \sum_{i=2}^m L_i\, p_i'(t)\, \varphi(t) v_j \,\mathrm{d}x\,\mathrm{d}t.$$

The linear hull of v_j is dense in V_θ. The functions $\varphi(t)v_k$ are dense in $\mathcal{V}_\theta$ and therefore admissible test functions (cf. [Naumann 2005a]). It remains to show $\theta(0) = \theta_0$ for the existence of a solution. Integration by parts yields:

$$\int_0^T (\theta_n'(t), \varphi(t) v_k)_{H_\theta} \,\mathrm{d}t = -\int_0^T (\theta_n(t), \varphi'(t) v_k)_{H_\theta} \,\mathrm{d}t + (\theta_{0n}, \varphi(0) v_k)_{H_\theta}$$

for $k = 1, \dots, m$ and

$$\int_0^T (\theta'(t), \varphi(t) v_k)_{H_\theta} \,\mathrm{d}t = -\int_0^T (\theta(t), \varphi'(t) v_k)_{H_\theta} \,\mathrm{d}t + (\theta(0), \varphi(0) v_k)_{H_\theta}$$

for $k \in \mathbb{N}$. Because of the weak star convergence, the passage to the limit for $n \to \infty$ gives

$$(\theta_{0n}, \varphi(0) v_k)_{H_\theta} \overset{n\to\infty}{\longrightarrow} (\theta(0), \varphi(0) v_k)_{H_\theta}, \qquad \forall\ k \in \mathbb{N}.$$

A density argument yields $\theta_{0,n} \rightharpoonup \theta(0)$ for $n \to \infty$ in H_θ. Because $\theta_{0,n} \to \theta_0$ for $n \to \infty$ in V_θ^*, it follows $\theta(0) = \theta_0$. Therefore, θ is a weak solution of the original problem with given properties. Because of the weak lower semi-continuity of the norms also the boundedness of the solution in the appropriate spaces holds.

(4) *Uniqueness:*
Let $\theta = \theta_1 - \theta_2$, where θ_1, θ_2 are two different solutions of Problem $(\mathbf{P}_{A,\theta})$. Since $\theta(0) = \theta_{10} - \theta_{20} = 0$, we have:

$$\int_0^T \left\langle \rho_0\, c_e\, \theta'(t), w(t)\right\rangle_{V_\theta^* V_\theta} \mathrm{d}t + \int_0^T \left\langle A_\theta \theta(t), w(t)\right\rangle_{V_\theta^* V_\theta} \mathrm{d}t = 0$$

f.a. $w \in \mathcal{V}_\theta$. Multiplying by $\theta(t)$, we get

$$\rho_0\, c_e \int_0^\vartheta \left\langle \theta'(t), \theta(t) \right\rangle_{V_\theta^* V_\theta} \mathrm{d}t + \int_0^\vartheta \left\langle A_\theta \theta(t), \theta(t) \right\rangle_{V_\theta^* V_\theta} \mathrm{d}t = 0.$$

Integrating form 0 to t, we obtain

$$\frac{\rho_0\, c_e}{2} \|\theta(\vartheta)\|^2_{H_\theta} + c \int_0^\vartheta \|\theta(s)\|^2_{V_\theta} \,\mathrm{d}s \leq 0$$

Hence, we obtain

$$\|\theta(\vartheta)\|^2_{H_\theta} = 0 \quad \text{f.a.a. } t \in S,$$

which implies the uniqueness, $\theta(\mathbf{x}, t) = 0$ and the lemma is proved.

□

5.3.35 Lemma (Continuous Dependence on Parameters $\mathbf{u}$, $\mathbf{p}$ and $\tilde{r}$)
Let Assumption 4.3.1$_{(A2)}$ be valid. Then there exists a constant $c > 0$ s.t. for two different solutions θ_i of Problem ($\mathbf{P}_{A,\theta}$) from Lemma 5.3.34 corresponding to the data $(\mathbf{u}_i, \mathbf{p}_i, \tilde{r}_i)$, $i = 1, 2$ holds

$$\text{(5.39)} \quad \|\theta_1 - \theta_2\|_{\mathcal{H}_\theta^\infty \cap \mathcal{V}_\theta} \leq c(h^{-1}) \Big\{ \|\mathbf{u}_1 - \mathbf{u}_2\|_{\mathcal{V}_u} + \|\mathbf{p}_1' - \mathbf{p}_2'\|_{\mathcal{H}_p} + \|\tilde{r}_1 - \tilde{r}_2\|_{\mathcal{H}_\sigma} \Big\}.$$

Proof. Let θ_1, θ_2 be two different solutions of Problem ($\mathbf{P}_\theta$) (resp. Problem ($\mathbf{P}_{\theta,h}$)) corresponding to initial values θ_{i0} and data $\mathbf{u}_i$, $\boldsymbol{\sigma}_i$, $\boldsymbol{\varepsilon}_{trip,i}$, $\varepsilon_{cp,i}$, $\mathbf{p}_i$, $\tilde{r}_i$, $i = 1, 2$. Subtracting the equations for θ_1 and θ_2 from each other and testing with the difference $\theta := \theta_1 - \theta_2$ gives for any $t \leq T$ (cf. Proposition 5.3.32)

$$\left\langle \rho_0\, c_e\, \theta'(t), \theta(t) \right\rangle_{V_\theta^* V_\theta} + \left\langle \mathbf{A}_\theta \theta(t), \theta(t) \right\rangle_{V_\theta^* V_\theta} + 3 \int_\Omega K_\alpha\, \theta_0\, \mathrm{div}((\mathcal{S}_h \mathbf{u})'(t))\, \theta(t)\, \mathrm{d}\mathbf{x}$$
$$= \int_\Omega \rho_0 \sum_{i=1}^N L_i\, p_i'(t)\, \theta(t)\, \mathrm{d}\mathbf{x} + \Big(\tilde{r}(t), \theta(t)\Big)_{H_\theta}.$$

The same estimates as in Proposition 5.3.32 conclude the proof. □

5.3.36 Lemma (Time Regularity)
Let Assumptions 4.3.1$_{(A2)}$ and 4.3.2$_{(A2)}$ be valid. Assume $(\mathbf{u}, \mathbf{p}, \tilde{r}) \in \mathfrak{V}_u \times \mathfrak{H}_p \times \mathcal{H}_\theta$ Then any solution of Problem ($\mathbf{P}_{A,\theta}$) satisfies $\theta \in \mathcal{V}_\theta^\infty \cap \mathfrak{H}_\theta$ and for $c_1, c_2 > 0$:

$$\|\theta'\|_{\mathcal{H}_\theta} + \|\theta\|_{\mathcal{V}_\theta^\infty} \leq c_1(h^{-1}) + c_2(h^{-1}) \Big\{ \|\mathbf{u}\|_{\mathcal{V}_u} + \|\boldsymbol{\varepsilon}_{cp}'\|_{\mathcal{H}_\sigma} + \|\mathbf{p}'\|_{\mathcal{H}_p} + \|\tilde{r}\|_{\mathcal{H}_\theta} \Big\}.$$

Proof. We use the Galerkin scheme from the proof of Lemma 5.3.34, where the initial conditions are chosen s.t. $\theta_n(0) \to \theta_{n0}$ in V_θ. Hence, taking $v = \theta'_m(t)$ in the Equation (5.31) and integrating over S, we obtain for $\varepsilon_i > 0$, $i = 1, \dots, 4$, $\sum_{i=1}^{3} \varepsilon_i < \frac{\rho_0 c_e}{2}$, $\varepsilon_4 < \frac{c}{2}$

$$\begin{aligned}\Big(\frac{\rho_0\, c_e}{2} - \sum_{i=1}^{3} \varepsilon_i\Big) \int_0^t \|\theta_n(s)\|^2_{H_\theta}\, ds + \Big(\frac{c}{2} - \varepsilon_4\Big) \|\theta_n(t)\|^2_{V_\theta} &\le \frac{9K_\alpha^2\theta_0^2}{4\varepsilon_1}\|\mathbf{u}'(t)\|^2_{\mathrm{V_u}} \\ &+ \frac{\rho_0^2\|L\|^2_\infty}{4\varepsilon_2}\|\mathbf{p}'(t)\|^2_{\mathrm{H_p}} + \frac{1}{4\varepsilon_3}\|r(t)\|^2_{H_\theta} + \frac{1}{4\varepsilon_4}\|\theta'_\Gamma(t)\|^2_{L^2(\Gamma)} \\ &+ \frac{1}{2}\|\theta'_\Gamma(t)\|^2_{L^2(S;L^2(\Gamma))} + \frac{1}{2}\int_0^t \|\theta_n(s)\|^2_{H_\theta}\, ds.\end{aligned}$$

Therefore, applying Gronwall's inequality we obtain the following estimate:

$$(\theta_n) \quad \text{is bounded in} \quad \mathcal{V}_\theta^\infty \quad \text{and} \quad (\theta'_n) \quad \text{is bounded in} \quad \mathcal{H}_\theta.$$

So the a-priori estimates follows, independently of n. The result follows by passing to the limit $n \to \infty$. □

5.3.9 Step 8: Analysis of Problem $(\mathbf{P}_{A,FP3})$

Using the foregoing lemmas of this section, we are now in the position to formulate the existence and uniqueness result for Problem $(\mathbf{P}_{A,FP3})$ or rather for the full Problem $(\mathbf{P}_A)$.

Proof. We first observe that the solution $\theta \in \mathcal{V}_\theta^\infty \cap \mathfrak{H}_\theta$ of (5.2) satisfies

$$\begin{aligned}\|\theta_1 - \theta_2\|_{\mathcal{H}_\theta^\infty} + \|\theta_1 - \theta_2\|_{\mathcal{V}_\theta} &\le c_1(h^{-1})\Big\{\|\mathbf{u}_1 - \mathbf{u}_2\|_{\mathcal{V}_\mathrm{u}} + \|\mathbf{p}_1 - \mathbf{p}_2\|_{\mathcal{H}_\mathrm{p}}\Big\}, \\ \|\theta'_1 - \theta'_2\|_{\mathcal{H}_\theta} + \|\theta_1 - \theta_2\|_{\mathcal{V}_\theta^\infty} &\le c_2(h^{-1})\Big\{\|\mathbf{u}_1 - \mathbf{u}_2\|_{\mathcal{V}_\mathrm{u}} + \|\mathbf{p}_1 - \mathbf{p}_2\|_{\mathcal{H}_\mathrm{p}}\Big\},\end{aligned}$$

where c_1 and c_2 are positive constants depending on the regularisation parameter h.

We use the Lemmas 5.3.16 and 5.3.27 and Banach's fixed-point theorem (cf. the version in [Amann and Escher 2006]) with the (complete) weighted-norm space

$$\mathbf{X} := \mathfrak{H}_\theta, \qquad \|\mathbf{x}\|^2_{\mathrm{X}} := \sup_{t \in \bar{S}} \Big\{\exp(-\lambda t) \int_0^t \|\mathbf{x}(s)\|^2_{\mathrm{H}_\theta}\, ds\Big\}, \qquad \lambda > 0.$$

Let $\boldsymbol{\varepsilon}_{trip}$ the unique solution to the Problem $(\mathbf{P}_{A,FP1})$ corresponding to Lemma 5.3.16 for given θ, $\mathbf{u}$, $\boldsymbol{\sigma}$ and $\boldsymbol{\varepsilon}_{cp}$. Moreover, let $(\mathbf{u}, \boldsymbol{\sigma}, \boldsymbol{\varepsilon}_{cp})$ the unique solution to the Problem $(\mathbf{P}_{A,FP2})$ corresponding to Lemma 5.3.27 for fixed θ and $\boldsymbol{\varepsilon}_{trip}$. For $\hat{\theta} \in \mathbf{X}$ given, we consider the linearised problem (5.2) for $\theta \in \mathbf{X}$. Finally, by Lemmas 5.3.34 and 5.3.36 there exists a unique solution $\theta \in \mathcal{V}_\theta \cap \mathfrak{H}_\theta \hookrightarrow \mathbf{X}$ for given $\mathbf{u}, \boldsymbol{\sigma}, \boldsymbol{\varepsilon}_{cp}$ and $\boldsymbol{\varepsilon}_{trip}$. We define a fixed-point operator as

$$\mathbf{T}_{A,FP3} : \mathbf{X} \to \mathbf{X}, \qquad \mathbf{T}_{A,FP3}(\hat{\theta}) = \theta$$

and consider solutions θ_1, θ_2 corresponding to different data $\hat{\theta}_1, \hat{\theta}_2$. By similar arguments as for Lemmas 5.3.16 and 5.3.27 (and using the estimates (5.16) and (5.27)) we obtain f.a.a. $t \in S$ the following estimate

$$\begin{aligned}
&\|\theta_1(t) - \theta_2(t)\|_{H_\theta} + \|\theta_1'(t) - \theta_2'(t)\|_{H_\theta} \\
&\quad \le c(h^{-1}) \left\{ \int_0^t \|\mathbf{u}_1(s) - \mathbf{u}_2(s)\|^2_{\mathrm{V_u}} \,\mathrm{d}s + \int_0^t \|\mathbf{p}_1(s) - \mathbf{p}_2(s)\|^2_{\mathrm{H_p}} \,\mathrm{d}s \right\} \\
&\quad \le c(h^{-1}) \left\{ \int_0^t \|\hat{\theta}_1(t) - \hat{\theta}_2(t)\|^2_{H_\theta} \,\mathrm{d}s + \int_0^t \|\hat{\theta}_1'(t) - \hat{\theta}_2'(t)\|_{H_\theta} \,\mathrm{d}s \right\} \\
&\quad \le c(h^{-1}) \int_0^t \exp(\lambda t) \,\mathrm{d}s \sup_{s \in [0,t]} \left\{ \exp(-\lambda t) \Big(\|\hat{\theta}_1(t) - \hat{\theta}_2(t)\|^2_{H_\theta} + \|\hat{\theta}_1'(t) - \hat{\theta}_2'(t)\|_{H_\theta} \Big) \right\} \\
&\quad \le c(h^{-1}) \frac{\exp(\lambda t)}{\lambda} \|\hat{\theta}_1 - \hat{\theta}_2\|_{\mathrm{X}}.
\end{aligned}$$

It follows

$$\|\theta_1 - \theta_2\|_{\mathrm{X}} \le \sqrt{\frac{c(h^{-1})}{\lambda}}\, \|\hat{\theta}_1 - \hat{\theta}_2\|_{\mathrm{X}}$$

and hence, for λ large enough, $\mathbf{T}_{A,FP3}$ is strictly contractive. The conditions for Banach's fixed-point theorem are fulfilled and Problem ($\mathbf{P}_A$) possesses a unique solution. □

5.4 Proof of Theorem 5.2.3

This section provides the proof of Theorem 5.2.3. Subsection 5.4.1 gives the outline of the proof, whereas the proof itself is analogous to the proof in Subsection 5.3.1. The main differences are discussed in Subsections 5.4.3 and 5.4.4.

In contrast to the preceding section, we incorporate the intrinsic dissipation $\boldsymbol{\sigma} : (\boldsymbol{\varepsilon}'_{trip} + \boldsymbol{\varepsilon}'_{cp})$ in this setting, but we have to disregard the dependence of the evolution equations on the parameter θ' instead. Again, we only deal with the case of constant $\mathbf{K}$, but the proof in the situation of a time-dependent $\mathbf{K}$ works similarly and in the case of a parameter-dependent $\mathbf{K}$, e.g. $\mathbf{K}$ depending on θ or $\boldsymbol{\varepsilon}_{cp}$, one can integrate this parameter in the fixed-point argumentation.

5.4.1 Outline of the Proof

The main idea of the proof is based on classical arguments of functional analysis concerning variational problems and fixed-point arguments, cf. Section 4.1. The proof will be done in several steps (cf. Figure 5.2 for a schematic representation of the proof). In order to prove the unique existence of a weak solution $(\mathbf{u}, \boldsymbol{\sigma}, \boldsymbol{\varepsilon}_{trip}, \boldsymbol{\varepsilon}_{cp}, \theta, \mathbf{p})$ of Problem $(\mathbf{P}_A)$, we apply the same strategy as in the preceding section (cf. Subsection 5.3.1).

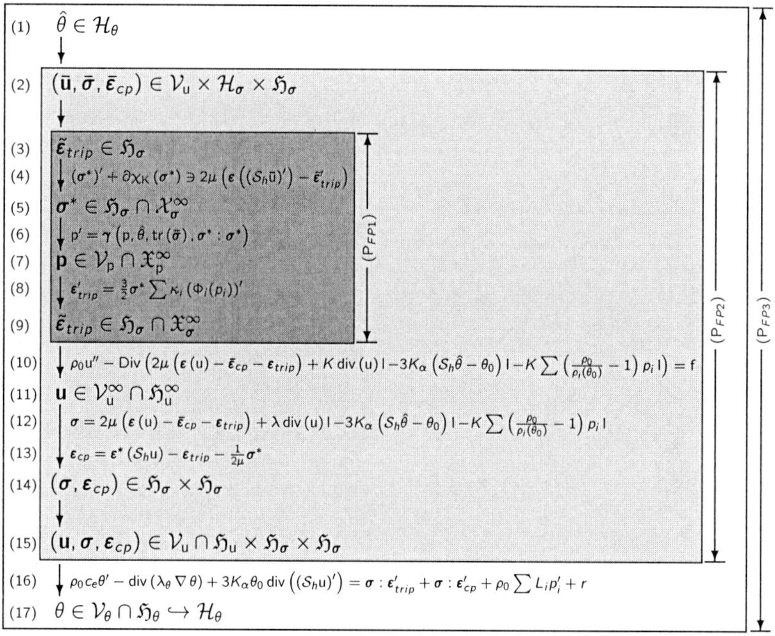

Figure 5.2: Scheme of the proof of existence for the Steklov regularised problem using Banach's fixed-point theorem for problems $(\mathbf{P}_{FPi})$, $i = 1, 2$ and Schauder's fixed-point theorem for problem $(\mathbf{P}_{FP3})$.

The solvability of Problem $(\mathbf{P}_{A,FP1})$ with given data $\hat{\theta}$, $\bar{\mathbf{u}}$, $\bar{\boldsymbol{\sigma}}$ and $\bar{\boldsymbol{\varepsilon}}_{cp}$ and the solvability of Problem $(\mathbf{P}_{A,FP2})$ for fixed $\hat{\theta}$ is obtained by using Banach's fixed-point theorem. In contrast to the preceding section, we investigate the existence of Problem $(\mathbf{P}_{A,FP3})$ or rather the existence the full Problem ($\mathbf{P}_A$) with Schauder's fixed-point theorem in this final step. Therefore, we only obtain the existence of at least one solution of the whole problem. The uniqueness of the solution has to be shown in a further step, certainly depending on more restrictive assumptions (in particular regarding the intrinsic dissipation term).

5.4.2 Step 1 – 6: Analysis of Problems ($\mathbf{P}_{A,\boldsymbol{\varepsilon}_{cp}}$) – ($\mathbf{P}_{A,FP2}$)

First, we remark that steps 1 – 6 of the proof are completely analogous to steps 1 – 6 of the proof of Theorem 5.2.1 and investigated in detail in Subsections 5.3.2 – 5.3.7.

5.4.3 Step 7: Analysis of Problem ($\mathbf{P}_{A,\theta}$)

We continue with Problem ($\mathbf{P}_{A,\theta}$) for given $(\mathbf{u}, \boldsymbol{\sigma}, \mathbf{p}, \boldsymbol{\varepsilon}_{trip}, \boldsymbol{\varepsilon}_{cp}) \in \mathfrak{H}_{\mathrm{u}} \times \mathcal{H}_{\sigma} \times \mathfrak{X}_{\mathrm{p}}^{\infty} \times \mathcal{H}_{\sigma} \times \mathcal{H}_{\sigma}$. Consider the (linearised) parabolic problem:

Problem ($\mathbf{P}_{A,\theta}$)
Find the temperature $\theta : \overline{\Omega}_T \to \mathbb{R}$ *s.t.*

(5.40)
$$\rho_0 c_e \frac{\partial \theta}{\partial t} - \operatorname{div}(\lambda_\theta \nabla \theta) + 3K_\alpha \theta_0 \operatorname{div}\left(\frac{\partial \mathcal{S}_h \mathbf{u}}{\partial t}\right) = \boldsymbol{\sigma} : (\boldsymbol{\varepsilon}'_{trip} + \boldsymbol{\varepsilon}'_{cp}) + \rho_0 \sum_{i=2}^{m} L_i p'_i + r \quad \text{in } \Omega_T$$

(5.41)
$$\theta(0) = \theta_0 \quad \text{in } \Omega \qquad \text{and} \qquad -\lambda_\theta \nabla \theta \nu = \delta(\theta - \theta_\Gamma) \quad \text{on } \partial\Omega$$

5.4.1 Definition (Weak Formulation)
Under the Assumption 4.3.1$_{(A2)}$ *the function* $\theta \in \mathcal{U}_\theta$ *is called a weak solution of the Problem* ($\mathbf{P}_{A,\theta}$), *if* $\theta(\mathbf{x}, 0) = \theta_0(\mathbf{x})$ *a.e. and*

$$(5.42)\quad \left\langle \rho_0\, c_e \frac{\partial}{\partial t}\theta(t), \vartheta \right\rangle_{V_\theta^* V_\theta} + \int_\Omega \lambda_\theta \nabla \theta(t) \nabla \vartheta \,\mathrm{d}\mathbf{x} + 3 \int_\Omega K_\alpha\, \theta_0 \operatorname{div}\left(\frac{\partial \mathcal{S}_h \mathbf{u}(t)}{\partial t}\right) \vartheta \,\mathrm{d}\mathbf{x}$$
$$= \int_\Omega \boldsymbol{\sigma}(t) : \boldsymbol{\varepsilon}'_{trip}(t)\, \vartheta \,\mathrm{d}\mathbf{x} + \int_\Omega \boldsymbol{\sigma}(t) : \boldsymbol{\varepsilon}'_{cp}(t)\, \vartheta \,\mathrm{d}\mathbf{x} + \int_\Omega \rho_0 \sum_{i=2}^{m} L_i\, p'_i(t)\, \vartheta \,\mathrm{d}\mathbf{x}$$
$$+ \int_\Omega r(t)\, \vartheta \,\mathrm{d}\mathbf{x} + \int_{\partial\Omega} \delta\, (\theta_\Gamma - \theta(t))\, \vartheta \,\mathrm{d}\sigma_x$$

f.a. $\vartheta \in V_\theta$, *f.a.a.* $t \in S$.

As in Subsection 5.3.8, we introduce the operator

$$A_\theta : V_\theta \to V_\theta^*, \qquad \langle A_\theta \theta, \vartheta \rangle_{V_\theta^* V_\theta} = \lambda_\theta \int_\Omega \nabla \theta : \nabla \vartheta \, \mathrm{dx} + \delta \int_\Gamma \theta \vartheta \, \mathrm{d}\sigma_x, \qquad \theta, \vartheta \in V_\theta.$$

and the function $\tilde{r} \in \mathcal{V}_\theta^*$ via

$$\langle \tilde{r}(t), v \rangle_{V_\theta^* V_\theta} := \int_\Omega r(t) v \, \mathbf{dx} + \delta \int_\Gamma \theta_\Gamma(t) v|_\Gamma \, \mathrm{d}\sigma_x, \qquad v \in V_\theta, \qquad \text{f.a.a. } t \in S.$$

5.4.2 Remark (Estimate for the Inelastic Dissipation)
Using the definitions $\boldsymbol{\sigma}_{vM} := \sqrt{\frac{3}{2}\boldsymbol{\sigma}^* : \boldsymbol{\sigma}^*}$ *and* $\Lambda := \frac{3}{2(R_0+R)} s'_{cp}$, $s'_{cp} := \sqrt{\frac{3}{2}\boldsymbol{\varepsilon}'_{cp} : \boldsymbol{\varepsilon}'_{cp}}$ *we get*

$$\boldsymbol{\sigma} : \boldsymbol{\varepsilon}'_{cp} = \boldsymbol{\sigma} : \Lambda \boldsymbol{\sigma}^* = \Lambda \boldsymbol{\sigma} : \boldsymbol{\sigma}^* = \Lambda |\boldsymbol{\sigma}^*|^2 \leq (R_0 + R) \mathbf{s}'_{cp}.$$

With the estimate $\|s'_p\|_{\mathrm{H}_\sigma} \leq \sqrt{\frac{2}{3}} \|\boldsymbol{\varepsilon}'_{cp}\|_{\mathrm{H}_\sigma}$ *it follows* $\boldsymbol{\sigma} : \boldsymbol{\varepsilon}'_{cp} \in \mathcal{H}_\sigma$ *for appropriate* R_0 *and* R*, cf. [Wolff et al. 2008b]. Moreover,*

$$\boldsymbol{\sigma} : \boldsymbol{\varepsilon}'_{trip} = \boldsymbol{\sigma}_{vM}^2 \sum_{i=1}^m \kappa_i \frac{\partial \Phi_i}{\partial p_i}(p) \max\{p'_i, 0\} \leq (R_0 + R)^2 M_\Phi \sum_{i=1}^m \kappa_i \max\{p'_i, 0\}.$$

Therefore, $\boldsymbol{\sigma} : \boldsymbol{\varepsilon}'_{trip} \in \mathcal{H}_\sigma$ *for constant (or rather bounded)* R_0, R, M_Φ, κ *and given* $\mathbf{p} \in \mathfrak{H}_\mathrm{p} \cap \mathfrak{X}_\mathrm{p}^\infty$*, cf. [Wolff et al. 2008b].*

5.4.3 Proposition (A-priori Estimate)
Let Assumption 4.3.1$_{(A2)}$ *be valid and* $\theta \in \mathcal{H}_\theta^\infty \cap \mathcal{V}_\theta$ *be a weak solution of the original Problem* $(\mathbf{P}_{A,\theta})$ *(without the Steklov regularisation for* $\mathbf{u}'$*). Then there exists* $c > 0$ *s.t.*

$$\|\theta\|_{\mathcal{H}_\theta^\infty} + \|\theta\|_{\mathcal{V}_\theta} \leq c \left\{ \|\mathbf{u}'\|_{\mathcal{V}_\mathrm{u}} + \|\boldsymbol{\varepsilon}'_{cp}\|_{\mathcal{H}_\sigma} + \|\mathbf{p}'\|_{\mathcal{H}_\mathrm{p}} + \|\tilde{r}\|_{\mathcal{V}_\theta^*} \right\}.$$

Proof. Taking $v = \theta(t)$ in the equation (5.31) , we obtain

$$\Big\langle \rho_0 \, c_e \, \theta'(t), \theta(t) \Big\rangle_{V_\theta^* V_\theta} + \Big\langle A_\theta \theta(t), \theta(t) \Big\rangle_{V_\theta^* V_\theta} + \underbrace{3 \int_\Omega K_\alpha \, \theta_0 \, \mathrm{div}(\mathbf{u}'(t)) \, \theta(t) \, \mathbf{dx}}_{=:\mathbb{I}_1}$$

$$= \underbrace{\int_\Omega \Big(\boldsymbol{\sigma}(t) : \big(\boldsymbol{\varepsilon}'_{trip} + \boldsymbol{\varepsilon}'_{cp} \big) \Big) \theta(t) \, \mathbf{dx}}_{=:\mathbb{I}_2} + \underbrace{\int_\Omega \rho_0 \sum_{i=1}^m L_i \, p'_i(t) \, \theta(t) \, \mathbf{dx}}_{=:\mathbb{I}_3} + \underbrace{\Big(\tilde{r}(t), \theta(t) \Big)_{H_\theta}}_{=:\mathbb{I}_4}.$$

The following estimates hold for arbitrary $\varepsilon > 0$:

$$|\mathbb{I}_1| \leq \frac{9 K_\alpha^2 \theta_0^2}{2} \|\mathbf{u}'(t)\|_{\mathrm{V_u}}^2 + \frac{1}{2} \|\theta\|_{H_\theta}^2,$$

$$|\mathbb{I}_2| \le \frac{\tilde{R}_0^2}{2}\|\boldsymbol{\varepsilon}'_{cp}(t)\|^2_{\mathrm{H}_\sigma} + \frac{R_0^2 M_\Phi^2 \|\kappa\|^2_\infty}{2}\|\mathbf{p}'(t)\|^2_{\mathrm{H}_\mathrm{p}} + \|\theta\|^2_{H_\theta},$$
$$|\mathbb{I}_3| \le \frac{\rho_0^2\|L\|^2_\infty}{2}\|\mathbf{p}'(t)\|^2_{\mathrm{H}_\mathrm{p}} + \frac{1}{2}\|\theta(t)\|^2_{H_\theta},$$
$$|\mathbb{I}_4| \le \frac{1}{2}\|r(t)\|^2_{H_\theta} + \frac{1}{2}\|\theta(t)\|^2_{H_\theta} + \frac{1}{4\varepsilon}\|\theta_\Gamma(t)\|^2_{L^2(\Gamma_T)} + \varepsilon\|\theta(t)\|^2_{V_\theta}.$$

Integrating equation over S and using the estimates $\mathbb{I}_i$, $i = 1, 2, 3, 4$ gives for $\varepsilon < c$:

$$\begin{aligned}\frac{\rho_0\, c_e}{2}\|\theta(t)\|^2_{H_\theta} + \left(c - \varepsilon\right)\int_0^t \|\theta(s)\|^2_{V_\theta}\,\mathrm{d}s \le \frac{9K_\alpha^2\theta_0^2}{2}\int_0^t \|\mathbf{u}'(s)\|^2_{\mathrm{V}_\mathrm{u}}\,\mathrm{d}s \\ + \frac{\tilde{R}_0^2}{2}\int_0^t \|\boldsymbol{\varepsilon}'_{cp}(s)\|^2_{\mathrm{H}_\sigma}\,\mathrm{d}s + \frac{R_0^2 M_\Phi^2\|\kappa\|^2_\infty + \rho_0^2\|L\|^2_\infty}{2}\int_0^t \|\mathbf{p}'(s)\|^2_{\mathrm{H}_\mathrm{p}}\,\mathrm{d}s \\ + \frac{1}{2}\int_0^t \|r(s)\|^2_{H_\theta}\,\mathrm{d}s + \frac{1}{4\varepsilon}\int_0^t \|\theta_\Gamma(s)\|^2_{L^2(\Gamma)}\,\mathrm{d}s + \frac{5}{2}\int_0^t \|\theta(s)\|^2_{H_\theta}\,\mathrm{d}s.\end{aligned}$$

Application of the Gronwall Lemma leads to:

$$\begin{aligned}\|\theta(t)\|^2_{H_\theta} + \int_0^t \|\theta(s)\|^2_{V_\theta}\,\mathrm{d}s \le c\left\{\int_0^t \|\mathbf{u}'(s)\|^2_{\mathrm{V}_\mathrm{u}}\,\mathrm{d}s + \int_0^t \|\boldsymbol{\varepsilon}'_{cp}(s)\|^2_{\mathrm{H}_\sigma}\,\mathrm{d}s \right.\\ \left. + \int_0^t \|\mathbf{p}'(s)\|^2_{\mathrm{H}_\mathrm{p}}\,\mathrm{d}s + \int_0^t \|r(s)\|^2_{H_\theta}\,\mathrm{d}s + \int_0^t \|\theta_\Gamma(s)\|^2_{L^2(\Gamma)}\,\mathrm{d}s\right\}.\end{aligned}$$

Hence,

$$\sup_{t\in\bar{S}}\|\theta(t)\|_{H_\theta} \le c, \qquad \int_0^T \|\theta(t)\|^2_{V_\theta}\,\mathrm{d}t \le c,$$

where c is a generic constant. □

5.4.4 Remark (A-priori Estimate for Problem ($\mathbf{P}_{A,\theta}$))
The a-priori estimate can also be proven for $\mathbf{u}' \in \mathcal{H}_\mathrm{u}$ if $\mathbf{u}'|_{\partial\Omega} = 0$ or $\nabla\theta\nu|_{\partial\Omega} = 0$.
For the Problem ($\mathbf{P}_{A,\theta}$) we use the approximation $\mathbf{u}' \approx \mathcal{D}_h\mathbf{u}$ with $\|\mathcal{D}_h\mathbf{u}\|_{\mathcal{V}_\mathrm{u}} \le \frac{2}{h}\|\mathbf{u}\|_{\mathcal{V}_\mathrm{u}}$ and therefore we obtain for $c_1, c_2 > 0$:

$$\|\theta\|_{\mathcal{H}_\theta^\infty} + \|\theta\|_{\mathcal{V}_\theta} \le c_1(h^{-1}) + c_2(h^{-1})\Big\{\|\mathbf{u}\|_{\mathcal{V}_\mathrm{u}} + \|\boldsymbol{\varepsilon}'_{cp}\|_{\mathcal{H}_\sigma} + \|\mathbf{p}'\|_{\mathcal{H}_\mathrm{p}} + \|\tilde{r}\|_{\mathcal{V}_\theta^*}\Big\}.$$

5.4.5 Lemma (Existence and Uniqueness)
Under Assumption 4.3.1$_{(A2)}$ the Problem ($\mathbf{P}_{A,\theta}$) admits a unique solution $\theta : \overline{\Omega}_T \to \mathbb{R}$, satisfying the following conditions:

$$\theta \in \mathcal{H}_\theta^\infty \cap \mathcal{V}_\theta, \qquad \theta' \in \mathcal{V}_\theta^*, \qquad \|\theta\|_{\mathcal{H}_\theta^\infty\cap\mathcal{V}_\theta} \le c.$$

Proof. The proof follows analogue to the proof of Lemma 5.3.34 with the help of the Galerkin method using the a-priori estimates in Remark 5.4.4. □

5.4.6 Lemma (Time Regularity)
Let Assumptions 4.3.1$_{(A2)}$ and 4.3.2$_{(A2)}$ be valid. Assume $(\mathbf{u}, \boldsymbol{\varepsilon}_{cp}, \mathbf{p}, \tilde{r}) \in \mathfrak{V}_\mathrm{u} \times \mathfrak{H}_\sigma \times \mathfrak{H}_\mathrm{p} \times \mathcal{H}_\theta$ *Then any solution of Problem* $(\mathbf{P}_{A,\theta})$ *satisfies* $\theta \in \mathcal{V}_\theta^\infty \cap \mathfrak{H}_\theta$ *and for* $c_1, c_2 > 0$

$$\|\theta'\|_{\mathcal{H}_\theta} + \|\theta\|_{\mathcal{V}_\theta^\infty} \leq c_1(h^{-1}) + c_2(h^{-1})\left\{\|\mathbf{u}\|_{\mathcal{V}_\mathrm{u}} + \|\boldsymbol{\varepsilon}'_{cp}\|_{\mathcal{H}_\sigma} + \|\mathbf{p}'\|_{\mathcal{H}_\mathrm{p}} + \|R\|_{\mathcal{H}_\theta}\right\}.$$

Proof. We use the Galerkin scheme from the proof of Lemma 5.4.5, where the initial conditions are chosen s.t. $\theta_n(0) \to \theta_{n0}$ in V_θ. Hence, taking $v = \theta'_m(t)$ in the Equation (5.31) and integrating over S, we obtain for $\varepsilon_i > 0$, $i = 1, \dots, 5$, $\sum_{i=1}^4 \varepsilon_i < \frac{\rho_0 c_e}{2}$, $\varepsilon_5 < \frac{c}{2}$

$$\begin{aligned}\left(\frac{\rho_0\, c_e}{2} - \sum_{i=1}^4 \varepsilon_i\right)\int_0^t \|\theta_n(s)\|^2_{H_\theta}\, ds + \left(\frac{c}{2} - \varepsilon_5\right)\|\theta_n(t)\|^2_{V_\theta} &\leq \frac{9K_\alpha^2\theta_0^2}{4\varepsilon_1}\|\mathbf{u}'(t)\|^2_{\mathrm{V_u}} \\ + \frac{\tilde{R}_0^2}{4\varepsilon_2}\|\boldsymbol{\varepsilon}'_{cp}(t)\|^2_{\mathrm{H}_\sigma} + \frac{R_0^2 M_\Phi^2\|\kappa\|^2_\infty + \rho_0^2\|L\|^2_\infty}{4\varepsilon_3}&\|\mathbf{p}'(t)\|^2_{\mathrm{H_p}} + \frac{1}{4\varepsilon_4}\|r(t)\|^2_{H_\theta} \\ + \frac{1}{4\varepsilon_5}\|\theta'_\Gamma(t)\|^2_{L^2(\Gamma)} + \frac{1}{2}\|\theta'_\Gamma(t)\|^2_{L^2(S;L^2(\Gamma))} &+ \frac{1}{2}\int_0^t \|\theta_n(s)\|^2_{H_\theta}\, ds.\end{aligned}$$

Therefore, applying Gronwall's inequality we obtain the following estimate:

$$(\theta_n) \quad \text{is bounded in} \quad \mathcal{V}_\theta^\infty \quad \text{and} \quad (\theta'_n) \quad \text{is bounded in} \quad \mathcal{H}_\theta.$$

So the a-priori estimates follows, independently of n. The result follows by passing to the limit $n \to \infty$. □

5.4.4 Step 8: Analysis of Problem $(\mathbf{P}_{A,FP3})$

We apply Schauder's fixed-point argument in this setting in order to prove the existence result.

Proof. Let $\boldsymbol{\varepsilon}_{trip}$ the unique solution to the Problem $(\mathbf{P}_{A,FP1})$ corresponding to Lemma 5.3.16 for given θ, $\mathbf{u}$, $\boldsymbol{\sigma}$ and $\boldsymbol{\varepsilon}_{cp}$. Moreover, let $(\mathbf{u}, \boldsymbol{\sigma}, \boldsymbol{\varepsilon}_{cp})$ the unique solution to the Problem $(\mathbf{P}_{A,FP2})$ corresponding to Lemma 5.3.27 for fixed θ and $\boldsymbol{\varepsilon}_{trip}$.
Let us define the set

$$\mathbf{M} := \left\{\theta \in \mathcal{V}_\theta^\infty \cap \mathfrak{H}_\theta : \|\theta\|_{\mathcal{V}_\theta^\infty} + \|\theta\|_{\mathfrak{H}_\theta} \leq R, (2.24)_3, (2.29)\right\} \subset \mathcal{H}_\theta$$

with a positive constant R to be defined in the preceding step. Define an operator

$$\mathbf{T}_{A,FP3} : \mathbf{M} \to \mathbf{M}, \qquad \hat{\theta} \mapsto \theta = \mathbf{T}_{A,FP3}(\hat{\theta})$$

where $\theta \in \mathcal{V}_\theta^\infty \cap \mathfrak{H}_\theta$ is the weak solution of the subproblem (5.2) and satisfies the estimate

$$\|\theta\|_{\mathcal{H}_\theta^\infty} + \|\theta\|_{\mathcal{V}_\theta} \leq c_1(h^{-1}) + c_2(h^{-1}) \left\{ \|\mathbf{u}\|_{\mathcal{V}_u} + \|\boldsymbol{\varepsilon}'_{cp}\|_{\mathcal{H}_\sigma} + \|\mathbf{p}'\|_{\mathcal{H}_p} + \|\tilde{r}\|_{\mathcal{V}_\theta^*} \right\},$$

where c_1 and c_2 are positive constants depending on the regularisation parameter h. We have to verify that

- $\mathbf{T}_{A,FP3} : \mathcal{H}_\theta \to \mathcal{H}_\theta$ is continuous and compact and
- for sufficiently large R (for fixed T) the operator $\mathbf{T}_{A,FP3}$ maps the set $\mathbf{M}$ into itself, i.e. the aim is to show $\mathbf{T}_{A,FP3}(\hat{\theta}) \in \mathbf{M}$ for $\hat{\theta} \in \mathbf{M}$.

By Subsection 5.4.3 we have the following estimates:

$$\begin{aligned}
\|\theta\|_{\mathcal{H}_\theta^\infty} + \|\theta\|_{\mathcal{V}_\theta} &\leq c_1 + c_2 \left\{ \|\mathbf{u}\|_{\mathcal{V}_u} + \|\mathbf{p}'\|_{\mathcal{H}_p} + \|\boldsymbol{\varepsilon}'_{cp}\|_{\mathcal{H}_\sigma} + \|\boldsymbol{\varepsilon}'_{trip}\|_{\mathcal{X}_\sigma^\infty} \|\boldsymbol{\sigma}\|_{\mathcal{H}_\sigma} \right\}, \\
\|\theta'\|_{\mathcal{H}_\theta} + \|\theta\|_{\mathcal{V}_\theta^\infty} &\leq c_1 + c_2 \left\{ \|\mathbf{u}\|_{\mathcal{V}_u} + \|\mathbf{p}'\|_{\mathcal{H}_p} + \|\boldsymbol{\varepsilon}'_{cp}\|_{\mathcal{H}_\sigma} + \|\boldsymbol{\varepsilon}'_{trip}\|_{\mathcal{X}_\sigma^\infty} \|\boldsymbol{\sigma}\|_{\mathcal{H}_\sigma} \right\}, \\
\|\mathbf{p}\|_{\mathcal{X}_p^\infty} &\leq c, \\
\|\boldsymbol{\varepsilon}_{cp}\|_{\mathfrak{H}_\sigma} &\leq c \left\{ \|\mathbf{u}\|_{\mathcal{V}_u} + \|\boldsymbol{\varepsilon}_{cp}\|_{\mathfrak{H}_\sigma} + \|\boldsymbol{\sigma}^*\|_{\mathfrak{H}_\sigma} \right\}, \\
\|\boldsymbol{\sigma}\|_{\mathcal{H}_\sigma} &\leq c \left\{ \|\mathbf{u}\|_{\mathcal{V}_u} + \|\hat{\theta}\|_{\mathcal{H}_\theta} + \|\mathbf{p}\|_{\mathcal{H}_p} + \|\boldsymbol{\varepsilon}_{cp}\|_{\mathcal{H}_\sigma} + \|\boldsymbol{\varepsilon}_{trip}\|_{\mathcal{H}_\sigma} \right\}, \\
\|\mathbf{u}'\|_{\mathcal{H}_u^\infty} + \|\mathbf{u}\|_{\mathcal{V}_u^\infty} &\leq c \left\{ + \|\hat{\theta}\|_{\mathcal{H}_\theta} + \|\mathbf{p}\|_{\mathfrak{H}_p} + \|\boldsymbol{\varepsilon}_{trip}\|_{\mathfrak{H}_\sigma} \right\}, \\
\|\boldsymbol{\varepsilon}_{trip}\|_{\mathfrak{H}_\sigma} &\leq c\, \|\boldsymbol{\sigma}^*\|_{\mathcal{X}_\sigma^\infty} \|\mathbf{p}\|_{\mathcal{X}_p^\infty}, \\
\|\boldsymbol{\sigma}^*\|_{\mathcal{X}_\sigma^\infty} &\leq c,
\end{aligned}$$

with positive constants c, c_1 and c_2. This implies the estimate:

$$\begin{aligned}
\|\theta\|_{\mathcal{V}_\theta^\infty} + \|\theta\|_{\mathfrak{H}_\theta} &\leq c_1 + c_2 \left\{ \|\mathbf{u}\|_{\mathcal{V}_u} + \|\mathbf{p}'\|_{\mathcal{H}_p} + \|\boldsymbol{\varepsilon}'_{cp}\|_{\mathcal{H}_\sigma} + \|\boldsymbol{\varepsilon}'_{trip}\|_{\mathcal{X}_\sigma^\infty} \|\boldsymbol{\sigma}\|_{\mathcal{H}_\sigma} \right\} \\
&\leq c_1 + c_2 \left\{ \|\hat{\theta}\|_{\mathcal{H}_\theta} + \|\boldsymbol{\varepsilon}_{cp}\|_{\mathfrak{H}_\sigma} + \|\boldsymbol{\varepsilon}_{trip}\|_{\mathfrak{H}_\sigma} + \|\boldsymbol{\sigma}\|_{\mathcal{H}_\sigma} \right\} \\
&\leq c_1 + c_2 \left\{ \|\hat{\theta}\|_{\mathcal{H}_\theta} + \|\boldsymbol{\varepsilon}_{trip}\|_{\mathfrak{H}_\sigma} + \|\boldsymbol{\sigma}^*\|_{\mathfrak{H}_\sigma} \right\} \\
&\leq c_1 + c_2 \, \|\hat{\theta}\|_{\mathcal{H}_\theta} \\
&\leq c_1 + c_2 \, R,
\end{aligned}$$

where c_1 and c_2 do not depend on R. Hence, choosing R sufficiently large, we have shown that the operator $\mathbf{T}_{A,FP3}$ maps the set $\mathbf{M}$ into itself. We observe that $\mathbf{M}$ is nonempty, closed, convex and compact in $\mathcal{H}_\theta$, cf. Lions-Aubin's compactness results in [Simon 1986]. To end the proof we have to show that $\mathbf{T}_{A,FP3}$ is continuous. Let $\hat{\theta}_n \to \hat{\theta}$

in $\mathcal{H}_\theta$. Then we have to prove that $\theta_n = \mathbf{T}_{A,FP3}(\hat{\theta}_n)$ converges to $\theta = \mathbf{T}_{A,FP3}(\hat{\theta})$ in $\mathcal{H}_\theta$. First we obtain from Lemma 5.3.10 that the sequence $(\mathbf{p}_n)_{n\in\mathbb{N}}$ of solutions of the ODEs

$$\mathbf{p}_n' = \boldsymbol{\gamma}(\mathbf{p}_n, \hat{\theta}_n, \operatorname{tr}(\boldsymbol{\sigma}_n), \boldsymbol{\sigma}_n^* : \boldsymbol{\sigma}_n^*) \text{ in } \Omega_T, \qquad \mathbf{p}_n(0) = \mathbf{0} \text{ in } \Omega$$

converges in the space $\mathcal{H}_\mathrm{p}^\infty \cap \mathfrak{H}_\mathrm{p}$ to the solution

$$\mathbf{p}' = \boldsymbol{\gamma}(\mathbf{p}, \hat{\theta}, \operatorname{tr}(\boldsymbol{\sigma}), \boldsymbol{\sigma}^* : \boldsymbol{\sigma}^*) \text{ in } \Omega_T, \qquad \mathbf{p}(0) = \mathbf{0} \text{ in } \Omega.$$

Moreover, the Lipschitz continuity of $\mathbf{T}_{A,FP2}$ yields $\mathbf{u}_n \to \mathbf{u}$ in $\mathcal{V}_\mathrm{u}^\infty \cap \mathfrak{H}_\mathrm{u}^\infty$, $\boldsymbol{\sigma}_n \to \boldsymbol{\sigma}$ in $\mathcal{H}_\sigma$ and $\boldsymbol{\varepsilon}_{cp,n} \to \boldsymbol{\varepsilon}_{cp}$ in $\mathfrak{H}_\sigma$, cf. Lemma 5.3.27. Furthermore, the Lipschitz continuity of $\mathbf{T}_{A,FP1}$ gives $\boldsymbol{\varepsilon}_{trip,n} \to \boldsymbol{\varepsilon}_{trip}$ in $\mathfrak{H}_\sigma$, cf. Lemma 5.3.16. In particular, we achieve with the help of Lebesgue's dominated convergence theorem

$$\|\boldsymbol{\sigma}_n : (\boldsymbol{\varepsilon}_{trip,n}' + \boldsymbol{\varepsilon}_{cp,n}') - \boldsymbol{\sigma} : (\boldsymbol{\varepsilon}_{trip}' + \boldsymbol{\varepsilon}_{cp}')\|_{\mathcal{H}_\sigma} \to 0.$$

By Lemma 5.3.35 we then obtain the convergence $\theta_n \to \theta$ in $\mathcal{V}_\theta^\infty \cap \mathfrak{H}_\theta \hookrightarrow \mathcal{H}_\theta$. The application of the Schauder fixed-point theorem completes the proof. □

5.5 Regularity Result

In this section we show some additional regularity of the weak solution of Problem $(\mathbf{P}_A)$ under suitable assumptions. The intention of this approach is to pass to the limit in the regularisation parameter. Therefore we use the results of the preceding sections. Unfortunately, the regularity result cannot be obtained for the fully coupled problem with our approach.

The lack of regularity of $\boldsymbol{\sigma}$ (resp. $\boldsymbol{\sigma}^*$) causes great difficulties in treating stress-dependent phase transformation behaviour, cf. the discussion in [Boettcher 2007]. Because the stress is proportional to the gradient of the displacement vector, we do not get the necessary estimates in order to use our fixed-point approach. A possibility to avoid these difficulties could be the usage of further regularisation methods, e.g. convolution with smooth functions, cf. [Boettcher 2007].

Therefore, we neglect stress-dependent transformation behaviour as discussed in the preceding sections and assume that the phase evolution is only dependent on the temperature, i.e. $\boldsymbol{\gamma} = \boldsymbol{\gamma}(\mathbf{p}, \theta)$. Moreover, we assume that the intrinsic dissipation vanishes, i.e. $\boldsymbol{\sigma} : \left(\boldsymbol{\varepsilon}_{trip}' + \boldsymbol{\varepsilon}_{cp}'\right) = 0$ and we consider the model of Tanaka for the saturation function in the relation for the TRIP strain, i.e. $\Phi = \mathbf{I}$. Furthermore, we assume simplified boundary conditions, i.e. $\theta|_{\partial\Omega} = 0$ or $\nabla\theta \cdot \nu|_{\partial\Omega} = 0$ and $\mathbf{u} = 0$ on $\partial\Omega$.

As before, we only consider the case of constant $\mathbf{K}$, but the proof in the situation of time- or parameter-dependent $\mathbf{K}$ works similarly.

First, we recall the existence and uniqueness result of weak solvability for Problem ($\mathbf{P}_A$) under these slightly modified assumptions.

5.5.1 Theorem (Existence and Uniqueness for the modified Problem ($\mathbf{P}_A$))
Let Assumption 4.3.1 be valid. Then the Problem ($\mathbf{P}_A$) possesses a unique weak solution $(\mathbf{u}, \theta, \mathbf{p}, \boldsymbol{\varepsilon}_{cp}, \boldsymbol{\varepsilon}_{trip}) \in \mathcal{V}_\mathrm{u} \cap \mathfrak{H}_\mathrm{u} \times \mathfrak{H}_\theta \times \mathfrak{H}_\mathrm{p} \times \mathfrak{H}_\sigma \times \mathfrak{H}_\sigma$.

The proof is analogous to the proof of Theorem 5.2.1 (resp. 5.2.3). We continue with studying additional regularity of (the modified) Problem ($\mathbf{P}_A$).

5.5.2 Theorem (Regularity for the modified Problem ($\mathbf{P}_A$))
In addition to the assumptions of Theorem 5.5.1 let Assumption 4.3.2 be valid. Then the solution $(\mathbf{u}, \theta, \mathbf{p}, \boldsymbol{\varepsilon}_{cp}, \boldsymbol{\varepsilon}_{trip})$ *of the modified Problem* ($\mathbf{P}_A$) *satisfies* $(\mathbf{u}, \theta, \mathbf{p}, \boldsymbol{\varepsilon}_{cp}, \boldsymbol{\varepsilon}_{trip}) \in \mathcal{W}_\mathrm{u} \cap \mathfrak{V}_\mathrm{u} \times \mathcal{W}_\theta \times \mathfrak{W}_\mathrm{p} \times \mathfrak{V}_\sigma \times \mathfrak{V}_\sigma$.

Previously, we have shown that the regularised Problem ($\mathbf{P}_A$) possesses a unique weak solution $(\mathbf{u}_h, \theta_h, \mathbf{p}_h, \boldsymbol{\varepsilon}_{cp,h}, \boldsymbol{\varepsilon}_{trip,h}) \in \mathcal{V}_\mathrm{u} \cap \mathfrak{H}_\mathrm{u} \times \mathfrak{H}_\theta \times \mathfrak{H}_\mathrm{p} \times \mathfrak{H}_\sigma \times \mathfrak{H}_\sigma$ f.a. $h > 0$ (we dropped the subscript h in our notation). Now we want to pass to the limit in the regularisation parameter h.

However, it is also possible to show the (unique) existence of the fully coupled Problem ($\mathbf{P}$) under these strong additional assumptions, using the analogous argumentation as in Section 5.2.

5.5.3 Remark (Passage to the Limit in the Regularisation Parameter)
In addition to the assumptions of Theorem 5.5.1 let Assumption 4.3.2 be valid.
Then $(\mathbf{u}_h, \theta_h, \mathbf{p}_h, \boldsymbol{\varepsilon}_{cp,h}, \boldsymbol{\varepsilon}_{trip,h}) \to (\mathbf{u}, \theta, \mathbf{p}, \boldsymbol{\varepsilon}_{cp}, \boldsymbol{\varepsilon}_{trip})$ *in* $\mathcal{V}_\mathrm{u} \times \mathcal{H}_\theta \times \mathcal{H}_\mathrm{p} \times \mathcal{H}_\sigma \times \mathcal{H}_\sigma$ *and* $(\mathbf{u}'_h, \mathbf{p}'_h, \boldsymbol{\varepsilon}'_{cp,h}, \boldsymbol{\varepsilon}'_{trip,h}) \to (\mathbf{u}', \mathbf{p}', \boldsymbol{\varepsilon}'_{cp}, \boldsymbol{\varepsilon}'_{trip})$ *in* $\mathcal{H}_\mathrm{u} \times \mathcal{H}_\mathrm{p} \times \mathcal{H}_\sigma \times \mathcal{H}_\sigma$ *for* $h \to 0$, *where* $(\mathbf{u}, \theta, \mathbf{p}, \boldsymbol{\varepsilon}_{cp}, \boldsymbol{\varepsilon}_{trip})$ *is the (unique weak) solution of the modified original Problem* ($\mathbf{P}$).

Proof. We roughly sketch the idea of the proof. From Lemmas 5.3.16 and 5.6.10 the boundedness of $\boldsymbol{\varepsilon}_{trip,h}$ and $\boldsymbol{\varepsilon}'_{trip,h}$ follows and the estimate

$$\|\boldsymbol{\varepsilon}_{trip,h} - \boldsymbol{\varepsilon}_{trip}\|_{\mathcal{H}_\sigma} + \|\boldsymbol{\varepsilon}'_{trip,h} - \boldsymbol{\varepsilon}'_{trip}\|_{\mathcal{H}_\sigma} \leq c\left\{\|\mathbf{p}'_h - \mathbf{p}'\|_{\mathcal{V}_u} + \|h^{-1}\mathcal{D}_h\mathbf{u} - \mathbf{u}'\|_{\mathcal{V}_u}\right\}$$

holds (cf. Subsection 5.3.4). Therefore, $\|\boldsymbol{\varepsilon}_{trip,h} - \boldsymbol{\varepsilon}_{trip}\|_{\mathcal{H}_\sigma} + \|\boldsymbol{\varepsilon}'_{trip,h} - \boldsymbol{\varepsilon}'_{trip}\|_{\mathcal{H}_\sigma} \to 0$ for $h \to 0$, using the properties in Section B.1. Moreover, Lemmas 5.3.27 and 5.6.14 give the boundedness of $(\mathbf{u}_h, \boldsymbol{\varepsilon}_{cp,h})$, $(\mathbf{u}'_h, \boldsymbol{\varepsilon}'_{cp,h})$ and the estimate (cf. Subsection 5.3.6)

$$\|\mathbf{u}_h - \mathbf{u}\|_{\mathcal{V}_u} + \|\mathbf{u}'_h - \mathbf{u}'\|_{\mathcal{H}_u} + \|\boldsymbol{\varepsilon}_{cp,h} - \boldsymbol{\varepsilon}_{cp}\|_{\mathfrak{H}_\sigma} \leq c\left\{\|\theta_h - \theta\|_{\mathfrak{H}_\theta} + \|\mathbf{p}_h - \mathbf{p}\|_{\mathfrak{H}_p}\right\}.$$

Hence, $\|\mathbf{u}_h - \mathbf{u}\|_{\mathcal{V}_u} + \|\mathbf{u}'_h - \mathbf{u}'\|_{\mathcal{H}_u} + \|\boldsymbol{\varepsilon}_{cp,h} - \boldsymbol{\varepsilon}_{cp}\|_{\mathfrak{H}_\sigma} \to 0$ for $h \to 0$. Furthermore, Theorems 5.2.1 and 5.5.2 give the boundedness of $(\mathbf{p}_h, \theta_h)$, $(\mathbf{p}'_h, \theta'_h)$ and the estimate (cf. Subsection 5.3.8)

$$\|\theta_h - \theta\|_{\mathcal{V}_\theta \cap \mathcal{H}_\theta} + \|\mathbf{p}_h - \mathbf{p}\|_{\mathfrak{H}_p} \leq 0.$$

Therefore, $\|\mathbf{p}_h - \mathbf{p}\|_{\mathfrak{H}_p} + \|\theta_h - \theta\|_{\mathcal{V}_\theta} \to 0$ for $h \to 0$. □

5.6 Proof of Theorem 5.5.2

This section provides the proof of Theorem 5.5.2. Subsection 5.6.1 describes the outline of the proof, whereas the proof itself is divided into Subsections 5.6.2 – 5.6.9.

5.6.1 Outline of the Proof

The main idea of the proof is based on classical arguments of functional analysis concerning variational problems and fixed-point arguments, cf. Section 4.1. The proof will be done in several steps (cf. Figure 5.3 for a schematic representation of the proof). In order to prove the unique existence of a weak solution $(\mathbf{u}, \boldsymbol{\sigma}, \boldsymbol{\varepsilon}_{trip}, \boldsymbol{\varepsilon}_{cp}, \theta, \mathbf{p})$ of Problem $(\mathbf{P}_A)$, we apply the same strategy as in Subsection 5.3.1.
First, we show regularity results for the subproblems summarised in Tables 4.2 and 4.4. Then, the subproblems will be extended sequentially by the other variables until the fully coupled problem is considered. In the sequel, we prove regularity results for the three fixed-point steps Problems $(\mathbf{P}_{A,FPi})$, $i = 1, 2, 3$ for Problem $(\mathbf{P}_A)$. The use of Banach's fixed-point theorem for the proof of the existence and uniqueness result for the three fixed-point steps Problems $(\mathbf{P}_{A,FPi})$, $i = 1, 2, 3$ gives us the opportunity to show additional regularity results.

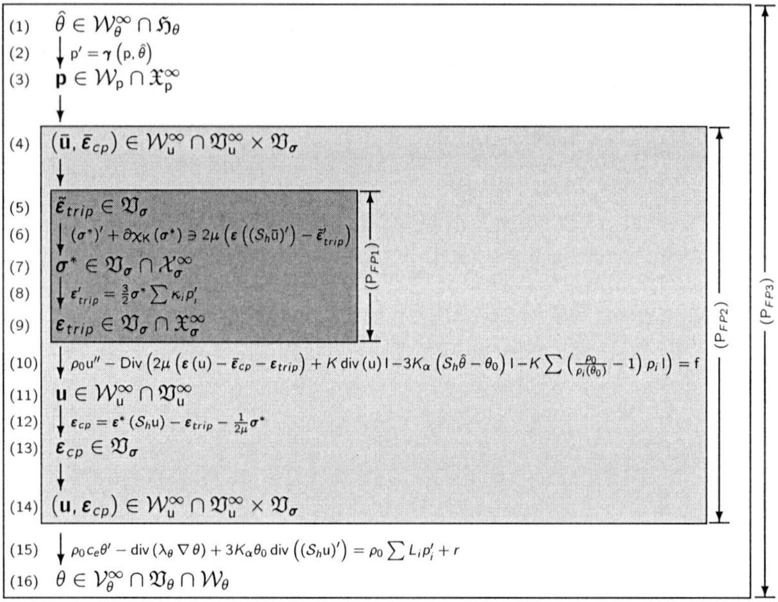

Figure 5.3: Scheme of the proof of additional regularity for the Steklov regularised problem using the Banach fixed-point theorem for problem ($\mathbf{P}_{FPi}$), $i = 1, 2, 3$.

Besides the statement that every contraction mapping on a complete metric space has a unique fixed point, the theorem provides the method of successive approximations: Let $\mathbf{X}$ be a complete metric space and $T : \mathbf{X} \to \mathbf{X}$ contractive. Then the sequence $\{\mathbf{x}_n\}_{n\in\mathbb{N}} \subset \mathbf{X}$, $\mathbf{x}_{n+1} := T(\mathbf{x}_n)$ that is constructed in the proof of the Banach fixed-point theorem not only converges to the fixed point $\bar{\mathbf{x}} \in \mathbf{X}$, it converges to $\bar{\mathbf{x}}$ (monotonically) for any initial $\mathbf{x}_0 \in \mathbf{X}$. Each successive term is closer to $\bar{\mathbf{x}}$ than its predecessor. Consequently, starting from any arbitrary point in $\mathbf{X}$, we can repeatedly apply the function T to the current approximation of $\bar{\mathbf{x}}$, obtaining a better approximation, and the approximations converge to $\bar{\mathbf{x}}$. This often provides a straightforward method for computing $\bar{\mathbf{x}}$, cf. e.g. [Amann and Escher 2006]

In order to show additional regularity (for the fixed-point operators $\mathbf{T}_{A,FPi}$, $i = 1, 2, 3$)

we choose an arbitrary $\mathbf{x}_0 \in \mathbf{Y}$ for a particular subspace $\mathbf{Y} \subset \mathbf{X}$ and prove $\|\mathbf{x}_n\|_{\mathrm{Y}} \leq$ const. f.a. $n \in \mathbb{N}$. Then there exists a subsequence $\{\mathbf{x}_{n_k}\}_{k\in\mathbb{N}} \subset \mathbf{Y}$ s.t. $\mathbf{x}_{n_k} \rightharpoonup \mathbf{z}$ for $k \to \infty$. Moreover, it holds $\mathbf{z} = \bar{\mathbf{x}}$, because $\mathbf{x}_n \to \bar{\mathbf{x}}$ in $\mathbf{X}$. Hence, $\bar{\mathbf{x}} \in \mathbf{Y}$, because of $\|\bar{\mathbf{x}}\|_{\mathrm{Y}} \leq \lim_{k\to\infty} \|\mathbf{x}_{n_k}\|_{\mathrm{Y}} \leq$ const.

Furthermore, the general idea to show the boundedness of the sequence of successive approximations is to use the Bielecki trick (or rather renormalisation trick) as follows:

$$
\begin{aligned}
\|\mathbf{x}_n(t)\|_{\mathrm{Y}}^2 &\leq c\Big(1 + \int_0^t \|\mathbf{x}_{n-1}(s)\|_{\mathrm{Y}}^2 \,\mathrm{d}s\Big) \\
&\leq c + c\int_0^t \|\mathbf{x}_{n-1}(s)\|_{\mathrm{Y}}^2 \exp(\lambda s)\exp(-\lambda s)\,\mathrm{d}s \\
&\leq c + c\|\mathbf{x}_{n-1}\|_{\mathrm{Y},\lambda}^2 \int_0^t \exp(\lambda s)\,\mathrm{d}s \\
&\leq c + c\|\mathbf{x}_{n-1}\|_{\mathrm{Y},\lambda}^2 \frac{1}{\lambda}\Big(\exp(\lambda t) - 1\Big) \\
&\leq c + \frac{1}{2}\|\mathbf{x}_{n-1}\|_{\mathrm{Y},\lambda}^2 \exp(\lambda t), \quad t \in S.
\end{aligned}
$$

Finally we get

$$
\|\mathbf{x}_n\|_{\mathrm{Y},\lambda}^2 \leq c + \frac{1}{2}\|\mathbf{x}_{n-1}\|_{\mathrm{Y},\lambda}^2 \leq \underbrace{\sum_{k=0}^{n} \frac{1}{2^k}}_{\leq 2} c + \frac{1}{2^n}\|\mathbf{x}_0\|_{\mathrm{Y},\lambda}^2.
$$

Details will be given in the next subsections.

5.6.2 Step 1: Analysis of Problem $(\mathbf{P}_{A,\mathbf{p}})$

We start with the regularity of Problem $(\mathbf{P}_{A,\mathrm{p}})$ for given $\theta \in \mathcal{V}_\theta$ as a fixed parameter or data (cf. the problem in Subsection 5.3.3).

5.6.1 Lemma (Spatial Regularity)
Let Assumptions 4.3.1$_{(A5)}$ and 4.3.2$_{(A3)}$ be valid. Assume $\theta \in \mathcal{V}_\theta$. Then $\nabla \mathbf{p} \in \mathcal{H}_{\mathrm{p}}^\infty$ and $\nabla \mathbf{p}' \in \mathcal{H}_{\mathrm{p}}$ s.t.

$$
\begin{aligned}
&\|\nabla \mathbf{p}(t)\|_{\mathrm{H_p}}^2 \leq c\int_0^t \|\nabla \theta(s)\|_{H_\theta}^2 \,\mathrm{d}s \quad \textit{and} \\
&\|\nabla \mathbf{p}'(t)\|_{\mathrm{H_p}}^2 \leq c\left(\|\nabla \theta(t)\|_{H_\theta}^2 + \int_0^t \|\nabla \theta(s)\|_{H_\theta}^2 \,\mathrm{d}s\right) \text{ f.a.a. } t \in S.
\end{aligned}
$$

Proof. The estimate for the gradient will be obtained from the initial value problem for $i = 1, \ldots, m$

$$\nabla p_i' = \sum_{k=1}^{m} \frac{\partial \gamma_i}{\partial p_k}(\mathbf{p}, \theta) \nabla p_k + \frac{\partial \gamma_i}{\partial \theta}(\mathbf{p}, \theta) \nabla \theta, \qquad \nabla p_i(0) = \nabla p_{0i}.$$

Integration gives

$$\nabla p_i(t) = \nabla p_{0i} + \int_0^t \sum_{k=1}^{m} \frac{\partial \gamma_i}{\partial p_k}(\mathbf{p}, \theta) \nabla p_k + \frac{\partial \gamma_i}{\partial \theta}(\mathbf{p}, \theta) \nabla \theta \, \mathrm{d}s, \; t \in S.$$

According to Gronwall's inequality the result follows immediately (cf. [Chelminski et al. 2007; Kern 2011] for details). □

5.6.2 Lemma (Time Regularity)
Let Assumptions 4.3.1$_{(A5)}$ and 4.3.2$_{(A3)}$ be valid. Assume $\theta \in \mathfrak{H}_\theta$. Then $\mathbf{p}'' \in \mathcal{H}_\mathrm{p}$ s.t.

$$\|\mathbf{p}''(t)\|_{\mathrm{H_p}}^2 \le c_1 + c_2 \int_0^t \|\theta'(s)\|_{H_\theta}^2 \, \mathrm{d}s, \; t \in S, \; c_1, c_2 > 0.$$

Proof. The estimate for the time derivative will be obtained from the following equation for $i = 1, \ldots, m$

$$p_i'' = \sum_{k=1}^{m} \frac{\partial \gamma_i}{\partial p_k}(\mathbf{p}, \theta) p_k' + \frac{\partial \gamma_i}{\partial \theta}(\mathbf{p}, \theta) \theta'.$$

According to the boundedness of $\boldsymbol{\gamma}$ the result follows immediately. □

Additional regularity of the temperature yields improved spatial regularity of $\mathbf{p}$. Analogously to Lemma 5.6.1 we get the following result.

5.6.3 Remark (Additional Spatial Regularity)
Let Assumptions 4.3.1$_{(A5)}$ and 4.3.2$_{(A3')}$ be valid. Assume $\theta \in \mathcal{W}_\theta$. Then it follows $\nabla \mathbf{p} \in \mathcal{V}_\mathrm{p}^\infty$, $\nabla \mathbf{p}' \in \mathcal{V}_\mathrm{p}$, $\nabla^2 \mathbf{p} \in \mathcal{H}_\mathrm{p}^\infty$ and $\nabla^2 \mathbf{p}' \in \mathcal{H}_\mathrm{p}$ via

$$\nabla^2 p_i' = \frac{\partial^2 \gamma_i}{\partial \theta^2} |\nabla \theta|^2 + \frac{\partial \gamma_i}{\partial \theta} \nabla^2 \theta + \sum_{j=1}^{m} \left[\frac{\partial^2 \gamma_i}{\partial p_j^2} |\nabla p_j|^2 + 2 \frac{\partial^2 \gamma_i}{\partial p_j \partial \theta} \nabla p_j \nabla \theta + \frac{\partial \gamma_i}{\partial p_j} \nabla^2 p_j \right]$$

with similar estimates. Analogously to Lemma 5.6.1 we obtain for $c > 0$ and $t \in S$:

$$\|\nabla^2 \mathbf{p}(t)\|_{\mathrm{H_p}}^2 \le c \left\{ \int_0^t \|\nabla \theta(s)\|_{H_\theta}^2 \, \mathrm{d}s + \int_0^t \|\nabla^2 \theta(s)\|_{H_\theta}^2 \, \mathrm{d}s \right\},$$

$$\|\nabla^2 \mathbf{p}'(t)\|_{\mathrm{H_p}}^2 \le c \Big\{ \|\nabla \theta(t)\|_{H_\theta}^2 + \|\nabla^2 \theta(t)\|_{H_\theta}^2 + \int_0^t \|\nabla \theta(s)\|_{H_\theta}^2 \, \mathrm{d}s + \int_0^t \|\nabla^2 \theta(s)\|_{H_\theta}^2 \, \mathrm{d}s \Big\}.$$

All in all it follows f.a.a. $t \in S$:

$$
\begin{aligned}
&\|\mathbf{p}(t)\|_{\mathsf{V_p}}^2 \le c \int_0^t \|\theta(s)\|_{V_\theta}^2 \,\mathrm{d}s, \\
&\|\mathbf{p}'(t)\|_{\mathsf{V_p}}^2 \le c \Big\{ \|\theta(t)\|_{V_\theta}^2 + \int_0^t \|\theta(s)\|_{V_\theta}^2 \,\mathrm{d}s \Big\}, \\
&\|\mathbf{p}(t)\|_{\mathsf{W_p}}^2 \le c \Big\{ \int_0^t \|\theta(s)\|_{V_\theta}^2 \,\mathrm{d}s + \int_0^t \|\theta(s)\|_{W_\theta}^2 \,\mathrm{d}s \Big\}, \\
&\|\mathbf{p}'(t)\|_{\mathsf{W_p}}^2 \le c \Big\{ \|\theta(t)\|_{V_\theta \cap W_\theta}^2 + \int_0^t \|\theta(s)\|_{V_\theta \cap W_\theta}^2 \,\mathrm{d}s \Big\}.
\end{aligned}
\tag{5.43}
$$

5.6.4 Remark (Additional Variables)
Let $\mathbf{H}_\xi := [L^2(\Omega)]^M$, $\mathbf{V}_\xi := [W^{1,2}(\Omega)]^M$, $\mathbf{W}_\xi := [W^{2,2}(\Omega)]^M$. *In the situation of Remark 5.3.7 we get analogously* f.a.a. $t \in S$:

$$
\begin{aligned}
\|\mathbf{p}''(t)\|_{\mathsf{H_p}}^2 &\le c \Big\{ 1 + \int_0^t \|\theta'(s)\|_{H_\theta} \,\mathrm{d}s + \int_0^t \|\boldsymbol{\xi}'(s)\|_{\mathsf{H}_\xi} \,\mathrm{d}s \Big\}, \\
\|\mathbf{p}(t)\|_{\mathsf{V_p}}^2 &\le c \Big\{ \int_0^t \|\theta(s)\|_{V_\theta}^2 \,\mathrm{d}s + \int_0^t \|\boldsymbol{\xi}(s)\|_{\mathsf{V}_\xi} \,\mathrm{d}s \Big\}, \\
\|\mathbf{p}'(t)\|_{\mathsf{V_p}}^2 &\le c \Big\{ \|\theta(t)\|_{V_\theta}^2 + \|\boldsymbol{\xi}(t)\|_{\mathsf{V}_\xi}^2 + \int_0^t \|\theta(s)\|_{V_\theta}^2 \,\mathrm{d}s + \int_0^t \|\boldsymbol{\xi}(s)\|_{\mathsf{V}_\xi}^2 \,\mathrm{d}s \Big\}, \\
\|\mathbf{p}(t)\|_{\mathsf{W_p}}^2 &\le c \Big\{ \int_0^t \|\theta(s)\|_{V_\theta \cap W_\theta}^2 \,\mathrm{d}s + \int_0^t \|\boldsymbol{\xi}(s)\|_{\mathsf{V}_\xi \cap \mathsf{W}_\xi}^2 \,\mathrm{d}s \Big\}, \\
\|\mathbf{p}'(t)\|_{\mathsf{W_p}}^2 &\le c \Big\{ \|\theta(t)\|_{V_\theta \cap W_\theta}^2 + \|\boldsymbol{\xi}(t)\|_{\mathsf{V}_\xi \cap \mathsf{W}_\xi}^2 + \int_0^t \|\theta(s)\|_{V_\theta \cap W_\theta}^2 \,\mathrm{d}s + \int_0^t \|\boldsymbol{\xi}(s)\|_{\mathsf{V}_\xi \cap \mathsf{W}_\xi}^2 \,\mathrm{d}s \Big\}.
\end{aligned}
$$

5.6.3 Step 2: Analysis of Problem $(\mathbf{P}_{A,\boldsymbol{\varepsilon}_{cp}})$ for constant K

We continue with the regularity of Problem $(\mathbf{P}_{A,\boldsymbol{\varepsilon}_{cp}})$ in case of constant $\mathbf{K}$ (cf. Subsection 5.3.2) for $(\mathbf{u}, \boldsymbol{\sigma}, \boldsymbol{\varepsilon}_{trip}) \in \mathcal{W}_\mathsf{u} \cap \mathfrak{V}_\mathsf{u} \times \mathcal{V}_\sigma \times \mathfrak{V}_\sigma$ considered as data. We define $F : \mathbb{R}^{3\times 3} \to \mathbb{R}$ via

$$
F(\boldsymbol{\sigma}) := \sqrt{\frac{3}{2}\boldsymbol{\sigma}^* : \boldsymbol{\sigma}^*} - (R_0 + R)
$$

for given constant $R_0, R \in \mathbb{R}^+$. The set of all admissible $\boldsymbol{\sigma}$ is convex (cf. [Han and Reddy 1999b]). Moreover, we define

$$
\mathbf{K}_F := \Big\{ \boldsymbol{\tau} \in \mathbb{R}^{3\times 3}_{\text{sym}}, \operatorname{tr}(\boldsymbol{\tau}) = 0 : F(\boldsymbol{\tau}) \le 0 \Big\}, \quad \mathbf{K} := \Big\{ \boldsymbol{\tau} \in \mathbf{V}_\sigma : \boldsymbol{\tau}(\mathbf{x}) \in \mathbf{K}_F \text{ f.a.a. } \mathbf{x} \in \Omega \Big\}.
$$

5.6.5 Lemma
Let Assumption $4.3.1_{(A3)}$ *be valid. Then* $\mathbf{K}$ *is a nonempty closed convex subset of* $\mathbf{V}_\sigma$.

Proof. The proof is analogous to the proof of Lemma 5.3.1. □

5.6.6 Lemma
Let Assumption 4.3.1$_{(A3)}$ be valid. Then

(1) *the indicator function χ_{K} is proper, convex and lower semi-continuous.*
(2) $\partial\chi_{\mathrm{K}} : \mathbf{V}_\sigma \to \mathcal{P}(\mathbf{V}_\sigma)$ *is maximal monotone.*
(3) $D(\chi_{\mathrm{K}}) = D(\partial\chi_{\mathrm{K}}) = \mathbf{K}$.

Proof. The proof is analogous to the proof of Lemma 5.3.2. □

5.6.7 Lemma (Regularity)
Let Assumption 4.3.1$_{(A3)}$ be valid and assume in addition $\mathbf{u}' \in \mathcal{V}_{\mathrm{u}}$, $\boldsymbol{\varepsilon}'_{trip} \in \mathcal{H}_\sigma$. Then the unique solution of Lemma 5.3.3 of Problem ($\mathbf{P}_{A,\boldsymbol{\varepsilon}_{cp}}$) yields $\boldsymbol{\sigma}^* \in C(\bar{S};\mathbf{V}_\sigma)\cap\mathfrak{V}_\sigma$ satisfying $\boldsymbol{\sigma}^*(t) \in \mathbf{K}$ f.a.a. $t \in S$ and

$$\|\boldsymbol{\sigma}^*(t) - \boldsymbol{\sigma}^*(0)\|_{\mathrm{V}_\sigma} \le c(h^{-1})\Big\{\int_0^t \|\mathbf{u}(s)\|_{\mathrm{W}_{\mathrm{u}}}\,\mathrm{ds} + \int_0^t \|\boldsymbol{\varepsilon}'_{trip}(s)\|_{\mathrm{V}_\sigma}\,\mathrm{ds}\Big\} \text{ f.a.a. } t \in S,$$

$$\|(\boldsymbol{\sigma}^*)'(t)\|_{\mathrm{V}_\sigma} \le c(h^{-1})\Big\{\|\mathbf{u}(t)\|_{\mathrm{W}_{\mathrm{u}}} + \|\boldsymbol{\varepsilon}'_{trip}(t)\|_{\mathrm{V}_\sigma}\Big\} \text{ f.a.a. } t \in S.$$

Proof. Using Lemma 5.6.5, Lemma 5.6.6 and applying Lemma C.2.1 proves the result. □

5.6.4 Step 3: Analysis of Problem ($\mathbf{P}_{A,\boldsymbol{\varepsilon}_{trip}}$)

We continue with the regularity of Problem ($\mathbf{P}_{A,\boldsymbol{\varepsilon}_{trip}}$) for either given $\boldsymbol{\sigma}^* \in \mathcal{V}_\sigma$ or (when using the identity (3.7), cf. Subsection 3.2.2) $(\mathbf{u}, \boldsymbol{\varepsilon}_{cp}) \in \mathcal{W}_{\mathrm{u}} \times \mathcal{V}_\sigma$ and $\mathbf{p} \in \mathcal{X}_{\mathrm{p}}^\infty$ with $\mathbf{p}' \in \mathcal{X}_{\mathrm{p}}^\infty$.

5.6.8 Lemma (Spatial Regularity)
Let Assumption 4.3.1$_{(A4)}$ be valid and assume in addition the model of Tanaka for the saturation function Φ, i.e. $\Phi = \mathbf{I}$. Moreover, assume $(\theta, \mathbf{u}, \boldsymbol{\sigma}) \in \mathcal{W}_\theta \times L^\infty(S;\mathbf{W}_{\mathrm{u}}) \times L^\infty(S;\mathbf{W}_\sigma)$. Then the solution of Lemma 5.3.13 fulfills the following estimate

$$\|\boldsymbol{\varepsilon}_{trip}(t)\|^2_{\mathrm{V}_\sigma} \le c_1 + c_2\int_0^t \Big\{\|\mathbf{u}(s)\|^{16}_{\mathrm{W}_{\mathrm{u}}} + \|\boldsymbol{\varepsilon}_{cp}(s)\|^{16}_{\mathrm{W}_{\mathrm{u}}} + \|\theta(s)\|^2_{W_\theta} + \|\theta(s)\|^4_{\mathrm{V}_\theta}\Big\}\,\mathrm{ds}$$

f.a.a. $t \in S$, *where the constants $c_1, c_2 > 0$ do not depend on the data.*

Proof. Using the multiplicator theorem, cf. Lemma A.5.6, we get

$$\|\boldsymbol{\varepsilon}'_{trip}(t)\|_{\mathrm{V}_\sigma} \le c\,\|\boldsymbol{\sigma}^*(t)\mathbf{p}'(t)\|_{\mathrm{V}_\sigma} \le c\,\|\boldsymbol{\sigma}^*(t)\|_{\mathrm{V}_\sigma}\,\|\mathbf{p}'(t)\|_{[W^{1,4}(\Omega)]^m}.$$

Integration gives

$$\|\boldsymbol{\varepsilon}_{trip}(t)\|_{\mathrm{V}_\sigma} \leq c \int_0^t \|\boldsymbol{\sigma}^*(s)\|_{\mathrm{V}_\sigma} \|\mathbf{p}'(s)\|_{[W^{1,4}(\Omega)]^m}\, \mathrm{d}s.$$

Using the Sobolev embedding $W^{1,6}(\Omega) \hookrightarrow W^{1,4}(\Omega)$, the interpolation inequality, cf. Lemma A.5.5, and the (generalised) Young inequality (cf. [Alt 2002; Růžička 2004]), one obtains

$$\begin{aligned}
\|\boldsymbol{\varepsilon}'_{trip}(t)\|^2_{\mathrm{V}_\sigma} &\leq c\, \|\boldsymbol{\sigma}^*(t)\|^2_{\mathrm{V}_\sigma} \|\mathbf{p}'(t)\|^2_{[W^{1,4}(\Omega)]^m} \\
&\leq c\, \|\boldsymbol{\sigma}^*(t)\|^2_{\mathrm{V}_\sigma} \|\mathbf{p}'(t)\|^2_{[W^{1,6}(\Omega)]^m} \\
&\leq c\, \|\boldsymbol{\sigma}^*(t)\|^2_{\mathrm{V}_\sigma} \|\mathbf{p}'(t)\|^{\frac{1}{2}}_{\mathrm{V}_p} \|\mathbf{p}'(t)\|^{\frac{3}{2}}_{\mathrm{W}_p} \\
&\leq c \left\{ \|\boldsymbol{\sigma}^*(t)\|^{16}_{\mathrm{V}_\sigma} + \|\mathbf{p}'(t)\|^4_{\mathrm{V}_p} + \|\mathbf{p}'(t)\|^2_{\mathrm{W}_p} \right\}.
\end{aligned}$$

Integration yields

$$\|\boldsymbol{\varepsilon}_{trip}(t)\|^2_{\mathrm{V}_\sigma} \leq c \left\{ \int_0^t \|\boldsymbol{\sigma}^*(s)\|^{16}_{\mathrm{V}_\sigma}\, \mathrm{d}s + \int_0^t \|\mathbf{p}'(s)\|^4_{\mathrm{V}_p}\, \mathrm{d}s + \int_0^t \|\mathbf{p}'(s)\|^2_{\mathrm{W}_p}\, \mathrm{d}s \right\}.$$

Using the estimates (5.43) for $\mathbf{p}'$ and again applying the general Hölder's (cf. [Cheung 2001; Michlin 1981]) and Young's inequality completes the proof. □

5.6.9 Remark (Regularity)
We discuss some additional aspects of the regularity result.

(1) Using the following estimate for the stress deviator

$$\|\boldsymbol{\sigma}^*(t)\|_{\mathrm{V}_\sigma} \leq c \left\{ \|\mathbf{u}(t)\|_{\mathrm{W}_u} + \|\boldsymbol{\varepsilon}_{trip}(t)\|_{\mathrm{V}_\sigma} + \|\boldsymbol{\varepsilon}_{cp}(t)\|_{\mathrm{V}_\sigma} \right\}$$

and applying the generalised Gronwall inequality, cf. Lemma A.5.1, we get the following estimate

$$\|\boldsymbol{\varepsilon}_{trip}(t)\|^2_{\mathrm{V}_\sigma} \leq c_1 + c_2 \int_0^t \left\{ \|\mathbf{u}(s)\|^{16}_{\mathrm{W}_u} + \|\boldsymbol{\varepsilon}_{cp}(s)\|^{16}_{\mathrm{V}_\sigma} + \|\theta(s)\|^2_{\mathrm{W}_\theta} + \|\theta(s)\|^4_{\mathrm{V}_\theta} \right\} \mathrm{d}s.$$

Analogously, we get

$$\begin{aligned}
\|\boldsymbol{\varepsilon}'_{trip}(t)\|^2_{\mathrm{V}_\sigma} \leq c_1 + c_2 \Big\{ &\|\mathbf{u}(t)\|^{16}_{\mathrm{W}_u} + \|\boldsymbol{\varepsilon}_{cp}(t)\|^{16}_{\mathrm{V}_\sigma} + \|\theta(t)\|^2_{\mathrm{W}_\theta} + \theta(t)\|^4_{\mathrm{V}_\theta} \\
&+ \int_0^t \|\theta(s)\|^2_{\mathrm{W}_\theta}\, \mathrm{d}s + \int_0^t \|\theta(s)\|^4_{\mathrm{V}_\theta}\, \mathrm{d}s \Big\}.
\end{aligned}$$

(2) Analogous to the proof of Lemma 5.6.1 one could calculate

$$\mathrm{Div}(\boldsymbol{\varepsilon}_{trip}) = \int_0^t \exp\Big(-\int_s^t \mathbf{p}'\,\mathrm{d}\tau\Big)\Big[\Big(\nabla\,\mathbf{p}'\Big)\boldsymbol{\varepsilon}(\mathbf{u}) - \mathbf{p}'\Big(\int_s^t \nabla\,\mathbf{p}'\,\mathrm{d}\tau\Big)\boldsymbol{\varepsilon}(\mathbf{u}) + \mathbf{p}'\,\mathrm{Div}(\boldsymbol{\varepsilon}(\mathbf{u}))\Big]\,\mathrm{d}s,$$

but the problem is, how to deal with the products $\Big(\nabla\,\mathbf{p}'\Big)\boldsymbol{\varepsilon}(\mathbf{u})$.

(3) Time regularity could be obtained similar to Lemma 5.6.2 (formally via differentiation of the Equation (3.8)*):*

$$\|\boldsymbol{\varepsilon}''_{trip}(t)\|_{\mathrm{H}_\sigma} \le c\Big\{\|\,(\boldsymbol{\sigma}^*)'\,(t)\|_{\mathrm{H}_\sigma}\|\mathbf{p}'(t)\|_{\mathrm{H}_p} + \|\boldsymbol{\sigma}^*(t)\|_{\mathrm{H}_\sigma}\|\mathbf{p}''(t)\|_{\mathrm{H}_p}\Big\},$$

$$\|\boldsymbol{\varepsilon}''_{trip}(t)\|_{\mathrm{V}_\sigma} \le c\Big\{\|\,(\boldsymbol{\sigma}^*)'\,(t)\|_{\mathrm{V}_\sigma}\|\mathbf{p}'(t)\|_{[W^{1,4}(\Omega)]^m} + \|\boldsymbol{\sigma}^*(t)\|_{\mathrm{V}_\sigma}\|\mathbf{p}''(t)\|_{[W^{1,4}(\Omega)]^m}\Big\}.$$

Because the second time derivative of $\mathbf{p}$ *does apriori not exist, in [Boettcher 2007] the Steklov average of* $\mathbf{p}''$ *is used in a similar situation.*

(4) For general Φ *the estimate*

$$\|\boldsymbol{\varepsilon}'_{trip}(t)\|_{\mathrm{V}_\sigma} \le c\,\|\boldsymbol{\sigma}^*(t)\|_{\mathrm{V}_\sigma}\left\|\frac{\partial\Phi}{\partial\mathbf{p}}(t)\right\|_{[W^{1,4}(\Omega)]^m}\|\mathbf{p}'(t)\|_{[W^{1,4}(\Omega)]^m}$$

holds f.a.a. $t \in S$. *Therefore,* $\frac{\partial\Phi}{\partial\mathrm{p}} \in [W^{1,4}(\Omega)]^m$ *is a minimal requirement, which would lead to local (in time) solutions for the fully coupled problem.*
We remark that similar estimates are used in the analysis of the Navier-Stokes equations.

5.6.5 Step 4: Analysis of Problem ($\mathbf{P}_{A,FP1}$)

In the next step, we investigate the regularity of Problem ($\mathbf{P}_{A,FP1}$) for $\hat{\theta}$, $\bar{\mathbf{u}}$, $\bar{\boldsymbol{\sigma}}$ and $\bar{\boldsymbol{\varepsilon}}_{cp}$ considered as given data. and obtain the following result.

5.6.10 Lemma (Regularity)
In addition to the assumptions of Theorem 5.5.1 let Assumption 4.3.2 be valid. Assume that $(\hat{\theta}, \bar{\mathbf{u}}, \bar{\boldsymbol{\sigma}}, \bar{\boldsymbol{\varepsilon}}_{cp}) \in \mathcal{W}_\theta \times \mathcal{W}_{\mathrm{u}} \times \mathcal{W}_\sigma \times \mathcal{W}_\sigma$. *Then the solution* $(\boldsymbol{\varepsilon}_{trip}, \boldsymbol{\sigma}^*)$ *of the modified Problem* ($\mathbf{P}_{A,FP1}$) *satisfies* $(\boldsymbol{\varepsilon}_{trip}, \boldsymbol{\sigma}^*) \in \mathfrak{V}_\sigma \times \mathfrak{V}_\sigma$.
Moreover, the following estimate holds

$$(5.44)\quad \|\boldsymbol{\varepsilon}_{trip}(t)\|^2_{\mathrm{V}_\sigma} \le c_1 + c_2\int_0^t \Big\{\|\bar{\mathbf{u}}(s)\|^{16}_{\mathrm{W}_{\mathrm{u}}} + \|\bar{\boldsymbol{\varepsilon}}_{cp}(s)\|^{16}_{\mathrm{W}_{\mathrm{u}}} + \|\hat{\theta}(s)\|^2_{\mathrm{W}_\theta} + \|\hat{\theta}(s)\|^4_{\mathrm{V}_\theta}\Big\}\,\mathrm{d}s$$

f.a.a. $t \in S$, *where the constants* $c_1, c_2 \ge 0$ *do not depend on the data.*

Proof. According to Subsection 5.6.1 it is sufficient to show the estimate (5.44). The proof of the estimate can be found in the preceding section. □

5.6.6 Step 5: Analysis of Problem ($\mathbf{P}_{A,\mathbf{u}}$)

We continue with the solvability of Problem ($\mathbf{P}_{A,u}$). Assume either $(\theta, \mathbf{p}, \boldsymbol{\varepsilon}_{trip}, \boldsymbol{\varepsilon}_{cp}) \in \mathfrak{V}_\theta \times \mathfrak{V}_p \cap \mathfrak{X}_p^\infty \times \mathfrak{V}_\sigma \times \mathfrak{V}_\sigma$ or $(\theta, \mathbf{p}, \boldsymbol{\varepsilon}_{trip}, \boldsymbol{\varepsilon}_{cp}) \in \mathcal{W}_\theta \times \mathcal{W}_p \cap \mathfrak{X}_p \times \mathfrak{V}_\sigma \times \mathfrak{V}_\sigma$ are given data.

5.6.11 Lemma (Regularity)
Let Assumptions 4.3.1$_{(A1)}$ and 4.3.2$_{(A1)}$ be valid and assume in addition $\mathbf{u} = 0$ on $\partial\Omega$. Then the solution of Problem ($\mathbf{P}_{A,u}$) corresponding to Lemma 5.3.25 fulfills

$$\|\mathbf{u}'\|_{\mathcal{V}_u^\infty} + \|\mathbf{u}\|_{\mathcal{W}_u^\infty} \le c_1 + c_2 \Big\{ \|\mathbf{f}\|_{\mathcal{V}_u} + \|\theta\|_{\mathcal{V}_\theta} + \|\theta'\|_{\mathcal{V}_\theta} + \|\mathbf{p}\|_{\mathcal{V}_p} + \|\mathbf{p}'\|_{\mathcal{V}_p} + \|\boldsymbol{\varepsilon}'_{trip}\|_{\mathcal{V}_\sigma} + \|\boldsymbol{\varepsilon}_{cp}\|_{\mathcal{V}_\sigma^\infty} + \|\boldsymbol{\varepsilon}'_{cp}\|_{\mathcal{V}_\sigma} \Big\}$$

and for $(\theta, \mathbf{p}) \in \mathcal{W}_\theta \times \mathcal{W}_p \cap \mathfrak{X}_p$

$$\|\mathbf{u}'\|_{\mathcal{V}_u^\infty} + \|\mathbf{u}\|_{\mathcal{W}_u^\infty} \le c_1 + c_2 \Big\{ \|\mathbf{f}\|_{\mathcal{V}_u} + \|\theta\|_{\mathcal{W}_\theta} + \|\mathbf{p}\|_{\mathcal{W}_p} + \|\boldsymbol{\varepsilon}'_{trip}\|_{\mathcal{V}_\sigma} + \|\boldsymbol{\varepsilon}_{cp}\|_{\mathcal{V}_\sigma} + \|\boldsymbol{\varepsilon}'_{cp}\|_{\mathcal{V}_\sigma} \Big\}.$$

Proof. We use the Galerkin scheme from the proof of Lemma 5.3.25. Next, we have to test the Galerkin equations (5.23) with $\mathbf{v} = -\Delta \mathbf{u}'_n$, but we only sketch the main idea. The formal derivation of the estimate is obtained by testing (5.18) with $\mathbf{v} = -\Delta \mathbf{u}'$:

$$\frac{\mathrm{d}}{\mathrm{d}t}\|\nabla \mathbf{u}'\|^2 + (\mathrm{Div}(\mathbf{A}_u(\mathbf{u})), \Delta \mathbf{u}') = -(\mathbf{f}, \Delta \mathbf{u}') + (\nabla(\theta - \theta_0), \Delta \mathbf{u}') + (\nabla \mathbf{p}, \Delta \mathbf{u}') + (\mathrm{Div}(\boldsymbol{\varepsilon}_{trip}), \Delta \mathbf{u}') + (\mathrm{Div}(\boldsymbol{\varepsilon}_{cp}), \Delta \mathbf{u}') + (\mathbf{u}'', \nabla \mathbf{u}' \nu)_{\partial\Omega}.$$

First we note that $\frac{\partial}{\partial t}\|\mathbf{u}(t)\|^2_{W_u} \le \mathrm{Div}(\mathbf{A}_u \mathbf{u}(t)), \Delta \mathbf{u}'(t))$, cf. [Pawlow and Zochowski 2002; Pawlow and Zajaczkowski 2005a,b]. Moreover, it holds $\nabla \mathbf{u}' \cdot \nu = 0$ on $\partial\Omega$ for zero Dirichlet boundary conditions. Using this inequality and Poincare's inequality we get

$$\begin{aligned}
\|\mathbf{u}'(t)\|^2_{V_u} + \|\mathbf{u}(t)\|^2_{W_u} &\le \|\mathbf{u}_0\|^2_{W_u} + \|\mathbf{u}_1\|^2_{V_u} + (\mathbf{u}''(t), \nabla \mathbf{u}'(t)\nu)_{\partial\Omega} - \int_0^t (\mathbf{f}'(s), \Delta \mathbf{u}(s))\,\mathrm{d}s \\
&+ (\mathbf{f}(t), \Delta \mathbf{u}(t)) + \int_0^t (\nabla \theta'(s), \Delta \mathbf{u}(s))\,\mathrm{d}s - (\nabla \theta(t), \Delta \mathbf{u}(t)) + \int_0^t (\nabla \mathbf{p}'(s), \Delta \mathbf{u}(s))\,\mathrm{d}s \\
&- (\nabla \mathbf{p}(t), \Delta \mathbf{u}(t)) + \int_0^t (\mathrm{Div}(\boldsymbol{\varepsilon}'_{cp}(s)), \Delta \mathbf{u}(s))\,\mathrm{d}s - (\mathrm{Div}(\boldsymbol{\varepsilon}_{cp}(s)), \Delta \mathbf{u}(s)) \\
&+ \int_0^t (\mathrm{Div}(\boldsymbol{\varepsilon}'_{trip}(s)), \Delta \mathbf{u}(s))\,\mathrm{d}s - (\mathrm{Div}(\boldsymbol{\varepsilon}_{trip}(s)), \Delta \mathbf{u}(s)).
\end{aligned}$$

Integrating over $\bar{S}$ we obtain the estimate

$$\|\mathbf{u}'(t)\|^2_{\mathrm{V_u}}+\|\mathbf{u}(t)\|^2_{\mathrm{W_u}} \leq c_1+c_2\left\{\|\mathbf{f}\|_{\mathcal{V}}+\|\theta\|_{\mathcal{W}_\theta}+\|\mathbf{p}\|_{\mathcal{W}_\mathrm{p}}+\|\boldsymbol{\varepsilon}'_{trip}\|_{\mathcal{V}_\sigma}+\|\boldsymbol{\varepsilon}_{cp}\|_{\mathcal{V}_\sigma}+\|\boldsymbol{\varepsilon}'_{cp}\|_{\mathcal{V}_\sigma}\right\}.$$

Hence, using the Gronwall inequality, we have

$$(\mathbf{u}_n) \quad \text{bounded in} \quad \mathcal{W}_\mathrm{u} \quad \text{and} \quad (\mathbf{u}'_n) \quad \text{bounded in} \quad \mathcal{V}_\mathrm{u}.$$

So, the a-priori estimate follows, independently of n. The result follows by passing to the limit $n \to \infty$.
The second inequality in Lemma 5.6.11 follows via a second integration by parts and the use of the zero Dirichlet boundary conditions. □

5.6.12 Remark (Time Regularity)
The standard methods, like (formally) using $\mathbf{u}''$ as test function, testing with the Steklov average $\mathcal{S}_h\mathbf{u}$, differentiation of the Galerkin equations w.r.t. time or using $-\Delta\mathbf{u}'$ as a test function for zero boundary conditions are not leading to the desired results. The main problem is, that the second time derivatives of $\boldsymbol{\varepsilon}_{trip}$ and $\boldsymbol{\varepsilon}_{cp}$ are a-priori not existent (cf. the discussion in [Boettcher 2007]; usage of the Steklov average of $\mathbf{p}$ in the evolution equation for $\boldsymbol{\varepsilon}_{trip}$).
The treatment as a system of evolution equations of first order in t is not leading to the desired results, too.

5.6.13 Remark (Spatial Regularity)
An idea to get additional spatial regularity of the weak solution is the usage of special (Galerkin) bases (cf. [Dautray and Lions 1992; Lions 1969]). A more general approach is related to the concept of fractional powers of operators.
In the concept of very weak solutions of hyperbolic evolution equations one uses fractional powers of the corresponding operator as test functions (cf. [Gilbert and Jacek 1994; Lions 1969; Martínez Carracedo and Sanz Alix 2001; Mustard 1998; Sohr 2001] for the concept of fractional powers). Let us assume Dirichlet boundary conditions. Taking (formally for the moment) the scalar product of the Equation (5.17) *with $\mathbf{A}_\mathrm{u}^\alpha\mathbf{u}'$, $\alpha \in [0,1]$ one obtains additional estimates for the solution $\mathbf{u}$. There is a strong connection to interpolation theory (cf. [Eck et al. 2005] for a crash course on interpolation theory or [Adams and Fournier 2003; Amann 2000, 1993; Bergh and Löfström 1976; Sohr 2001; Tartar 2000; von Wahl 1985, 1981] for details) because $D(\mathbf{A}_\mathrm{u}^{\frac{1+\alpha}{2}}) = [\mathbf{H}_\mathrm{u}, \mathbf{V}_\mathrm{u}]_\alpha$, where $[\mathbf{H}_\mathrm{u}, \mathbf{V}_\mathrm{u}]_\alpha$ is the intermediate space between $\mathbf{H}_\mathrm{u}$ and $\mathbf{V}_\mathrm{u}$; $\mathbf{A}_\mathrm{u}$ is a self-adjoint, strictly positive operator defined by the variational framework and with compact resolvent. Using spectral decomposition $(\lambda_k, \mathbf{v}_k)_{k\in\mathbb{N}}$ of the operator $\mathbf{A}_\mathrm{u}$ (with $\mathbf{v}_k$ an orthonormal basis*

of eigenvectors of $\mathbf{A}_u$*, the eigenvalues* λ_k *of* $\mathbf{A}_u$ *being s.t.* $0 < \lambda_1 \leq \ldots \lambda_k \leq \ldots$ *and counted in their multiplicity) we set (f.a.* $s \in \mathbb{R}$*)*

$$D(\mathbf{A}_u^s) = \left\{ \mathbf{u} = \sum_{k=1}^{\infty} \mu_k \mathbf{v}_k, \sum_{k=1}^{\infty} \lambda_k^{2s} |\mu_k|^2 < +\infty \right\}, \qquad \|\mathbf{u}\|_{D(\mathbf{A}_u^s)} = \left(\sum_{k=1}^{\infty} \lambda_k^{2s} |\mu_k|^2 \right)^{\frac{1}{2}},$$

which is a Hilbert space. The original problem can be reduced by setting $\mathbf{u}(t) = \sum_k \mu_k(t) \mathbf{v}_k$*,* $u_0 = \sum_k \mu_{0k} \mathbf{v}_k$*,* $u_1 = \sum_k \mu_{1k} \mathbf{v}_k$ *and* $\mathbf{F}(t) = \sum_k \nu_k(t) \mathbf{v}_k$ *to the solution of problems in* $\mu_k(t)$*,* $t \in S$*:*

$$\mu_k''(t) + \lambda_k \mu_k(t) = \nu_k(t), \qquad \mu_k(0) = \mu_{0k}, \qquad \mu_k'(0) = \mu_{1k}.$$

We refer to [Dautray and Lions 1992] for details in case of a standard parabolic evolution problem or to [Lions and Magenes 1973b]. Moreover, we recommend [Lions and Magenes 1973a,b,c] for general theory of evolution equations.
As an example, for $\mathbf{u}_0 \in D(\mathbf{A}_u^{\frac{1+\alpha}{2}})$*,* $\mathbf{u}_1 \in D(\mathbf{A}_u^{\frac{\alpha}{2}})$ *and either* $\mathbf{F} \in L^2(S; D(\mathbf{A}_u^{\frac{\alpha}{2}}))$ *or* $\mathbf{F}, \mathbf{F}' \in L^2(S; D(\mathbf{A}_u^{\frac{\alpha-1}{2}}))$*, we would obtain* $\mathbf{u} \in L^\infty(S; D(\mathbf{A}_u^{\frac{1+\alpha}{2}}))$ *with* $\mathbf{u}' \in L^\infty(S; D(\mathbf{A}_u^{\frac{\alpha}{2}}))$ *in the particular case.*

5.6.7 Step 6: Analysis of Problem $(\mathbf{P}_{A,FP2})$

We continue with studying the solvability of Problem $(\mathbf{P}_{A,FP2})$ for fixed data $(\hat{\theta}, \mathbf{p})$.

5.6.14 Lemma (Regularity)
In addition to the assumptions of Theorem 5.5.1 let Assumption 4.3.2 be valid. Assume the solution $(\mathbf{u}, \boldsymbol{\varepsilon}_{cp})$ *of (the modified) Problem* $(\mathbf{P}_{A,FP2})$ *satisfies* $(\mathbf{u}, \boldsymbol{\varepsilon}_{cp}) \in \mathcal{W}_u^\infty \cap \mathfrak{V}_u^\infty \times \mathfrak{V}_\sigma$*.*
Moreover, the following estimate holds f.a.a. $t \in S$

$$\text{(5.45)} \quad \|\mathbf{u}'(t)\|_{V_u}^2 + \|\mathbf{u}(t)\|_{W_u}^2 + \|\boldsymbol{\varepsilon}_{cp}(t)\|_{V_\sigma}^2 + \|\boldsymbol{\varepsilon}_{cp}'(t)\|_{V_\sigma}^2 \leq c_1 + c_2 \int_0^t \left\{ \|\hat{\theta}(s)\|_{W_\theta}^2 + \|\hat{\theta}(s)\|_{V_\theta}^4 + \|\mathbf{p}\|_{V_p}^2 \right\} ds,$$

where the constants $c_1, c_2 \geq 0$ *do not depend on the data.*

Proof. Again, in the argumentation of Subsection 5.3.1 it is sufficient to show the estimate (5.45). First, we collect the following regularity results for the subproblems. Let $(\theta, \mathbf{p}, \boldsymbol{\varepsilon}_{trip}, \boldsymbol{\varepsilon}_{cp}) \in \mathcal{W}_\theta \times \mathfrak{V}_p \times \mathfrak{V}_\sigma \times \mathfrak{V}_\sigma$. The solution $\mathbf{u}$ of modified problem (5.1) satisfies

$$\|\mathbf{u}'(t)\|^2_{\mathsf{V}_u} + \|\mathbf{u}(t)\|^2_{\mathsf{W}_u} \le c\Big\{\|\mathbf{u}_0\|^2_{\mathsf{W}_u} + \|\mathbf{u}_1\|^2_{\mathsf{V}_u} + \int_0^t \|\mathbf{f}(s)\|^2_{\mathsf{V}_u}\,\mathrm{d}s + \int_0^t \|\theta(s)\|^2_{\mathcal{W}_\theta}\,\mathrm{d}s$$
$$+ \int_0^t \|\mathbf{p}(s)\|^2_{\mathsf{V}_p}\,\mathrm{d}s + \int_0^t \|\boldsymbol{\varepsilon}'_{trip}(s)\|^2_{\mathsf{V}_\sigma}\,\mathrm{d}s + \int_0^t \|\boldsymbol{\varepsilon}_{cp}(s)\|^2_{\mathsf{V}_\sigma}\,\mathrm{d}s + \int_0^t \|\boldsymbol{\varepsilon}'_{cp}(s)\|^2_{\mathsf{V}_\sigma}\,\mathrm{d}s\Big\},$$

where $c_1, c_2 \ge 0$ (cf. Subsection 5.3.6). Moreover, let $(\mathbf{u}, \boldsymbol{\varepsilon}_{trip}) \in \mathcal{W}_u \times \mathfrak{V}_\sigma$. The solution $\boldsymbol{\varepsilon}_{cp}$ of modified problem (5.6) satisfies

$$\|\boldsymbol{\varepsilon}_{cp}\|_{\mathcal{V}_\sigma} \le c_1\Big\{\|\mathbf{u}\|_{\mathcal{W}_u} + \|\boldsymbol{\varepsilon}_{trip}\|_{\mathcal{V}_\sigma}\Big\} \quad \text{and} \quad \|\boldsymbol{\varepsilon}'_{cp}\|_{\mathcal{V}_\sigma} \le c_2\Big\{\|(\mathcal{S}_h\mathbf{u})'\|_{\mathcal{W}_u} + \|\boldsymbol{\varepsilon}'_{trip}\|_{\mathcal{V}_\sigma}\Big\},$$

where $c_1, c_2 \ge 0$ (cf. Subsection 5.6.4).

Using the preceding estimates and the boundedness of $\mathbf{u}_0$, $\mathbf{u}_1$ and $\mathbf{f}$ gives

$$\|\mathbf{u}'(t)\|^2_{\mathsf{V}_u} + \|\mathbf{u}(t)\|^2_{\mathsf{W}_u} + \|\boldsymbol{\varepsilon}_{cp}(t)\|^2_{\mathsf{V}_\sigma} + \|\boldsymbol{\varepsilon}'_{cp}(t)\|^2_{\mathsf{V}_\sigma} \le c_1 + c_2\int_0^t \Big\{\|\theta(s)\|^2_{\mathcal{W}_\theta} + \|\hat\theta(s)\|^4_{\mathsf{V}_\theta}$$
$$+ \|\mathbf{p}(s)\|^2_{\mathsf{V}_p} + \|\mathbf{u}(s)\|^2_{\mathsf{W}_u} + \|\mathcal{S}_h\mathbf{u}(s)\|^2_{\mathsf{W}_u} + \|\mathbf{u}(s)\|^{16}_{\mathsf{W}_u} + \|\boldsymbol{\varepsilon}_{cp}(s)\|^{16}_{\mathsf{W}_u}\Big\}\,\mathrm{d}s,$$

where $c_1, c_2 \ge 0$, depending on h, and $(\theta, \mathbf{p})$, are given. Applying the result of Lemma 5.6.10 and using the generalised Gronwall inequality, cf. Lemma A.5.1 completes the proof. □

5.6.8 Step 7: Analysis of Problem ($\mathbf{P}_\theta$)

We continue with Problem ($\mathbf{P}_{A,\theta}$) for given $(\mathbf{u}, \mathbf{p}, \boldsymbol{\varepsilon}_{cp}) \in \mathcal{V}_u \times \mathfrak{X}^\infty_p \times \mathfrak{H}_\sigma$.

5.6.15 Remark (Additional Time Regularity)
Under Assumptions 4.3.1$_{(A2)}$ and 4.3.2$_{(A2)}$, $\mathbf{u}'' \in \mathcal{V}_u^*$, $\mathbf{u}''|_{\partial\Omega} = 0$ or $\nabla\theta'\nu\big|_{\partial\Omega} = 0$ and additionally vanishing mechanical dissipation, it is possible to prove the boundedness of θ' in $\mathcal{H}^\infty_\theta \cap \mathcal{V}_\theta$ via differentiation of the Galerkin equations and using θ' as a test function.

5.6.16 Lemma (Spatial Regularity)
Let Assumptions 4.3.1$_{(A2)}$ and 4.3.2$_{(A2')}$ be valid and assume in addition zero (Dirichlet or Neumann) boundary conditions, i.e. $\theta|_{\partial\Omega} = 0$ or $\nabla\theta\cdot\nu|_{\partial\Omega} = 0$. Then any solution of Problem ($\mathbf{P}_{A,\theta}$) satisfies $\theta \in \mathcal{V}^\infty_\theta \cap \mathcal{W}_\theta$ and (for $c_1, c_2 > 0$)

$$\|\theta\|_{\mathcal{V}^\infty_\theta\cap\mathcal{W}_\theta} \le c_1(h^{-1}) + c_2(h^{-1})\Big\{\|\mathbf{u}\|_{\mathcal{V}_u} + \|\boldsymbol{\varepsilon}'_{cp}\|_{\mathcal{H}_\sigma} + \|\mathbf{p}'\|_{\mathcal{H}_p} + \|\tilde r\|_{\mathcal{V}_\theta}\Big\}.$$

Proof. We use the Galerkin scheme from the proof of Lemma 5.3.34, where the initial conditions are chosen s.t. $\theta_n(0) \to \theta_{n0}$ in V_θ. First we note that the Ladyzhenskaya-Sobolevski inequality $\|\theta(t)\|_{W_\theta} \leq \|\Delta\theta(t)\|_{H_\theta}$ holds (cf. [Dautray and Lions 1992; Pawlow and Zochowski 2002; Pawlow and Zajaczkowski 2005a]). Next, taking $v = -\Delta\theta_n(t)$ in the equation (5.2) and using the hypothesis of zero boundary conditions, we obtain

$$\frac{\rho_0\, c_e}{2}\frac{\mathrm{d}}{\mathrm{d}t}\| \nabla \theta_n(t)\|^2_{H_\theta} + \frac{\lambda_\theta}{2}\|\Delta\theta_n(t)\|^2_{H_\theta} + \frac{\rho_0\, c_e}{2}\frac{\mathrm{d}}{\mathrm{d}t}\|\theta(t)\|^2_{V_\theta}$$
$$= \int_{\partial\Omega} \theta_n'(t)\theta_\Gamma(t)\,\mathrm{d}\sigma_x - \int_\Omega \hat{r}(t)\Delta\theta_n(t)\,\mathrm{d}\mathbf{x}$$

with

$$\hat{r}(t) := -3K_\alpha\,\theta_0\,\mathrm{div}(\mathbf{u}'(t)) + \Big(\boldsymbol{\sigma}(t) : \big(\boldsymbol{\varepsilon}'_{trip} + \boldsymbol{\varepsilon}_{cp}\big)\Big) + \rho_0 \sum_{i=1}^{N} L_i\, p_i'(t) + r(t).$$

Now, integrating from 0 to t and using the Ladyzhenskaya-Sobolevski inequality as well as Young's inequality, we obtain the estimate

$$\frac{\rho_0\, c_e}{2}\| \nabla \theta_n(t)\|^2_{H_\theta} + \Big(\frac{\lambda_\theta}{2} - \varepsilon_1\Big)\int_0^t \|\theta(s)\|^2_{W_\theta}\,\mathrm{d}s + \Big(\frac{\rho_0\, c_e}{2} - \varepsilon_2\Big)\|\theta_n(t)\|^2_{V_\theta}$$
$$\leq \frac{1}{4\varepsilon_2}\|\theta_\Gamma(t)\|^2_{L^2(\Gamma)} + \frac{1}{4\varepsilon_2}\int_0^t \|\theta_\Gamma'(s)\|^2_{L^2(\Gamma)}\,\mathrm{d}s + C(\varepsilon_1^{-1})\int_0^t \|\hat{r}(s)\|^2_{H_\theta}\,\mathrm{d}s$$

with $0 < \varepsilon_1 < \frac{\lambda_\theta}{2}$ and $0 < \varepsilon_2 < \frac{\rho_0\, c_e}{2}$. Hence, using Poincaré's and Gronwall's inequality, we get

$$(\theta_n) \quad \text{is bounded in} \quad \mathcal{V}_\theta^\infty \cap \mathcal{W}_\theta.$$

So the a-priori estimates follows, independently of n. The result follows by passing to the limit $n \to \infty$. □

5.6.9 Step 8: Analysis of Problem $(\mathbf{P}_{A,FP3})$

Using the foregoing lemmas in this section, we are now in the position to formulate a regularity result for (the modified) Problem $(\mathbf{P}_{A,FP3})$ resp. the full (modified) Problem $(\mathbf{P}_A)$.

Proof. First, we observe that the solution $\mathbf{p}$ of the modified problem (5.3) satisfies

$$\|\mathbf{p}'(t)\|_{V_p} \leq c\Big\{\|\theta(t)\|^2_{V_\theta} + \int_0^t \|\theta(s)\|^2_{V_\theta}\,\mathrm{d}s\Big\},$$

$$\|\mathbf{p}'(t)\|_{\mathcal{W}_p} \leq c\left\{\|\theta(t)\|^2_{V_\theta\cap\mathcal{W}_\theta} + \int_0^t \|\theta(s)\|^2_{V_\theta\cap\mathcal{W}_\theta}\,\mathrm{d}s\right\},$$

with $c \geq 0$ and given θ (cf. Subsection 5.6.2 for details). Moreover, the solution of the heat equation satisfies

$$\|\theta(t)\|^2_{V_\theta} + \int_0^t \|\theta(s)\|^2_{\mathcal{W}_\theta}\,\mathrm{d}s \leq c\left\{\|\theta_0\|^2_{V_\theta} + \int_0^t \|(\mathcal{S}_h\mathbf{u})'(s)\|^2_{V_u}\,\mathrm{d}s \right.$$
$$\left. + \int_0^t \|\theta_\Gamma(s)\|^2_{W_\theta}\,\mathrm{d}s + \int_0^t \|\mathbf{p}'(s)\|^2_{H_p}\,\mathrm{d}s + \int_0^t \|r(s)\|^2_{V_\theta}\,\mathrm{d}s\right\}$$

with $c \geq 0$ and given data $\mathbf{u}$ (cf. Subsection 5.6.8). Using the boundedness of θ_0, θ_Γ and r as well as the results of Lemmas 5.6.10 and 5.6.14 gives

$$\|\theta(t)\|^2_{V_\theta} + \int_0^t \|\theta(s)\|^2_{\mathcal{W}_\theta}\,\mathrm{d}s \leq c_1 + c_2\left\{\int_0^t \|\theta(s)\|^2_{\mathcal{W}_\theta}\,\mathrm{d}s + \int_0^t \|\theta(s)\|^4_{V_\theta}\,\mathrm{d}s\right\}$$

with $c_1, c_2 \geq 0$. Using the generalized Gronwall inequality and following the argumentation in Subsection 5.6.1 completes the proof. □

5.7 Further Aspects and Remarks*

In this section some further aspects and remarks of the Problem ($\mathbf{P}_A$) are provided (and also applicable for Problems ($\mathbf{P}_{VE}$) and ($\mathbf{P}_{QS}$)). Subsection 5.7.1 discusses an application of the subproblem of phase transitions to (concrete) phase transformations in steel. In Subsections 5.7.2 and 5.7.3, Problem ($\mathbf{P}_{A,\boldsymbol{\varepsilon}_{cp}}$) is investigated for time- and parameter-dependent $\mathbf{K}$. Finally, subsection 5.7.4 presents some additional properties of the temperature.

5.7.1 Application of Problem ($\mathbf{P}_{A,\mathbf{p}}$) to Phase Transformations in Steel*

In this work, Equations (5.9) (resp. the right-hand sides of Equations (5.9)) are not amplified and not specified concretely for individual phase transitions. References are [Wolff et al. 2006a, 2007a,c], in which general and specific approaches are discussed for phase-change models. For the mathematical analysis of the overall problem it is essential that for a given $\theta \in \mathcal{H}_\theta$ the Cauchy problem (5.9), (5.10) has exactly one solution (cf. Assumption 4.3.1 for exact conditions). This Cauchy problem itself is solved in Lemma 5.3.8 (cf. [Hüßler 2007] for details). In general, the right-hand sides in (5.9) may also depend on the stress and other (internal) variables (cf. Remark 5.3.7 and [Wolff et al. 2007a]).

Consider the following system of differential equations (cf. Section 2.4)

$$\frac{\partial p_i}{\partial t}(\mathbf{x}, t) = -\sum_{j=1}^{m} a_{ij} H(p_i) H(\bar{p}_{ij} - p_j) G_{ij}(\theta) + \sum_{j=1}^{m} a_{ji} H(p_j) H(\bar{p}_{ji} - p_i) G_{ji}(\theta) \tag{5.46}$$

$$p_i(\mathbf{x}, 0) = p_{0i}(\mathbf{x}) \tag{5.47}$$

for $i = 1, \dots, m$, $(\mathbf{x}, t) \in \Omega_T$, where

$$a_{ij} = a_{ij}(\mathbf{p}(\mathbf{x}, t), \theta(\mathbf{x}, t)) \tag{5.48}$$

for $i, j = 1, \dots, m$. The system (5.46) – (5.48) describes the phase transformation in steel (cf. Section 2.4).

5.7.1 Assumption
Assume that

(A.a) The Heaviside function is regularised via

$$H_\varepsilon(s) := \begin{cases} 0, & s \leq 0 \\ s/\varepsilon, & 0 < s < \varepsilon \\ 1, & s \geq \varepsilon \end{cases} \qquad \text{with} \qquad \varepsilon > 0.$$

(A.b) a_{ij} *fulfills Assumption 4.3.1*$_{(A5),(A6)}$, *i.e. f.a.* $i, j = 1, \dots, m$

- *Let* $a_{ij} : \mathbb{R}^m \times \mathbb{R} \to \mathbb{R}$ *be a Lebesgue measurable function and bounded function, i.e. there exist a non-negative function* $h \in C(\mathbb{R}^m)$ *and two constants* $c_1, c_2 \geq 0$ *s.t.*

$$\|a_{ij}(\mathbf{p}, \theta)\|_\infty \leq c_1 + c_2 h(\mathbf{p}) \qquad \text{f.a.} \qquad \mathbf{p} \in \mathbb{R}^N, \qquad \theta \in \mathbb{R}.$$

- *Moreover, there exists a constant* $L_p > 0$ *s.t.*

$$\|a_{ij}(\mathbf{p}, \theta) - a_{ij}(\mathbf{q}, \theta)\|_\infty \leq L_p \|\mathbf{p} - \mathbf{q}\|_\infty \quad \text{f.a.} \quad \mathbf{p}, \mathbf{q} \in \mathbb{R}^m, \quad \theta \in \mathbb{R}.$$

(A.c) Let $G_{ij}(\theta(\cdot, \cdot)) \in L^\infty(\Omega_T)$ *f.a.* $\theta \in \mathcal{H}_\theta$.

(A.d) $\mathbf{p}_0 \in \mathcal{X}_p^\infty$

and $a_{ij} \geq 0$, $G_{ij} \geq 0$ *f.a.* $\mathbf{p} \in \mathbf{X}_p$ *and* $\theta \in \mathcal{H}_\theta$.

5.7.2 Remark (Balance Condition)
Obviously, we get from (5.46)

$$\sum_{i=1}^{m} \frac{\partial p_i}{\partial t}(\mathbf{x}, t) = 0, \qquad (\mathbf{x}, t) \in \Omega_T,$$

i.e. the sum of the evolution of the phase fractions does not change.

5.7.3 Lemma (Existence and Uniqueness)
Let Assumption 5.7.1 be valid. Assume that $\theta \in \mathcal{H}_\theta$. Then the 'special' initial value problem (5.46), (5.47) *has a unique solution $\mathbf{p} \in \mathfrak{H}_\mathrm{p} \cap \mathcal{X}_\mathrm{p}^\infty$ with $\frac{\partial \mathrm{p}}{\partial t} \in \mathcal{X}_\mathrm{p}^\infty$.*

Proof. One can show that the Assumption $4.3.1_{(A5),(A6)}$ holds and apply Lemma 5.3.8 (cf. [Hüßler 2007] for details). □

5.7.4 Lemma (Balance Condition and Non-negativity Condition)
Let assumption 5.7.1 hold. In addition, the initial value fulfills a balance condition and a non-negativity condition, i.e.

$$\sum_{i=1}^{m} p_{0i}(\mathbf{x}) = 1, \qquad p_{0i}(\mathbf{x}) \geq 0, \qquad i = 1, \ldots, m.$$

Then the solution of (5.46), (5.47) *satisfies*

$$\sum_{i=1}^{m} p_i(\mathbf{x}, t) = 1, \qquad 0 \leq p_i(\mathbf{x}, t) \leq 1, \qquad i = 1, \ldots, m$$

f.a. $t \in S$ and f.a.a. $\mathbf{x} \in \Omega$.

Proof. The balance condition follows from Remark 5.7.2. The non-negativity condition is proven in [Hüßler 2007]. It follows as a consequence of a general situation in reaction kinetics (cf. [Volpert and Chudyaev 1975]). □

5.7.5 Remark (Balance Condition)
A similar situation for $m = 2, 3$ is proven by means of classical comparison criterion for ODE in [Chelminski et al. 2007; Fasano and Primicerio 1996; Fasano et al. 2007; Panizzi 2010].

5.7.6 Lemma (Continuous Dependence on the Parameter θ)
Let Assumption 5.7.1 hold and assume

(A.e) There exists a constant $L_{\theta,a} > 0$ s.t.

$$\|a_{ij}(\mathbf{p}, \theta) - a_{ij}(\mathbf{p}, \vartheta)\|_\infty \leq L_{\theta,a} |\theta - \vartheta| \qquad \text{f.a.} \qquad \mathbf{p} \in \mathbb{R}^m, \qquad \theta, \vartheta \in \mathbb{R}.$$

(A.f) There exists a constant $L_{\theta,G} > 0$ s.t.

$$\|G_{ij}(\theta) - G_{ij}(\vartheta)\|_\infty \leq L_{\theta,G} |\theta - \vartheta| \qquad \text{f.a.} \qquad \theta, \vartheta \in \mathbb{R}.$$

Let $\mathbf{p}_i$ be solutions from Lemma 5.3.8 corresponding to θ_i, $i = 1, 2$. Then

$$\sup_{t \in \bar{S}} \|\mathbf{p}_1(t) - \mathbf{p}_2(t)\|^2_{\mathrm{H}_p} \leq c \, \|\theta_1 - \theta_2\|^2_{\mathcal{H}_\theta} \qquad \text{and} \qquad \|\mathbf{p}_1' - \mathbf{p}_2'\|^2_{\mathcal{H}_p} \leq c \, \|\theta_1 - \theta_2\|^2_{\mathcal{H}_\theta}.$$

Proof. Apply Lemma 5.3.10 (cf. [Hüßler 2007]). □

Analogous regularity results like in Lemma 5.6.1, 5.6.2 and Remark 5.6.3 can be shown for appropriate conditions on a_{ij} and G_{ij}. In the following we discuss a simple example.

5.7.7 Example (Two-Phase Transformation)
Consider only the cooling of a steel specimen from high temperature (austenitic phase with phase fraction p_A) to the martensitic phase with phase fraction p_M. A prototypical model (and a simple transformation law) to describe this behaviour is given by the so-called Leblond-Devaux model (cf. Example 2.4.2)

$$p_A'(\mathbf{x},t) = -p_M'(\mathbf{x},t), \qquad p_A(x,0)=1 \tag{5.49}$$

$$p_M'(\mathbf{x},t) = \mu(\theta(\mathbf{x},t))\big(\bar{p}_M(\theta(\mathbf{x},t)) - p_M(\mathbf{x},t)\big), \qquad p_M(x,0)=0 \tag{5.50}$$

where $\bar{p}_M$ represents the maximal martensitic phase fraction that can be attained and $\theta \in \mathcal{H}_\theta$ the given temperature. We use the following Koistinen-Marburger ansatz

$$\bar{p}_M(\theta) = 1 - \exp\left(\frac{\theta-\theta_{ms}}{\theta_{m0}}\right), \quad \theta \in [\theta_{mf}, \theta_{ms}], \quad \text{and} \quad \bar{p}_M(\theta) = 0, \quad \theta \in [\theta_{ms}, \theta_0],$$

where $\mu(\theta) = a(\theta_0-\theta)^2$ (we choose $a = 8.8$, cf. [Dachkovski and Böhm 2004b]).

(1) **Existence and Uniqueness** *We remark, that $\mu, \bar{p} \in C^1([\theta_{mf},\theta_0])$ satisfying $\mu(\theta) \in [0, a\theta_0^2]$ and $\bar{p}(\theta) \in [0,1]$ f.a. $\theta \in [\theta_{mf},\theta_0]$. Rewriting the problem as*

$$z'(t) = F(z(t),t) \quad \text{in } S, \qquad z(0)=0$$

with $F(z,\cdot) = \mu(\theta(\cdot))\,(\bar{p}(\theta)(\cdot) - z)$, one can easily show that the hypothesis of the existence theorem of Carathéodory holds. The continuity regarding z is obvious and the measurability follows as a consequence of the measurability of θ. According to the Carathéodory Theorem (cf. [Roubíček 2005; Zeidler 1990b]) the problem (5.49),(5.50) has a solution $p_M \in W^{1,1}(S;L^2(\Omega))$. The Lipschitz continuity of F yields the uniqueness of the solution.

(2) **Regularity** *We calculate the partial derivatives*

$$\frac{\partial\gamma}{\partial\theta} = \mu'(\theta)(\bar{p}(\theta)-p) + \mu(\theta)\bar{p}'(\theta) \quad \text{and}$$
$$\frac{\partial^2\gamma}{\partial\theta^2} = \mu''(\theta)(\bar{p}(\theta)-p) + 2\mu'(\theta)\bar{p}'(\theta) + \mu(\theta)\bar{p}''(\theta).$$

For $\mu(\theta) = \alpha(\theta_0-\theta)^2$ holds $\mu'(\theta) = -2\alpha(\theta_0-\theta)$ and $\mu'' = 2\alpha$ where $\alpha = \text{const}$. Hence,

$$\bar{p}' = -\frac{1}{\theta_{M0}}\exp\left(\frac{\theta-\theta_{MS}}{\theta_{M0}}\right) = \frac{1}{\theta_{M0}}(\bar{p}-1) \quad \text{and}$$

$$\bar{p}'' = -\frac{1}{\theta_{M0}^2}\exp\left(\frac{\theta-\theta_{MS}}{\theta_{M0}}\right) = \frac{1}{\theta_{M0}^2}(\bar{p}-1)$$

for $\theta \in [\theta_{M0}, \theta_{MS}]$. Thus, the hypothesis of Lemma 5.6.2 are fulfilled and we obtain additional regularity of the solution.

(3) **Balance Condition** *Obviously, $p_A' + p_M' = 0$ and therefore integration by parts w.r.t. time yields $p_A + p_M = 1$.*

In this case, the solution reads as

$$p_M(t) = \exp\Big(\int_0^t \mu(t)\,\mathrm{d}s\Big)\int_0^t \mu(t)\bar{p}(t)\exp\Big(\int_0^s \mu(\tau)\,\mathrm{d}\tau\Big)\,\mathrm{d}s \quad \textit{and}$$
$$p_A(t) = 1 - \int_0^t \mu(s)\bar{p}(s)\,\mathrm{d}s + \int_0^t \mu(s)p_M(s)\,\mathrm{d}s.$$

5.7.8 Example (Transformation-induced Plasticity)
Assume the same situation as in Example 5.7.7 and use Tanaka's ansatz for the saturation function. Then the problem (5.11),(5.12) *reads as*

$$\boldsymbol{\varepsilon}_{trip}' = \frac{3}{2}\boldsymbol{\sigma}^*\left(\kappa_A p_A' + \kappa_M p_M'\right) = \frac{3}{2}\boldsymbol{\sigma}^*\left(\kappa_M - \kappa_A\right)p_M', \qquad \boldsymbol{\varepsilon}_{trip}(0) = \mathbf{0},$$

where $\boldsymbol{\sigma}^* = 2\mu\left(\boldsymbol{\varepsilon}^*(\mathbf{u}) - \boldsymbol{\varepsilon}_{trip} - \boldsymbol{\varepsilon}_{cp}\right)$. Existence of a unique solution is obvious and yields the solution

$$\boldsymbol{\varepsilon}_{trip}(t) = \int_0^t 3\mu\left(\kappa_M - \kappa_A\right)p_M' \exp\Big(-\int_s^t 3\mu\left(\kappa_M-\kappa_A\right)p_M'\,\mathrm{d}\tau\Big)\Big(\boldsymbol{\varepsilon}^*(\mathbf{u}) - \boldsymbol{\varepsilon}_{cp}\Big)\,\mathrm{d}s.$$

5.7.2 Analysis of Problem ($\mathbf{P}_{cp}$) for time-dependent $\mathbf{K}^*$

Due to the general time-dependence of R_0, R, the set of admissible stresses varies in time, when considering a time-dependent process (cf. e.g. [Han and Reddy 1999b]). In the application problem we are concerned with cooling or quenching processes. Therefore, we assume a growing yield radius with the time. We define $F : \mathbb{R}^{3\times3}\times\mathbb{R}\times\mathbb{R}^3 \to \mathbb{R}$ via

$$F(\boldsymbol{\sigma}, t, \mathbf{x}) := \sqrt{\frac{3}{2}\boldsymbol{\sigma}^* : \boldsymbol{\sigma}^*} - (R_0 + R(t,\mathbf{x})),$$
$$\mathbf{K}_F(t,\mathbf{x}) := \Big\{\boldsymbol{\tau} \in \mathbb{R}^{3\times3}_{\mathrm{sym}}, \mathrm{tr}(\boldsymbol{\tau}) = 0 : F(\boldsymbol{\tau}, t, \mathbf{x}) \le 0\Big\},$$
$$\mathbf{K}(t) := \Big\{\boldsymbol{\tau} \in \mathbf{H}_\sigma : \boldsymbol{\tau}(\mathbf{x}) \in \mathbf{K}_F(t,\mathbf{x}) \text{ f.a.a. } \mathbf{x} \in \Omega\Big\},$$

where R is defined, e.g., as in Equation (2.66) or (2.67).

Let $\chi_{K(t)} : \mathbf{H}_\sigma \to \mathbb{R}\cup\{+\infty\}$ be the indicator function on $\mathbf{K}(t)$, i.e. $\chi_{K(t)}(\mathbf{u}) = 0$ if $\mathbf{u} \in \mathbf{K}(t)$ and $\chi_{K(t)}(\mathbf{u}) = +\infty$ if $\mathbf{u} \notin \mathbf{K}(t)$ f.a. $t \in S$.

5.7.9 Lemma
Let Assumption 4.3.1$_{(A3)}$ be valid. Then $\mathbf{K}(t)$ *is a nonempty closed convex subset of* $\mathbf{H}_\sigma$ *(resp.* $\mathbf{V}_\sigma$ *for* $\mathbf{K}(t) := \{\boldsymbol{\tau} \in \mathbf{V}_\sigma : \boldsymbol{\tau}(\mathbf{x}) \in \mathbf{K}_F(t, \mathbf{x})$ *f.a.a.* $\mathbf{x} \in \Omega\}$*) f.a.* $t \in S$.

Proof. The proof is analogous to the proof of Lemma 5.3.1. □

5.7.10 Lemma
Let Assumption 4.3.1$_{(A3)}$ be valid. Then

(1) *the indicator function* $\chi_{\mathrm{K}(t)}$ *is proper, convex and lower semi-continuous f.a.* $t \in S$.
(2) $\partial\chi_{\mathrm{K}(t)} : \mathbf{H}_\sigma \to \mathcal{P}(\mathbf{H}_\sigma)$ *is maximal monotone f.a.* $t \in S$.
(3) $D(\chi_{\mathrm{K}(t)}) = D(\partial\chi_{\mathrm{K}(t)}) = \mathbf{K}(t)$ *f.a.* $t \in S$.

Proof. The proof is analogous to the proof of Lemma 5.3.2. □

5.7.11 Lemma (Existence and Uniqueness)
Let Assumption 4.3.1$_{(A3)}$ be valid. Assume $(\mathbf{u}, \boldsymbol{\varepsilon}_{trip}) \in \mathcal{V}_{\mathrm{u}} \times \mathfrak{H}_{\boldsymbol{\sigma}}$. *Then there exist a unique solution* $\boldsymbol{\sigma}^* \in C(\bar{S}; \mathbf{H}_\sigma) \cap \mathfrak{H}_\sigma$ *s.t. Problem* $(\mathbf{P}_{A,\boldsymbol{\varepsilon}_{cp}})$ *is fulfilled and* $\boldsymbol{\sigma}^*(0) = \boldsymbol{\sigma}_0^*$, $\boldsymbol{\sigma}^*(t) \in \mathbf{K}(t)$ *f.a.a.* $t \in S$, $t \to \chi_{\mathrm{K}(t)}(\boldsymbol{\sigma}^*(t))$ *bounded and*

$$\|\boldsymbol{\sigma}^*(t) - \boldsymbol{\sigma}^*(0)\|_{\mathrm{H}_\sigma} \le c(h^{-1}) \left\{ \int_0^t \|\mathbf{u}(s)\|_{\mathrm{V_u}} \, \mathrm{d}s + \int_0^t \|\boldsymbol{\varepsilon}'_{trip}(s)\|_{\mathrm{H}_\sigma} \, \mathrm{d}s \right\} \text{ f.a.a. } t \in S,$$
$$\|(\boldsymbol{\sigma}^*)'(t)\|_{\mathrm{H}_\sigma} \le c(h^{-1}) \left\{ \|\mathbf{u}(t)\|_{\mathrm{V_u}} + \|\boldsymbol{\varepsilon}'_{trip}(t)\|_{\mathrm{H}_\sigma} \right\} \text{ f.a.a. } t \in S.$$

Proof. We use Lemma 5.7.9 and Lemma 5.7.10. We define the projection on $\mathbf{K}(t)$ via $\mathbb{P}_{\mathrm{K}(t)} : \mathbf{H}_\sigma \to \mathbf{K}(t)$ with

$$\|\boldsymbol{\sigma} - \mathbb{P}_{\mathrm{K}(t)}\boldsymbol{\sigma}\|_{\mathrm{H}_\sigma} = \operatorname{dist}(\boldsymbol{\sigma}, \mathbf{K}(t)) = \inf_{\boldsymbol{\tau} \in \mathrm{K}(t)} \|\boldsymbol{\sigma} - \boldsymbol{\tau}\|_{\mathrm{H}_\sigma} \qquad \text{f.a.} \qquad \boldsymbol{\sigma} \in \mathbf{H}_\sigma.$$

It is clear that $\mathbb{P}^2_{\mathrm{K}(t)} = \mathbb{P}_{\mathrm{K}(t)}$, since $\mathbb{P}_{\mathrm{K}(t)} = \mathbf{I}$ on $\mathbf{K}(t)$. $\mathbb{P}_{\mathrm{K}(t)}$ is not linear, because then the image of $\mathbf{K}(t)$ would have to be a linear subspace of $\mathbf{H}_\sigma$, which is, of course, not the case in general. If $\mathbf{K}(s) \subset \mathbf{K}(t)$ for $t \ge s$, then $R(t) \ge R(s)$, i.e. $\boldsymbol{\sigma} \in \mathbf{K}(s) \implies \boldsymbol{\sigma} \in \mathbf{K}(t)$. Moreover, $\mathbb{P}_{\mathrm{K}(t)}\boldsymbol{\sigma} = \boldsymbol{\sigma}$ for $\boldsymbol{\sigma} \in \mathbf{K}(t)$ and $\mathbb{P}_{\mathrm{K}(s)}\boldsymbol{\sigma} = \boldsymbol{\sigma}$ for $\boldsymbol{\sigma} \in \mathbf{K}(s)$. Therefore we get by geometrical observation

$$\begin{aligned}
\mathrm{h}(\mathbf{K}(s), \mathbf{K}(t)) &= \max\Big\{ \sup_{\boldsymbol{\tau} \in \mathrm{K}(s)} \|\boldsymbol{\tau} - \mathbb{P}_{\mathrm{K}(t)}\boldsymbol{\tau}\|, \sup_{\boldsymbol{\tau} \in \mathrm{K}(t)} \|\boldsymbol{\tau} - \mathbb{P}_{\mathrm{K}(s)}\boldsymbol{\tau}\| \Big\} \\
&\le \sup_{\boldsymbol{\tau} \in \mathrm{K}(s) \cup \mathrm{K}(t)} \|\mathbb{P}_{\mathrm{K}(s)}\boldsymbol{\tau} - \mathbb{P}_{\mathrm{K}(t)}\boldsymbol{\tau}\| \\
&\le \sup_{\boldsymbol{\tau} \in \mathrm{K}(s) \cup \mathrm{K}(t)} \Big\| |\mathbb{P}_{\mathrm{K}(s)}\boldsymbol{\tau}|^2 - |\mathbb{P}_{\mathrm{K}(t)}\boldsymbol{\tau}|^2 \Big\|
\end{aligned}$$

$$
\begin{aligned}
&\leq \frac{2}{3}\Big[(R_0 + R(s))^2 - (R_0 + R(t))^2\Big] \\
&\leq \frac{2}{3}\|R(s) - R(t)\| \Big[\|2R_0 + R(s) + R(t)\|\Big] \\
&\leq c\|R(s) - R(t)\| \\
&\leq c \int_t^s \frac{\mathrm{d}}{\mathrm{d}t}\|R(\tau)\| \,\mathrm{d}\tau
\end{aligned}
$$

Hence, the prerequisites of Lemma C.2.11 are fulfilled and applying Lemma C.2.11 yields the existence and uniqueness result.

□

5.7.12 Lemma (Continuous Dependence on Parameters $\mathbf{u}$ and $\boldsymbol{\varepsilon}_{trip}$)
Let Assumption 4.3.1$_{(A3)}$ be valid and assume in addition $\boldsymbol{\sigma}^*(0) \in \mathbf{K}(0)$. Then two different solutions $\boldsymbol{\sigma}_i^*$ corresponding to Lemma 5.7.11 of Problem ($\mathbf{P}_{A,\boldsymbol{\varepsilon}_{cp}}$) corresponding to the data ($\mathbf{u}_i, \boldsymbol{\varepsilon}_{trip,i}$), $i = 1, 2$ fulfill the estimate

$$
\begin{aligned}
&\|\boldsymbol{\sigma}_1^*(t) - \boldsymbol{\sigma}_2^*(t)\|_{\mathrm{H}_\sigma} \leq \|\boldsymbol{\sigma}_1^*(0) - \boldsymbol{\sigma}_2^*(0)\|_{\mathrm{H}_\sigma} \\
&+ c(h^{-1})\Big\{ \int_0^t \|\mathbf{u}_1(s) - \mathbf{u}_2(s)\|_{\mathrm{V}_u} \,\mathrm{d}s + \int_0^t \|\boldsymbol{\varepsilon}'_{trip,1}(s) - \boldsymbol{\varepsilon}'_{trip,2}(s)\|_{\mathrm{H}_\sigma} \,\mathrm{d}s \Big\} \text{ f.a.a. } t \in S.
\end{aligned}
$$

Proof. Applying Remark C.2.15 gives the result. □

5.7.13 Lemma (Regularity)
Let Assumption 4.3.1$_{(A3)}$ be valid and assume in addition $(\mathbf{u}, \boldsymbol{\varepsilon}_{trip}) \in \mathcal{W}_u \times \mathfrak{V}_\sigma$. Then the solution corresponding to Lemma 5.7.11 of Problem ($\mathbf{P}_{A,\boldsymbol{\varepsilon}_{cp}}$) fulfills $\boldsymbol{\sigma}^* \in C(S; \mathbf{V}_\sigma) \cap \mathfrak{V}_\sigma$ and

$$
\begin{aligned}
\|\boldsymbol{\sigma}^*(t) - \boldsymbol{\sigma}^*(0)\|_{\mathrm{V}_\sigma} &\leq c(h^{-1})\Big\{ \int_0^t \|\mathbf{u}(s)\|_{\mathrm{W}_u} \,\mathrm{d}s + \int_0^t \|\boldsymbol{\varepsilon}'_{trip}(s)\|_{\mathrm{V}_\sigma} \,\mathrm{d}s \Big\} \text{ f.a.a. } t \in S, \\
\|(\boldsymbol{\sigma}^*)'(t)\|_{\mathrm{V}_\sigma} &\leq c(h^{-1})\Big\{ \|\mathbf{u}(t)\|_{\mathrm{W}_u} + \|\boldsymbol{\varepsilon}'_{trip}(t)\|_{\mathrm{V}_\sigma} \Big\} \text{ f.a.a. } t \in S.
\end{aligned}
$$

Proof. Using Lemma 5.7.9, Lemma 5.7.10 and applying Lemma C.2.11 proves the result. □

5.7.3 Analysis of Problem ($\mathbf{P}_{cp}$) for parameter-dependent $\mathbf{K}^*$

We define $F : \mathbb{R}^{3\times 3} \times \mathbb{R} \to \mathbb{R}$ via

$$
F(\boldsymbol{\tau}, \theta) := \sqrt{\frac{3}{2}\boldsymbol{\sigma}^* : \boldsymbol{\sigma}^*} - (R_0 + R(\theta)),
$$

$$\mathbf{K}_F(\theta) := \Big\{\boldsymbol{\tau} \in \mathbb{R}^{3\times 3}_{\text{sym}}, \operatorname{tr}(\boldsymbol{\tau}) = 0 : F(\boldsymbol{\tau}, \theta) \leq 0\Big\},$$
$$\mathbf{K}(\theta) := \Big\{\boldsymbol{\tau} \in \mathbf{H}_\sigma : \boldsymbol{\tau}(\mathbf{x}) \in \mathbf{K}_F(\theta) \text{ f.a.a. } \mathbf{x} \in \Omega\Big\},$$

where R is defined, e.g., as in Remark 2.6.2 or 2.6.3 and θ is a given parameter, e.g. the temperature. In [Babadjian et al. 2011; Chelminski and Racke 2006] the yield function depends explicitly on the temperature. We assume in our application problem cooling or quenching processes and therefore assume a decreasing temperature, i.e. a growing yield radius.

Let $\chi_{\mathbf{K}(\theta)} : \mathbf{H}_\sigma \to \mathbb{R} \cup \{+\infty\}$ be the indicator function on $\mathbf{K}(\theta)$, i.e. $\chi_{\mathbf{K}(\theta)}(\mathbf{u}) = 0$ if $\mathbf{u} \in \mathbf{K}(\theta)$ and $\chi_{\mathbf{K}(\theta)}(\mathbf{u}) = +\infty$ if $\mathbf{u} \notin \mathbf{K}(\theta)$ for a given parameter θ.

5.7.14 Remark
Lemmas 5.7.9 and 5.7.10 apply analogously to this case.

5.7.15 Lemma (Existence and Uniqueness)
Let Assumption 4.3.1$_{(A3)}$ be valid and assume in addition $\mathbf{u} \in \mathcal{V}_{\mathrm{u}}$, $\boldsymbol{\varepsilon}'_{trip} \in \mathcal{H}_\sigma$, $\theta \in \mathfrak{H}_\theta$ and $\boldsymbol{\sigma}^*(0) \in \mathbf{K}(0)$. Then, there exist a unique solution $\boldsymbol{\sigma}^* \in C(\bar{S}; \mathbf{H}_\sigma)$ s.t. Problem $(\mathbf{P}_{A,\boldsymbol{\varepsilon}_{cp}})$ is fulfilled and $\boldsymbol{\sigma}^*(0) = \boldsymbol{\sigma}^*_0$, $\boldsymbol{\sigma}^*(t) \in \mathbf{K}(t)$ f.a.a. $t \in S$, $(\boldsymbol{\sigma}^*)' \in \mathcal{V}_\sigma$, $t \to \chi_{\mathbf{K}(t)}(\boldsymbol{\sigma}^*(t))$ is bounded and

$$\|\boldsymbol{\sigma}^*(t) - \boldsymbol{\sigma}^*(0)\|_{\mathrm{H}_\sigma} \leq c(h^{-1})\Big\{\int_0^t \|\mathbf{u}(s)\|_{\mathrm{V}_{\mathrm{u}}}\,\mathrm{d}s + \int_0^t \|\boldsymbol{\varepsilon}'_{trip}(s)\|_{\mathrm{H}_\sigma}\,\mathrm{d}s\Big\} \text{ f.a.a. } t \in S,$$
$$\|(\boldsymbol{\sigma}^*)'(t)\|_{\mathrm{H}_\sigma} \leq c(h^{-1})\Big\{\|\mathbf{u}(t)\|_{\mathrm{V}_{\mathrm{u}}} + \|\boldsymbol{\varepsilon}'_{trip}(t)\|_{\mathrm{H}_\sigma}\Big\} \text{ f.a.a. } t \in S.$$

Proof. We use Lemma 5.7.9, Lemma 5.7.10 and apply Lemma C.2.19. The proof of Lemma C.2.11 yields

$$\begin{aligned} \mathrm{h}(\mathbf{K}(s,\theta), \mathbf{K}(t,\theta)) &\leq c|R(\theta(t)) - R(\theta(s))| \\ &\leq c|\theta(t) - \theta(s)| \\ &\leq c\int_s^t \|\theta'(\tau)\|_{H_\theta}\,\mathrm{d}\tau \end{aligned}$$

In order to verify the prerequisites of Lemma C.2.19, we have to show: $\mathbf{K}(t,\theta_n) \stackrel{M}{\to} \mathbf{K}(t,\theta)$ for $\theta_n \to \theta$ in the Mosco sense (cf. Appendix C.2.3 for the term 'Mosco convergence'), i.e.

(1) $\forall \boldsymbol{\sigma}^*_n \in \mathbf{K}(t,\theta_n)$, $\boldsymbol{\sigma}^*_n \to \boldsymbol{\sigma}^* \implies \boldsymbol{\sigma}^* \in \mathbf{K}(t,\theta)$ and
(2) $\forall \boldsymbol{\sigma}^* \in \mathbf{K}(t,\theta)\ \exists \boldsymbol{\sigma}^*_n \in \mathbf{K}(t,\theta_n)$ s.t. $\boldsymbol{\sigma}^*_n \to \boldsymbol{\sigma}^*$.

The first statement follows because of the continuity of R analogous to the closure of $\mathbf{K}$, cf. Lemma 5.3.1. Second, let $\boldsymbol{\sigma}^* \in \mathbf{K}(t,\theta)$ and choose $\boldsymbol{\sigma}^*_n = \boldsymbol{\sigma}^* + \sqrt{\frac{2}{3}}(R(\theta_n) - R(\theta))\,\mathbf{I}$. In order to show the convergence, we show:

$$\begin{aligned}\sqrt{\frac{2}{3}\boldsymbol{\sigma}^*_n : \boldsymbol{\sigma}^*_n} &= \sqrt{\frac{2}{3}\Big(\boldsymbol{\sigma}^* + \sqrt{\frac{2}{3}}(R(\theta_n) - R(\theta))\,\mathbf{I}\Big) : \Big(\boldsymbol{\sigma}^* + \sqrt{\frac{2}{3}}(R(\theta_n) - R(\theta))\,\mathbf{I}\Big)} \\ &= \sqrt{\frac{2}{3}\boldsymbol{\sigma}^* : \boldsymbol{\sigma}^* + 2\sqrt{\frac{2}{3}}\,\mathrm{tr}(\boldsymbol{\sigma}^*)(R(\theta_n) - R(\theta)))\,\mathbf{I} + \frac{2}{3}(R(\theta_n) - R(\theta))^2\,\mathbf{I}} \\ &= \sqrt{\frac{2}{3}\boldsymbol{\sigma}^* : \boldsymbol{\sigma}^* + \frac{2}{3}(R(\theta_n) - R(\theta))^2\,\mathbf{I}} \to \sqrt{\frac{2}{3}\boldsymbol{\sigma}^* : \boldsymbol{\sigma}^*}\end{aligned}$$

for $\theta_n \to \theta$. This completes the proof. □

5.7.16 Lemma (Continuous Dependence on Parameters $\mathbf{u}$, $\boldsymbol{\varepsilon}_{trip}$ and θ)
Let Assumption 4.3.1$_{(A3)}$ be valid and assume in addition $\mathbf{u} \in \mathcal{V}_{\mathrm{u}}$, $\boldsymbol{\varepsilon}'_{trip} \in \mathcal{H}_{\boldsymbol{\sigma}}$, $\theta \in \mathfrak{H}_\theta$ and $\boldsymbol{\sigma}^*(0) \in \mathbf{K}(0)$. Then two different solutions $\boldsymbol{\sigma}^*_i$ of Problem ($\mathbf{P}_{A,\boldsymbol{\varepsilon}_{cp}}$) from Lemma 5.7.15 corresponding to the data $(\mathbf{u}_i, \boldsymbol{\varepsilon}_{trip,i})$, $i = 1,2$ fulfill

$$\begin{aligned}\|\boldsymbol{\sigma}^*_1(t) - \boldsymbol{\sigma}^*_2(t)\|_{\mathrm{H}_\sigma} \le \|\boldsymbol{\sigma}^*_1(0) - \boldsymbol{\sigma}^*_2(0)\|_{\mathrm{H}_\sigma} + c(h^{-1})\Big\{\int_0^t \|\mathbf{u}_1(s) - \mathbf{u}_2(s)\|_{\mathrm{V}_\mathrm{u}}\,\mathrm{d}s \\ + \int_0^t \|\boldsymbol{\varepsilon}'_{trip,1}(s) - \boldsymbol{\varepsilon}'_{trip,2}(s)\|_{\mathrm{H}_\sigma}\,\mathrm{d}s + \int_0^t \|\theta_1(s) - \theta_2(s)\|_{H_\theta}\,\mathrm{d}s\Big\} \text{ f.a.a. } t \in S.\end{aligned}$$

Proof. Applying Remark C.2.21 gives the result. □

5.7.17 Lemma (Regularity)
Let Assumption 4.3.1$_{(A3)}$ hold and assume in addition $\mathbf{u} \in \mathcal{V}_{\mathrm{u}}$, $\boldsymbol{\varepsilon}'_{trip} \in \mathcal{H}_{\boldsymbol{\sigma}}$, $\theta \in \mathfrak{H}_\theta$ and $\boldsymbol{\sigma}^*(0) \in \mathbf{K}(0)$. Then for a solution of Problem ($\mathbf{P}_{A,\boldsymbol{\varepsilon}_{cp}}$) corresponding to Lemma 5.7.15 holds

$$\begin{aligned}\|\boldsymbol{\sigma}^*(t) - \boldsymbol{\sigma}^*(0)\|_{\mathrm{V}_\sigma} \le c(h^{-1})\Big\{\int_0^t \|\mathbf{u}(s)\|_{\mathrm{W}_\mathrm{u}}\,\mathrm{d}s + \int_0^t \|\boldsymbol{\varepsilon}'_{trip}(s)\|_{\mathrm{V}_\sigma}\,\mathrm{d}s\Big\} \text{ f.a.a. } t \in S, \\ \|(\boldsymbol{\sigma}^*)'(t)\|_{\mathrm{V}_\sigma} \le c(h^{-1})\Big\{\|\mathbf{u}(t)\|_{\mathrm{W}_\mathrm{u}} + \|\boldsymbol{\varepsilon}'_{trip}(t)\|_{\mathrm{V}_\sigma}\Big\} \text{ f.a.a. } t \in S.\end{aligned}$$

Proof. Using Lemma 5.7.9, Lemma 5.7.10 and applying Lemma C.2.19 gives the result. □

5.7.4 Additional Properties of the Temperature*

In this subsection we present some additional properties of the temperature. In the following, let θ be the solution of Problem ($\mathbf{P}_{A,\theta}$) corresponding to Lemma 5.3.34.

5.7.18 Remark (Positivity of the Temperature)
For cooling or quenching processes without inverse (or rather reverse) phase transformation, like in the case of heating or tempering, it holds $\sum_{i=2}^{m} L_i p_i' > 0,\ m \in \mathbb{N}$, *i.e. the the latent heat source term is non-negative*[1]. *Furthermore, we assume (besides Assumption 4.3.1*$_{(A2)}$*)* $r > 0,\ \theta_\Gamma > 0,\ \theta_0 > 0$ *and* $\boldsymbol{\sigma} : \boldsymbol{\varepsilon}'_{trip} > 0$ *as well as* $\boldsymbol{\sigma} : \boldsymbol{\varepsilon}'_{cp} > 0$ *which is conform with the remaining dissipation inequality* (2.62). *In this case, the temperature* θ *is non-negative.*
The idea of proving the non-negativity of the temperature is to show $\|\theta^-\|_{\mathcal{V}_\theta} \leq 0$. *Then it follows* $\theta \geq 0$ *a.e. Multiplying Equation (5.31) with* $-\theta^-$ *gives*

$$\begin{aligned}\frac{\mathrm{d}}{\mathrm{d}t}\int_\Omega \frac{\rho_0\, c_e}{2}|\theta^-(t)|^2\,\mathrm{d}\mathbf{x} + \lambda_\theta \int_\Omega |\nabla\theta^-(t)|^2\,\mathrm{d}\mathbf{x} + \int_{\partial\Omega}|\theta^-(t)|^2\,\mathrm{d}\sigma_x \\ = \int_\Omega \beta(\theta(t))\,\mathrm{div}(\mathbf{u}'(t))\theta^-(t)\,\mathrm{d}\mathbf{x} - \underbrace{\int_\Omega \boldsymbol{\sigma} : (\boldsymbol{\varepsilon}'_{trip} + \boldsymbol{\varepsilon}'_{cp})\theta^-(t)\,\mathrm{d}\mathbf{x}}_{\geq 0} \\ - \underbrace{\rho_0 \int_\Omega \sum L_i p_i'(t)\theta^-(t)\,\mathrm{d}\mathbf{x}}_{\geq 0} - \underbrace{\int_\Omega r(t)\theta^-(t)\,\mathrm{d}\mathbf{x}}_{\geq 0} - \underbrace{\int_{\partial\Omega}\theta_\Gamma(t)\theta^-(t)\,\mathrm{d}\sigma_x}_{\geq 0}.\end{aligned}$$

Hence

$$\|\theta^-(t)\|^2_{H_\theta} + \int_0^t \|\theta^-(s)\|^2_{V_\theta}\,\mathrm{d}s \leq \int_0^t \|\theta^-(s)\|^2_{H_\theta}\|\,\mathrm{div}(\mathbf{u}'(s))\|^2_{H_\theta}\,\mathrm{d}s.$$

Applying the Gronwall inequality completes the proof.

5.7.19 Remark (Asymptotic Behaviour of the Temperature)
Let Assumption 4.3.1$_{(A2)}$ *be valid and assume in addition zero Dirichlet boundary conditions. Multiplying Equation* (5.31) *with* θ *and integrating gives*

$$\|\theta(t)\|^2_{H_\theta} + \int_0^t \|\nabla\theta(s)\|^2_{H_\theta}\,\mathrm{d}s \leq c\Big\{\int_0^t \|\hat{r}(s)\|^2_{H_\theta}\,\mathrm{d}s + \int_0^t \|\theta(s)\|^2_{H_\theta}\,\mathrm{d}s + \|\theta_0\|^2_{H_\theta}\Big\}.$$

Using Poincare's and Gronwall's inequality we get

$$\|\theta(t)\|^2_{H_\theta} \leq \left(\|\hat{r}\|^2_{\mathcal{H}_\theta} + \|\theta_0\|^2_{H_\theta}\right)\exp(-ct).$$

Let θ_∞ *be the solution of the corresponding elliptic problem* $-\Delta\theta = \hat{r}$. *Put* $\vartheta := \theta - \theta_\infty$ *and consider*

$$\vartheta' - \Delta\vartheta = 0, \qquad \vartheta|_{\partial\Omega} = 0.$$

[1] private communications, M. Wolff, 2011

Then, multiplication with ϑ yields

$$\frac{\partial}{\partial t}\|\vartheta(t)\|^2_{H_\theta} = -c\|\nabla\vartheta(t)\|^2_{H_\theta} \leq -c\|\vartheta(t)\|^2_{H_\theta}.$$

Therefore $\frac{\partial}{\partial t}(\|\vartheta(t)\|^2_{H_\theta} exp(ct)) \leq 0$. *Integration gives* $\|\vartheta(t)\|^2_{H_\theta} \leq \|\vartheta_0\|^2_{H_\theta}\exp(-ct)$. *Hence,* $\|\theta - \theta_\infty\|_{\mathcal{H}_\theta} \to 0$ *for* $t \to \infty$ *(cf. [Eck et al. 2008] for details).*

5.7.20 Remark (Maximum Principle for the Temperature)
The inelastic constitutive equation implies that the temperature θ is bounded from above by some positive constant M_θ. Thus, to have a control for the function θ we slightly modify the right-hand side of the heat equation

$$\operatorname{div}(\mathbf{u}'(x,t))\theta(x,t) = \operatorname{div}(\mathbf{u}'(x,t))\beta(\theta(x,t)) \quad \textit{and} \quad \hat{r}(\mathbf{x},t) = \hat{r}(\mathbf{x},t)\beta(\theta(x,t))$$

where β is defined by

$$\beta(\theta) = \begin{cases} 1 & \textit{for } \theta < M_\theta \\ [0,1] & \textit{for } \theta = M_\theta \\ 0 & \textit{for } \theta > M_\theta \end{cases}.$$

Of course, this switch function is not needed if the divergence term, the dissipation and the right-hand side of the heat equation either vanish or are positive.
Let us fix the positive time interval S and assume in addition to Assumption 4.3.1$_{(A2)}$ that $(\theta, \mathbf{u}) \in V^\infty_\theta \times \mathcal{H}_u$ satisfy Equation (5.29) in the weak sense. If the temperature data θ_Γ and θ_0 satisfy

$$\theta_\Gamma(x,t) \leq M_\theta, \qquad (x,t) \in \Omega_T \qquad \theta_0(x) \leq M_\theta, \qquad x \in \Omega$$

then the temperature θ cannot cross the critical value M_θ:

$$\theta(x,t) \leq M_\theta, \qquad (x,t) \in \Omega_T$$

Proof. The proof is standard, but we sketch it here for the convenience of the reader, cf. [Chelminski and Racke 2006] for the same idea. The trick is to show $(\theta - M_\theta)_+(x,t) = 0$ a.e. in Ω_T. Testing Equation (5.31) with $(\theta - M_\theta)_+$ gives

$$\begin{aligned}
\frac{\mathrm{d}}{\mathrm{d}t}\int_\Omega \frac{\rho_0 c_e}{2}(\theta - M_\theta)^2_+ \,\mathrm{d}x &= \int_\Omega (\theta - M_\theta)_+ \rho_0 c_e \theta' \,\mathrm{d}x \\
&= \int_\Omega (\theta - M_\theta)_+ \lambda_\theta \Delta\theta \,\mathrm{d}x - \int_\Omega \hat{r}(\mathbf{x},t)\beta(\theta(x,t))(\theta - M_\theta)_+ \,\mathrm{d}x \\
&= -\int_\Omega \operatorname{sign}\left((\theta - M_\theta)_+\right)\lambda_\theta|\nabla\theta|^2 \,\mathrm{d}x + \int_{\partial\Omega}(\theta - M_\theta)_+ \lambda_\theta \frac{\partial\theta}{\partial\nu}\,\mathrm{d}s
\end{aligned}$$

$$\leq \int_{\partial\Omega} (\theta - M_\theta)_+ \, \delta \, (\theta_\Gamma - \theta) \, \mathrm{ds} \leq 0.$$

Using $\beta(\theta)(\theta - M_\theta)_+ \equiv 0$ and integrate the last inequality in time, we conclude

$$\frac{\rho_0 \, c_e}{2} \int_\Omega (\theta - M_\theta)_+^2 \, \mathrm{dx} \leq \frac{\rho_0 \, c_e}{2} \int_\Omega (\theta_0 - M_\theta)_+^2 \, \mathrm{dx} = 0.$$

□

5.7.21 Remark (References for the Heat Equation)
The initial-boundary value problem for the heat equation is discussed in a mathematical way in [Nakao 1972; Rincon et al. 2005]. More general, the mathematical investigation of parabolic evolution equations can be found in [Boccardo et al. 1999; Eck 2004; Escher 1989; Glitzky and Hünlich 2006; Ladyženskaya et al. 1968; Rehberg 2009], where [Eck 2004] contains a (compact) summary of regularity results for parabolic differential equations of second order.

Chapter 6

Analysis of Problem ($\mathbf{P}_{VE}$)

In this chapter we investigate Problem ($\mathbf{P}_{VE}$) introduced in Subsection 3.3.2. We call this problem 'Visco-elastic Regularisation', because we use $h\boldsymbol{\varepsilon}\left(\frac{\partial \mathbf{u}}{\partial t}\right)$ as an additional regularising term in the stress tensor $\boldsymbol{\sigma}$. This corresponds to a slightly different medium with a small viscosity (or rather damping force in the context of friction).

In order to prove existence and uniqueness of a weak solution of the Problem ($\mathbf{P}_{VE}$) we assume in this chapter $h > 0$ sufficiently small. Although the solution of the regularised problem depends on the regularisation parameter h, we omit this detail in the notation (unless otherwise expressly noted). The passage to the limit of the comprehensive model to the original one is not obtained a-priori.

The weak formulation of Problem ($\mathbf{P}_{VE}$) is given in Section 6.1. The existence and uniqueness result is described in Section 6.2. and the corresponding proof is given in Section 6.3. Section 6.4 provides further aspects and remarks.

6.1 Weak Formulation of the Problem

The weak formulation is as follows:

6.1.1 Definition (Weak Formulation of Problem ($\mathbf{P}_{VE}$))
Under the Assumption 4.3.1 a quintuple $(\mathbf{u}, \theta, \mathbf{p}, \boldsymbol{\varepsilon}_{cp}, \boldsymbol{\varepsilon}_{trip}) \in \mathcal{U}_{\mathrm{u}} \cap \mathfrak{V}_{\mathrm{u}} \times \mathcal{U}_\theta \times \mathfrak{X}_{\mathrm{p}}^\infty \times \mathfrak{H}_\sigma \times \mathfrak{H}_\sigma$ *is called a weak solution of the Problem* ($\mathbf{P}_{VE}$), *if*

$$(6.1)\quad \left\langle \rho_0 \frac{\partial^2}{\partial t^2}\mathbf{u}(t), \mathbf{v}\right\rangle_{V_u^* V_u} + 2\int_\Omega \mu\, \boldsymbol{\varepsilon}(\mathbf{u}(t)) : \boldsymbol{\varepsilon}(\mathbf{v})\, \mathrm{d}\mathbf{x} + \int_\Omega \lambda\, \mathrm{div}(\mathbf{u}(t))\, \mathrm{div}(\mathbf{v})\, \mathrm{d}\mathbf{x}$$

$$+ h \int_\Omega \boldsymbol{\varepsilon}(\mathbf{u}'(t)) : \boldsymbol{\varepsilon}(\mathbf{v})\,\mathrm{d}\mathbf{x} = 3 \int_\Omega K_\alpha\,(\theta(t) - \theta_0)\,\mathrm{div}(\mathbf{v})\,\mathrm{d}\mathbf{x}$$

$$+ \int_\Omega K \sum_{i=1}^{m} \left(\frac{\rho_0}{\rho_i(\theta_0)} - 1 \right) p_i(t)\,\mathrm{div}(\mathbf{v})\,\mathrm{d}\mathbf{x} + 2 \int_\Omega \mu\, \boldsymbol{\varepsilon}_{trip}(t) : \nabla \mathbf{v}\,\mathrm{d}\mathbf{x}$$

$$+ 2 \int_\Omega \mu\, \boldsymbol{\varepsilon}_{cp}(t) : \nabla \mathbf{v}\,\mathrm{d}\mathbf{x} + \int_\Omega \mathbf{f}(t)\,\mathbf{v}\,\mathrm{d}\mathbf{x}$$

f.a. $\mathbf{v} \in \mathbf{V}_{\mathrm{u}}$, *f.a.a.* $t \in S$ *and* $\mathbf{u}(\mathbf{x}, 0) = \mathbf{u}_0(\mathbf{x})$ *a.e.*, $\mathbf{u}'(\mathbf{x}, 0) = \mathbf{u}_1(\mathbf{x})$ *a.e.*,

$$\text{(6.2)} \quad \left\langle \rho_0\, c_e \frac{\partial}{\partial t} \theta(t), \vartheta \right\rangle_{V_\theta^* V_\theta} + \int_\Omega \lambda_\theta\, \nabla \theta(t)\, \nabla \vartheta\,\mathrm{d}\mathbf{x} + 3 \int_\Omega K_\alpha\, \theta_0\,\mathrm{div}\left(\frac{\partial \mathbf{u}(t)}{\partial t} \right) \vartheta\,\mathrm{d}\mathbf{x}$$

$$+ \int_{\partial\Omega} \delta\, \theta(t)\, \vartheta\,\mathrm{d}\sigma_x = \int_\Omega \boldsymbol{\sigma}(t) : \boldsymbol{\varepsilon}'_{trip}(t)\, \vartheta\,\mathrm{d}\mathbf{x} + \int_\Omega \boldsymbol{\sigma}(t) : \boldsymbol{\varepsilon}'_{cp}(t)\, \vartheta\,\mathrm{d}\mathbf{x}$$

$$+ \int_\Omega \rho_0 \sum_{i=2}^{m} L_i\, p_i'(t)\, \vartheta\,\mathrm{d}\mathbf{x} + \int_\Omega r(t)\, \vartheta\,\mathrm{d}\mathbf{x} + \int_{\partial\Omega} \delta\, \theta_\Gamma(t)\, \vartheta\,\mathrm{d}\sigma_x$$

f.a. $\vartheta \in V_\theta$, *f.a.a.* $t \in S$ *and* $\theta(\mathbf{x}, 0) = \theta_0(\mathbf{x})$ *a.e.*,

$$\text{(6.3)} \qquad \frac{\partial \mathbf{p}}{\partial t}(\mathbf{x}, t) = \boldsymbol{\gamma}(\mathbf{p}(\mathbf{x}, t), \theta(\mathbf{x}, t), \theta'(\mathbf{x}, t), \mathrm{tr}(\boldsymbol{\sigma}(\mathbf{x}, t)), \boldsymbol{\sigma}^*(\mathbf{x}, t) : \boldsymbol{\sigma}^*(\mathbf{x}, t))$$

f.a.a. $(\mathbf{x}, t) \in \Omega_T$, $\mathbf{p}(\mathbf{x}, 0) = \mathbf{p}_0(\mathbf{x})$ *a.e.*,

$$\text{(6.4)} \qquad \boldsymbol{\varepsilon}'_{trip}(\mathbf{x}, t) = \frac{3}{2} \boldsymbol{\sigma}^*(\mathbf{x}, t) \sum_{i=1}^{m} \kappa_i \frac{\partial \Phi_i}{\partial p_i}(p_i(\mathbf{x}, t)) \max\{p_i'(\mathbf{x}, t), 0\}$$

$$\text{(6.5)} \qquad \boldsymbol{\varepsilon}_{cp}(\mathbf{x}, t) = \boldsymbol{\varepsilon}^*(\mathbf{u}(\mathbf{x}, t)) - \boldsymbol{\varepsilon}_{trip}(\mathbf{x}, t) - \frac{1}{2\mu} \boldsymbol{\sigma}^*(\mathbf{x}, t)$$

f.a.a. $(\mathbf{x}, t) \in \Omega_T$ *and* $\boldsymbol{\varepsilon}_{trip}(\mathbf{x}, 0) = \mathbf{0}$ *a.e.*, $\boldsymbol{\varepsilon}_{cp}(\mathbf{x}, 0) = \mathbf{0}$ *a.e.*,

$$\text{(6.6)} \qquad (\boldsymbol{\sigma}^*)'(t) + \partial \chi_{\mathrm{K}}(\boldsymbol{\sigma}^*(t)) \ni 2\mu \left(\boldsymbol{\varepsilon}^*(\mathbf{u}'(t)) - \boldsymbol{\varepsilon}'_{trip}(t) \right)$$

f.a.a. $t \in S$, $\boldsymbol{\sigma}^*(0) = \boldsymbol{\sigma}_0^* \in \mathbf{K}$, $\mathbf{K} := \{\boldsymbol{\tau} \in \mathbf{H}_\sigma : \boldsymbol{\tau}(\mathbf{x}) \in \mathbf{K}_F$ *f.a.a.* $\mathbf{x} \in \Omega\}$,
$\mathbf{K}_F := \left\{ \boldsymbol{\tau} \in \mathbb{R}^{3\times 3}_{sym}, \mathrm{tr}(\boldsymbol{\tau}) = 0 : F(\boldsymbol{\tau}) \le 0 \right\}$, $F(\boldsymbol{\sigma}) := \sqrt{\tfrac{3}{2} \boldsymbol{\sigma}^* : \boldsymbol{\sigma}^*} - (R_0 + R)$.

6.2 Existence and Uniqueness Results

In this section, we summarise the existence and uniqueness results for Problem $(\mathbf{P}_{VE})$ and discuss some additional remarks.

6.2.1 Theorem (Existence and Uniqueness for Problem $(\mathbf{P}_{VE})$)
Let Assumptions 4.3.1 be valid and assume in addition that the intrinsic dissipation vanishes, i.e. $\boldsymbol{\sigma} : (\boldsymbol{\varepsilon}'_{trip} + \boldsymbol{\varepsilon}'_{cp}) = 0$. *Then the Problem* $(\mathbf{P}_{VE})$ *possesses a unique weak solution* $(\mathbf{u}, \theta, \mathbf{p}, \boldsymbol{\varepsilon}_{cp}, \boldsymbol{\varepsilon}_{trip}) \in \mathfrak{V}_{\mathsf{u}} \times \mathfrak{H}_\theta \times \mathfrak{H}_{\mathsf{p}} \times \mathfrak{H}_\sigma \times \mathfrak{H}_\sigma$.

6.2.2 Remark
We mention some remarks:

(1) *Again, we only consider the case of constant* $\mathbf{K}$ *in the proof of Theorem 6.2.1, but the proof in the situation of time- or parameter-dependent* $\mathbf{K}$ *works similarly.*

(2) *We also obtain (without going into details at this point) the continuous dependence of the solution on the parameters using suitable assumptions, i.e. it holds*

$$\begin{aligned} &\|\theta_1-\theta_2\|_{\mathcal{V}_\theta^\infty\cap\mathfrak{H}_\theta}+\|\mathbf{u}_1-\mathbf{u}_2\|_{\mathcal{V}_{\mathsf{u}}^\infty\cap\mathfrak{V}_{\mathsf{u}}^\infty}+\|\mathbf{p}_1-\mathbf{p}_2\|_{\mathfrak{H}_{\mathsf{p}}}+\|\boldsymbol{\sigma}_1-\boldsymbol{\sigma}_2\|_{\mathcal{H}_\sigma}+\|\boldsymbol{\varepsilon}_{trip,1}-\boldsymbol{\varepsilon}_{trip,2}\|_{\mathfrak{H}_\sigma} \\ &\quad + \|\boldsymbol{\varepsilon}_{cp,1}-\boldsymbol{\varepsilon}_{cp,2}\|_{\mathfrak{H}_\sigma} \le c\Big\{\|\mathbf{u}_{0,1}-\mathbf{u}_{0,2}\|_{\mathsf{V}_{\mathsf{u}}}+\|\mathbf{u}_{1,1}-\mathbf{u}_{1,2}\|_{\mathsf{H}_{\mathsf{u}}}+\|\mathbf{f}_1-\mathbf{f}_2\|_{\mathcal{H}_{\mathsf{u}}} \\ &\qquad + \|\theta_{0,1}-\theta_{0,2}\|_{H_\theta}+\|\theta_{\Gamma,1}-\theta_{\Gamma,2}\|_{V_\theta}+\|r_1-r_2\|_{H_\theta}+\|\mathbf{p}_{0,1}-\mathbf{p}_{0,2}\|_{\mathsf{H}_{\mathsf{p}}}\Big\}. \end{aligned}$$

(3) *Note that the embedding* $\mathcal{V}_\theta\cap\mathfrak{H}_\theta \hookrightarrow \mathbf{X}$ *is not compact. Therefore we have chosen a technique of proving the existence that does not need any compactness arguments. The proof of the Problem* $(\mathbf{P}_{FP3})$ *does not work with the Schauder fixed-point argument in this setting, because of this lack of compactness.*
In contrast to the Problem $(\mathbf{P}_A)$ *the proof of the fixed-point property of the fixed-point operator* $\mathbf{T}_{FP3}$ *via the Schauder fixed-point argument does not work in this setting without any further simplifications because of lack of regularity in the coupling of the heat equation and the equation for the displacement.*

(4) *A-posteriori estimates provide no further information since these are not independent of the regularisation parameter.*

(5) *Additional regularity of the weak solution of Problem* $(\mathbf{P}_{VE})$ *under suitable assumptions in order to pass to the limit in the regularisation parameter can be shown in correspondence to the Problem* $(\mathbf{P}_A)$. *Unfortunately, the regularity result cannot*

be obtained for the fully coupled problem with our approach. Using the additional assumptions

- *independence of the phase transitions on the time change of the temperature,*
- *negligence of stress-dependent transformation behaviour as discussed in Section 5.5 and negligence of the intrinsic dissipation,*
- *application of the model of Tanaka for the saturation function in the relation for the TRIP strain and*
- *simplified boundary conditions,*

gives additional regularity of the solution and therefore one can prove the passage to the limit in the regularisation parameter. The further course of action is described in Subsection 5.3.1 and carried out analogously to Section 5.5.
However, it is also possible to show the (unique) existence of the fully coupled Problem (**P**) *under these strong additional assumptions.*

6.2.3 Theorem (Existence for Problem ($\mathbf{P}_{VE}$))
Let Assumptions 4.3.1 be valid and assume in addition that the evolution equations for the phase transitions do not dependent on the time derivative of temperature, but on the temperature itself. Then the Problem ($\mathbf{P}_{VE}$) *possesses at least one weak solution* $(\mathbf{u}, \theta, \mathbf{p}, \boldsymbol{\varepsilon}_{cp}, \boldsymbol{\varepsilon}_{trip}) \in \mathfrak{V}_{\mathrm{u}} \times \mathfrak{H}_\theta \times \mathfrak{H}_{\mathrm{p}} \times \mathfrak{H}_\sigma \times \mathfrak{H}_\sigma$.

Proof. The proof is analogous to the proof of Theorem 5.2.3 using the analysis of the corresponding subproblems in the proof of Theorem 6.2.1. □

6.2.4 Remark (Uniqueness for Problem ($\mathbf{P}_{VE}$))
In addition to the assumptions of Theorem 6.2.3 we assume that either the intrinsic dissipation vanishes, i.e. $\boldsymbol{\sigma} : (\boldsymbol{\varepsilon}'_{trip} + \boldsymbol{\varepsilon}'_{cp}) = 0$ *(cf. the setting in Theorem 6.2.1) or that at least one solution possesses better regularity than obtained in the proof of Theorem 6.2.3, e.g.* $\boldsymbol{\sigma} \in \mathbf{X}^\infty_\sigma$ *and* $\boldsymbol{\varepsilon}_{trip}, \boldsymbol{\varepsilon}_{cp} \in \mathfrak{X}^\infty_\sigma$*. Then the weak solution* $(\mathbf{u}, \theta, \mathbf{p}, \boldsymbol{\varepsilon}_{cp}, \boldsymbol{\varepsilon}_{trip}) \in \mathfrak{V}_{\mathrm{u}} \times \mathfrak{H}_\theta \times \mathfrak{H}_{\mathrm{p}} \times \mathfrak{H}_\sigma \times \mathfrak{H}_\sigma$ *of Problem* ($\mathbf{P}_{VE}$) *corresponding to Theorem 6.2.3 is unique.*

6.3 Proof of Theorem 6.2.1

This section provides the proof of Theorem 6.2.1. Subsection 6.3.1 gives the outline of the proof, whereas the proof itself is analogous to the proof in Subsection 5.3.1. The main differences are discussed in Subsections 6.3.2, 6.3.5 and 6.3.7.

6.3.1 Outline of the Proof

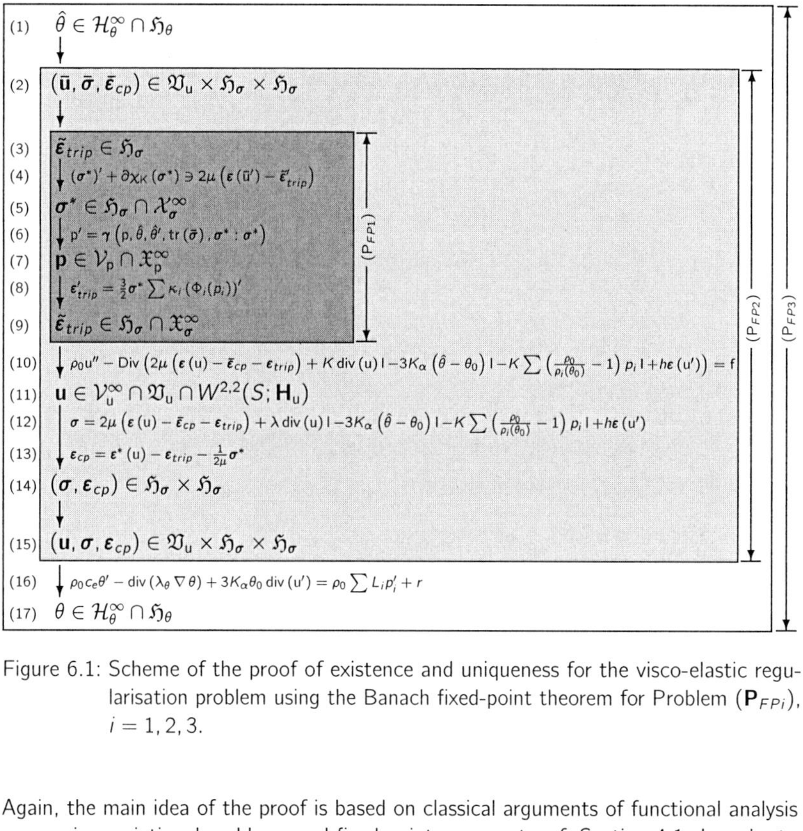

Figure 6.1: Scheme of the proof of existence and uniqueness for the visco-elastic regularisation problem using the Banach fixed-point theorem for Problem ($\mathbf{P}_{FPi}$), $i = 1, 2, 3$.

Again, the main idea of the proof is based on classical arguments of functional analysis concerning variational problems and fixed-point arguments, cf. Section 4.1. In order to prove the unique existence of a weak solution $(\mathbf{u}, \boldsymbol{\sigma}, \boldsymbol{\varepsilon}_{trip}, \boldsymbol{\varepsilon}_{cp}, \theta, \mathbf{p})$ of Problem ($\mathbf{P}_{VE}$), we apply the same strategy as described in Subsection 5.3.1. The proof will be done in several steps (cf. Figure 6.1 for a schematic representation of the proof). The subproblems (as summarised in Tables 4.2 and 4.4) will be extended sequentially until the fully coupled problem is considered. The unique existence of a solution of the three fixed-point steps for Problem ($\mathbf{P}_{VE}$) is shown by using the Banach fixed-point theorem.

6.3.2 Step 1: Analysis of Problem ($\mathbf{P}_{VE,\boldsymbol{\varepsilon}_{cp}}$) for constant K

First, we start with the solvability of Problem ($\mathbf{P}_{VE,\boldsymbol{\varepsilon}_{cp}}$) in case of constant $\mathbf{K}$ for $(\mathbf{u}, \boldsymbol{\sigma}, \boldsymbol{\varepsilon}_{trip}) \in \mathfrak{V}_{\mathrm{u}} \times \mathfrak{H}_{\boldsymbol{\sigma}} \times \mathfrak{H}_{\boldsymbol{\sigma}}$ considered as data. We define $F : \mathbb{R}^{3\times3} \to \mathbb{R}$ via

$$F(\boldsymbol{\sigma}) := \sqrt{\frac{3}{2}\boldsymbol{\sigma}^* : \boldsymbol{\sigma}^*} - (R_0 + R)$$

for given constant $R_0, R \in \mathbb{R}^+$. Moreover, we define

$$\mathbf{K}_F := \left\{\boldsymbol{\tau} \in \mathbb{R}^{3\times3}_{\mathrm{sym}}, \operatorname{tr}(\boldsymbol{\tau}) = 0 : F(\boldsymbol{\tau}) \le 0\right\}, \quad \mathbf{K} := \left\{\boldsymbol{\tau} \in \mathbf{H}_{\boldsymbol{\sigma}} : \boldsymbol{\tau}(\mathbf{x}) \in \mathbf{K}_F \text{ f.a.a. } \mathbf{x} \in \Omega\right\}.$$

6.3.1 Lemma
Let Assumption 4.3.1$_{(A3)}$ be valid. Then $\mathbf{K}$ *is a nonempty closed convex subset of* $\mathbf{H}_{\boldsymbol{\sigma}}$. *Moreover,*

(1) the indicator function χ_{K} *is proper, convex and lower semi-continuous.*
(2) $\partial\chi_{\mathrm{K}} : \mathbf{H}_{\boldsymbol{\sigma}} \to \mathcal{P}(\mathbf{H}_{\boldsymbol{\sigma}})$ *is maximal monotone.*
(3) $D(\chi_{\mathrm{K}}) = D(\partial\chi_{\mathrm{K}}) = \mathbf{K}$.

Proof. The proof is analogous to the proofs of Lemma 5.3.1 and 5.3.2. □

6.3.2 Lemma (Existence and Uniqueness)
Let Assumption 4.3.1$_{(A3)}$ be valid. Assume $(\mathbf{u}, \boldsymbol{\varepsilon}_{trip}) \in \mathfrak{V}_{\mathrm{u}} \times \mathfrak{H}_{\boldsymbol{\sigma}}$. *Then there exists a unique solution* $\boldsymbol{\sigma}^* \in C(\bar{S}; \mathbf{H}_{\boldsymbol{\sigma}}) \cap \mathfrak{H}_{\boldsymbol{\sigma}}$ *of Problem* ($\mathbf{P}_{VE,\boldsymbol{\varepsilon}_{cp}}$) *satisfying* $\boldsymbol{\sigma}^*(t) \in \mathbf{K}$ f.a.a. $t \in S$ *and*

$$\|\boldsymbol{\sigma}^*(t) - \boldsymbol{\sigma}^*(0)\|_{\mathrm{H}_{\sigma}} \le c\left\{\int_0^t \|\mathbf{u}'(s)\|_{\mathrm{V}_{\mathrm{u}}}\,\mathrm{d}s + \int_0^t \|\boldsymbol{\varepsilon}'_{trip}(s)\|_{\mathrm{H}_{\sigma}}\,\mathrm{d}s\right\} \text{ f.a.a. } t \in S,$$
$$\|(\boldsymbol{\sigma}^*)'(t)\|_{\mathrm{H}_{\sigma}} \le c\left\{\|\mathbf{u}'(t)\|_{\mathrm{V}_{\mathrm{u}}} + \|\boldsymbol{\varepsilon}'_{trip}(t)\|_{\mathrm{H}_{\sigma}}\right\} \text{ f.a.a. } t \in S.$$

Proof. Using Lemma 6.3.1 and applying Lemma C.2.1 proves the result. □

6.3.3 Remark (Boundedness of the Solution)
Let Assumption 4.3.1$_{(A3)}$ be valid and assume in addition $\boldsymbol{\sigma}^*(t) \in \mathbf{K}$ f.a.a. $t \in S$. *Then* $F(\boldsymbol{\sigma}) \le 0$, *which means*

$$|\boldsymbol{\sigma}^*(t, \mathbf{x})|^2 \le \frac{3}{2}\left(R_0 + R\right)^2 < \infty$$

f.a.a. $t \in S$ *and* f.a. $\mathbf{x} \in \Omega$. *Therefore,* $\boldsymbol{\sigma}^* \in \mathcal{X}^{\infty}_{\boldsymbol{\sigma}}$.

6.3.4 Remark
For $\mathbf{u}' \in \mathcal{H}_{\mathrm{u}}$ *and* $\boldsymbol{\varepsilon}'_{trip} \in \mathcal{V}^*_{\boldsymbol{\sigma}}$ *there exists a solution* $\boldsymbol{\sigma} \in \mathcal{H}_{\boldsymbol{\sigma}}$ *(cf. Remark C.2.3).*

6.3.5 Lemma (Continuous Dependence on Parameters $\mathbf{u}$ and $\boldsymbol{\varepsilon}_{trip}$)
Let Assumption 4.3.1$_{(A3)}$ be valid. Then it follows for two different solutions $\boldsymbol{\sigma}_i^*$ of Problem $(\mathbf{P}_{VE,\boldsymbol{\varepsilon}_{cp}})$ in the sense of Lemma 6.3.2 corresponding to the data $(\mathbf{u}_i, \boldsymbol{\varepsilon}_{trip,i})$, $i = 1,2$

$$\|\boldsymbol{\sigma}_1^*(t) - \boldsymbol{\sigma}_2^*(t)\|_{\mathrm{H}_\sigma} \leq \|\boldsymbol{\sigma}_1^*(0) - \boldsymbol{\sigma}_2^*(0)\|_{\mathrm{H}_\sigma} + c\left\{\int_0^t \|\mathbf{u}_1'(s) - \mathbf{u}_2'(s)\|_{\mathrm{V}_u}\,\mathrm{d}s + \int_0^t \|\boldsymbol{\varepsilon}_{trip,1}'(s) - \boldsymbol{\varepsilon}_{trip,2}'\|_{\mathrm{H}_\sigma}(s)\,\mathrm{d}s\right\} \text{ f.a.a. } t \in S.$$

Proof. Applying Remark C.2.4 gives the result. □

6.3.3 Step 2 – 3: Analysis of Problem $(\mathbf{P}_{VE,\mathbf{p}})$ – $(\mathbf{P}_{VE,\boldsymbol{\varepsilon}_{trip}})$

Next, we remark that steps 2 and 3 of the proof are completely analogous to steps 2 and 3 of the proof of Theorem 5.2.1 and investigated in detail in Subsections 5.3.3 – 5.3.4.

6.3.4 Step 4: Analysis of Problem $(\mathbf{P}_{VE,FP1})$

We continue with the solvability of Problem $(\mathbf{P}_{VE,FP1})$ for $\hat{\theta}$, $\bar{\mathbf{u}}$, $\bar{\boldsymbol{\sigma}}$ and $\bar{\boldsymbol{\varepsilon}}_{cp}$ considered as data.

6.3.6 Lemma (Existence and Uniqueness)
Let Assumption 4.3.1 be valid. Assume $(\hat{\theta}, \bar{\mathbf{u}}, \bar{\boldsymbol{\sigma}}, \bar{\boldsymbol{\varepsilon}}_{cp}) \in \mathfrak{H}_\theta \times \mathfrak{V}_u \times \mathcal{H}_\sigma \times \mathcal{H}_\sigma$. Then there exists a weak solution $(\boldsymbol{\varepsilon}_{trip}, \boldsymbol{\sigma}^*, \mathbf{p}) \in \mathfrak{H}_\sigma \times \mathfrak{H}_\sigma \times \mathfrak{H}_p$ of Problem $(\mathbf{P}_{VE,FP1})$.
Moreover, let $(\boldsymbol{\varepsilon}_{trip,i}, \boldsymbol{\sigma}_i^*, \mathbf{p}_i)$ be solutions w.r.t. the data $(\hat{\theta}_i, \bar{\mathbf{u}}_i, \bar{\boldsymbol{\sigma}}_i, \bar{\boldsymbol{\varepsilon}}_{cp,i})$, $i = 1,2$. Then the following estimate holds:

$$\begin{aligned}(6.7)\quad &\|\boldsymbol{\varepsilon}_{trip,1}(t) - \boldsymbol{\varepsilon}_{trip,2}(t)\|_{\mathrm{H}_\sigma} + \|\boldsymbol{\varepsilon}_{trip,1}'(t) - \boldsymbol{\varepsilon}_{trip,2}'(t)\|_{\mathrm{H}_\sigma} + \|\boldsymbol{\sigma}_1^*(t) - \boldsymbol{\sigma}_2^*(t)\|_{\mathrm{H}_\sigma} \\ &+ \|\mathbf{p}_1(t) - \mathbf{p}_2(t)\|_{\mathrm{H}_p} + \|\mathbf{p}_1'(t) - \mathbf{p}_2'(t)\|_{\mathrm{H}_p} \leq c\Big\{\|\bar{\boldsymbol{\sigma}}_1(t) - \bar{\boldsymbol{\sigma}}_2(t)\|_{\mathrm{H}_\sigma} \\ &+ \int_0^t \|\bar{\boldsymbol{\sigma}}_1(s) - \bar{\boldsymbol{\sigma}}_2(s)\|_{\mathrm{H}_\sigma}\,\mathrm{d}s + \|\hat{\theta}_1(t) - \hat{\theta}_2(t)\|_{\mathrm{H}_\sigma} + \int_0^t \|\hat{\theta}_1(s) - \hat{\theta}_2(s)\|_{\mathrm{H}_\sigma}\,\mathrm{d}s + \\ &\|\hat{\theta}_1'(t) - \hat{\theta}_2'(t)\|_{\mathrm{H}_\sigma} + \int_0^t \|\hat{\theta}_1'(s) - \hat{\theta}_2'(s)\|_{\mathrm{H}_\sigma}\,\mathrm{d}s + \int_0^t \|\bar{\mathbf{u}}_1'(s) - \bar{\mathbf{u}}_2'(s)\|_{\mathrm{V}_u}\,\mathrm{d}s\Big\}\end{aligned}$$

f.a.a. $t \in S$, *where the positive constant c does not depend on the data.*

Proof. We use a standard fixed-point argument. Before we start with the proof let us give the following results for the subproblems.

Let $(\mathbf{u}, \boldsymbol{\varepsilon}_{trip}) \in \mathfrak{V}_u \times \mathfrak{H}_\sigma$. The problem (6.6) possesses a unique solution $\boldsymbol{\sigma}^* \in \mathfrak{H}_\sigma \cap \mathcal{X}_\sigma^\infty$ s.t.

$$\|\boldsymbol{\sigma}_1^*(t) - \boldsymbol{\sigma}_2^*(t)\|_{\mathrm{H}_\sigma} \le c\left\{\|\boldsymbol{\varepsilon}'_{trip,1} - \boldsymbol{\varepsilon}'_{trip,2}\|_{\mathcal{H}_\sigma} + \|\mathbf{u}'_1 - \mathbf{u}'_2\|_{\mathcal{V}_u}\right\},$$

where c is a positive constant.
Let $(\theta, \boldsymbol{\sigma}) \in \mathfrak{H}_\theta \times \mathcal{H}_\sigma$. The problem (6.3) possesses a unique solution $\mathbf{p} \in \mathfrak{H}_p \cap \mathcal{X}_p^\infty$ s.t.

$$\|\mathbf{p}_1(t) - \mathbf{p}_2(t)\|_{\mathrm{H}_p} \le c_1\left\{\|\theta_1 - \theta_2\|_{\mathfrak{H}_\theta} + \|\theta'_1 - \theta'_2\|_{\mathfrak{H}_\theta} + \|\boldsymbol{\sigma}_1 - \boldsymbol{\sigma}_2\|_{\mathcal{H}_\sigma} + \|\boldsymbol{\sigma}_1^* + \boldsymbol{\sigma}_2^*\|_{\mathcal{H}_\sigma}\right\},$$
$$\|\mathbf{p}'_1 - \mathbf{p}'_2\|_{\mathcal{H}_p} \le c_2\left\{\|\theta_1 - \theta_2\|_{\mathfrak{H}_\theta} + \|\theta'_1 - \theta'_2\|_{\mathfrak{H}_\theta} + \|\boldsymbol{\sigma}_1 - \boldsymbol{\sigma}_2\|_{\mathcal{H}_\sigma} + \|\boldsymbol{\sigma}_1^* + \boldsymbol{\sigma}_2^*\|_{\mathcal{H}_\sigma}\right\}$$

f.a.a. $t \in S$, where c_1 and c_2 are positive constants. Details can be found in Subsection 5.3.3.
Let $(\mathbf{p}, \boldsymbol{\sigma}) \in \mathfrak{H}_p \times \mathcal{H}_\sigma$. The problem (6.4) possesses a unique solution $\boldsymbol{\varepsilon}_{trip} \in \mathfrak{H}_\sigma \cap \mathcal{X}_\sigma^\infty$ s.t.

$$\|\boldsymbol{\varepsilon}_{trip,1}(t) - \boldsymbol{\varepsilon}_{trip,2}(t)\|_{\mathrm{H}_\sigma} \le c_1\left\{\|\boldsymbol{\sigma}_1^* - \boldsymbol{\sigma}_2^*\|_{\mathcal{H}_\sigma} + \|\mathbf{p}_1 - \mathbf{p}_2\|_{\mathfrak{H}_p}\right\},$$
$$\|\boldsymbol{\varepsilon}_{trip,1} - \boldsymbol{\varepsilon}_{trip,2}\|_{\mathcal{H}_\sigma} \le c_2\left\{\|\boldsymbol{\sigma}_1^* - \boldsymbol{\sigma}_2^*\|_{\mathcal{H}_\sigma} + \|\mathbf{p}_1 - \mathbf{p}_2\|_{\mathfrak{H}_p}\right\}$$

f.a.a. $t \in S$, where c_1 and c_2 are positive constants, cf. Subsection 5.3.4 for details.

We use Banach's fixed-point theorem with the (complete) weighted-norm space

$$\mathbf{X} := \mathfrak{H}_\sigma, \quad \|\mathbf{x}\|_X := \|\mathbf{x}\|_{\mathcal{H}_\sigma,\lambda} + \|\mathbf{x}'\|_{\mathcal{H}_\sigma,\lambda}, \quad \|\mathbf{x}\|^2_{\mathcal{H}_\sigma,\lambda} := \sup_{t\in\bar{S}}\left\{\exp(-\lambda t)\int_0^t \|v(s)\|^2_{\mathrm{H}_\sigma}\,ds\right\}$$

with $\lambda > 0$. For $\tilde{\boldsymbol{\varepsilon}}_{trip} \in \mathbf{X}$ given, we consider the linearised problem (6.6), (6.3), (6.4) for $\boldsymbol{\varepsilon}_{trip}$ in which the parameters $\hat{\theta}$, $\bar{\mathbf{u}}$, $\bar{\boldsymbol{\sigma}}$ and $\bar{\boldsymbol{\varepsilon}}_{cp}$ are fixed. By Lemma 5.3.13 there exists a unique solution $\tilde{\varepsilon}_{trip} \in \mathcal{V}_\sigma \hookrightarrow \mathbf{X}$. We define a fixed-point operator as

$$\mathbf{T}_{VE,FP1} : \mathbf{X} \to \mathbf{X}, \qquad \mathbf{T}_{VE,FP1}(\tilde{\varepsilon}_{trip}) = \boldsymbol{\varepsilon}_{trip}$$

and consider solutions $\boldsymbol{\varepsilon}_{trip,1}, \boldsymbol{\varepsilon}_{trip,2}$ corresponding to different data $\tilde{\boldsymbol{\varepsilon}}_{trip,1}, \tilde{\boldsymbol{\varepsilon}}_{trip,2}$. Using the arguments from above we obtain f.a.a. $t \in S$ the following estimate

$$\begin{aligned}&\|\boldsymbol{\varepsilon}_{trip,1}(t) - \boldsymbol{\varepsilon}_{trip,2}(t)\|_{\mathrm{H}_\sigma} + \|\boldsymbol{\varepsilon}'_{trip,1}(t) - \boldsymbol{\varepsilon}'_{trip,2}(t)\|_{\mathrm{H}_\sigma}\\ &\quad \le c\Big\{\|\boldsymbol{\sigma}_1^*(t) - \boldsymbol{\sigma}_2^*(t)\|_{\mathrm{H}_\sigma} + \|\mathbf{p}_1(t) - \mathbf{p}_2(t)\|_{\mathrm{H}_p} + \|\mathbf{p}'_1(t) - \mathbf{p}'_2(t)\|_{\mathrm{H}_p}\end{aligned}$$

$$
\begin{aligned}
&+ \int_0^t \left(\|\boldsymbol{\sigma}_1^*(s) - \boldsymbol{\sigma}_2^*(s)\|_{\mathrm{H}_\sigma} + \|\mathbf{p}_1(s) - \mathbf{p}_2(s)\|_{\mathrm{H}_\mathrm{p}} + \|\mathbf{p}_1'(s) - \mathbf{p}_2'(s)\|_{\mathrm{H}_\mathrm{p}}\right) \mathrm{d}s\Big\} \\
\leq\; & c\Big\{\|\boldsymbol{\sigma}_1^*(t) - \boldsymbol{\sigma}_2^*(t)\|_{\mathrm{H}_\sigma} + \int_0^t \|\boldsymbol{\sigma}_1^*(s) - \boldsymbol{\sigma}_2^*(s)\|_{\mathrm{H}_\sigma} \mathrm{d}s\Big\} \\
\leq\; & c \int_0^t \|\tilde{\boldsymbol{\varepsilon}}'_{trip,1}(s) - \tilde{\boldsymbol{\varepsilon}}'_{trip,2}(s)\|_{\mathrm{H}_\sigma} \mathrm{d}s \\
\leq\; & c\Big\{ \int_0^t \left(\|\tilde{\boldsymbol{\varepsilon}}_{trip,1}(s) - \tilde{\boldsymbol{\varepsilon}}_{trip,2}(s)\|_{\mathrm{H}_\sigma} + \|\tilde{\boldsymbol{\varepsilon}}'_{trip,1}(s) - \tilde{\boldsymbol{\varepsilon}}'_{trip,2}(s)\|_{\mathrm{H}_\sigma}\right) \mathrm{d}s\Big\} \\
\leq\; & c \int_0^t \exp(\lambda t)\, \mathrm{d}s \sup_{s\in\bar{S}} \Big\{ \exp(-\lambda t)\Big(\|\tilde{\boldsymbol{\varepsilon}}_{trip,1}(s) - \tilde{\boldsymbol{\varepsilon}}_{trip,2}(s)\|^2_{\mathrm{H}_\sigma} \\
&+ \|\tilde{\boldsymbol{\varepsilon}}'_{trip,1}(s) - \tilde{\boldsymbol{\varepsilon}}'_{trip,2}(s)\|_{\mathrm{H}_\sigma}\Big)\Big\} \\
\leq\; & c\,\frac{\exp(\lambda t)}{\lambda}\, \|\tilde{\boldsymbol{\varepsilon}}_{trip,1} - \tilde{\boldsymbol{\varepsilon}}_{trip,2}\|_{\mathrm{X}}.
\end{aligned}
$$

It follows

$$
\|\boldsymbol{\varepsilon}_{trip,1} - \boldsymbol{\varepsilon}_{trip,2}\|_{\mathrm{X}} \leq \sqrt{\frac{c}{\lambda}} \|\tilde{\boldsymbol{\varepsilon}}_{trip,1} - \tilde{\boldsymbol{\varepsilon}}_{trip,2}\|_{\mathrm{X}}
$$

and hence, for λ large enough, $\mathbf{T}_{VE,FP1}$ is strictly contractive. The conditions for Banach's fixed-point theorem are fulfilled and Problem $(\mathbf{P}_{VE,FP1})$ possesses a unique solution. Moreover, the estimate (6.7) holds. □

6.3.5 Step 5: Analysis of Problem $(\mathbf{P}_{VE,\mathbf{u}})$

We continue with the analysis of Problem $(\mathbf{P}_{VE,\mathrm{u}})$ for given data $(\theta, \mathbf{p}, \boldsymbol{\varepsilon}_{trip}, \boldsymbol{\varepsilon}_{cp}) \in \mathcal{H}_\theta \times \mathfrak{X}_\mathrm{p}^\infty \times \mathfrak{H}_\sigma \times \mathfrak{H}_\sigma$. Consider the following problem:

Problem $(\mathbf{P}_{VE,\mathrm{u}})$
Find the displacement field $\mathbf{u} : \overline{\Omega}_T \to \mathbb{R}^3$ *and the stress field* $\boldsymbol{\sigma} : \overline{\Omega}_T \to \mathbb{R}^{3\times 3}_{sym}$ *s.t.*

(6.8)

$$
\begin{aligned}
&\rho_0 \frac{\partial^2 \mathbf{u}}{\partial t^2} - 2\,\mathrm{Div}(\mu \boldsymbol{\varepsilon}(\mathbf{u})) - \mathrm{grad}(\lambda\, \mathrm{div}(\mathbf{u})) - h\,\mathrm{Div}\left(\boldsymbol{\varepsilon}\left(\frac{\partial \mathbf{u}}{\partial t}\right)\right) + 3\,\mathrm{grad}(K_\alpha(\theta - \theta_0)) \\
&\quad + \mathrm{grad}\left(K \sum_{i=1}^m \left(\frac{\rho_0}{\rho_i(\theta_0)} - 1\right) p_i\right) + 2\,\mathrm{Div}(\mu \boldsymbol{\varepsilon}_{trip}) + 2\,\mathrm{Div}(\mu \boldsymbol{\varepsilon}_{cp}) = \mathbf{f} \quad \textit{in} \quad \Omega_T
\end{aligned}
$$

including conditions (2.10), (2.11) *as well as initial values* $(2.24)_1$, $(2.24)_2$ *and boundary conditions*

$$\mathbf{u} = 0 \quad on \quad \Gamma_1, \qquad\qquad \Big(\sigma + h\varepsilon(\mathbf{u}')\Big) \cdot \nu = \mathbf{0} \quad on \quad \Gamma_2.$$

In this subsection we deal with the parabolic or visco-elastic regularisation, sometimes also called penalty method or method of vanishing viscosity in the literature, cf. e.g. [Lions and Magenes 1973a].

It will be seen that adding viscosity to the model leads to a substantial increase in the regularity or the smoothness of the solution, and this allows further analysis.

6.3.7 Definition (Weak Formulation)
Under the Assumption $4.3.1_{(A1)}$ *a function* $\mathbf{u} \in \mathcal{U}_u \cap \mathfrak{V}_u$ *is called a weak solution of the Problem* ($\mathbf{P}_{VE,u}$), *if*

$$\begin{aligned}(6.9)\quad \Big\langle \rho_0 \frac{\partial^2}{\partial t^2}\mathbf{u}(t), \mathbf{v}\Big\rangle_{V_u^* V_u} &+ 2\int_\Omega \mu\, \boldsymbol{\varepsilon}(\mathbf{u}(t)) : \boldsymbol{\varepsilon}(\mathbf{v})\,\mathrm{d}\mathbf{x} + \int_\Omega \lambda\, \mathrm{div}(\mathbf{u}(t))\, \mathrm{div}(\mathbf{v})\,\mathrm{d}\mathbf{x} \\ &+ h\int_\Omega \boldsymbol{\varepsilon}(\mathbf{u}'(t)) : \boldsymbol{\varepsilon}(\mathbf{v})\,\mathrm{d}\mathbf{x} = 3\int_\Omega K_\alpha\, (\theta(t) - \theta_0)\, \mathrm{div}(\mathbf{v})\,\mathrm{d}\mathbf{x} \\ &+ \int_\Omega K \sum_{i=1}^m \Big(\frac{\rho_0}{\rho_i(\theta_0)} - 1\Big) p_i(t)\, \mathrm{div}(\mathbf{v})\,\mathrm{d}\mathbf{x} + 2\int_\Omega \mu\, \boldsymbol{\varepsilon}_{trip}(t) : \nabla \mathbf{v}\,\mathrm{d}\mathbf{x} \\ &+ 2\int_\Omega \mu\, \boldsymbol{\varepsilon}_{cp}(t) : \nabla \mathbf{v}\,\mathrm{d}\mathbf{x} + \int_\Omega \mathbf{f}(t)\, \mathbf{v}\,\mathrm{d}\mathbf{x}\end{aligned}$$

f.a. $\mathbf{v} \in \mathbf{V}_u$, *f.a.a.* $t \in S$ *and* $\mathbf{u}(\mathbf{x}, 0) = \mathbf{u}_0(\mathbf{x})$ *a.e.,* $\mathbf{u}'(\mathbf{x}, 0) = \mathbf{u}_1(\mathbf{x})$ *a.e.*

6.3.8 Proposition (A-priori Estimates)
Let Assumption $4.3.1_{(A1)}$ *be valid and let* $\mathbf{u} \in \mathfrak{V}_u$ *be the weak solution of Problem* ($\mathbf{P}_{VE,u}$). *Then there exist constants* $c_1, c_2 > 0$ *s.t.*

(6.10)

$$\|\mathbf{u}'\|_{\mathcal{H}_u^\infty} + \sqrt{h}\|\mathbf{u}'\|_{\mathcal{V}_u} + \|\mathbf{u}\|_{\mathcal{V}_u^\infty} \le c_1 + c_2 \Big\{\|\mathbf{f}\|_{\mathcal{H}_u} + \|\theta\|_{\mathcal{H}_\theta} + \|\mathbf{p}\|_{\mathcal{H}_p} + \|\boldsymbol{\varepsilon}_{trip}\|_{\mathcal{H}_\sigma} + \|\boldsymbol{\varepsilon}_{cp}\|_{\mathcal{H}_\sigma}\Big\}.$$

In addition we get

$$\|\sigma\|_{\mathcal{H}_\sigma^\infty} \le c_1 + c_2 \Big\{\|\mathbf{u}\|_{\mathcal{V}_u^\infty} + \|\theta\|_{\mathcal{H}_\theta^\infty)} + \|\mathbf{p}\|_{\mathcal{H}_p^\infty)} + \|\boldsymbol{\varepsilon}_{trip}\|_{\mathcal{H}_\sigma^\infty} + \|\boldsymbol{\varepsilon}_{cp}\|_{\mathcal{H}_\sigma^\infty}\Big\}.$$

Proof. The proof is analogous to the proof of Lemma 5.3.24. Using $\mathbf{u}'$ as a test function for the first estimates and $\mathbf{u}''$ for the second yields the statement. □

6.3.9 Lemma (Existence and Uniqueness)
Let Assumption 4.3.1$_{(A1)}$ be valid.

(1) *Then,* f.a. $h > 0$, *Problem* ($\mathbf{P}_{VE,\mathrm{u}}$) *has a unique weak solution* $\mathbf{u}_h$, *which satisfies* $\mathbf{u}_h \in C(\bar{S};\mathbf{V}_\mathrm{u}) \cap \mathcal{V}^\infty_\mathrm{u}$, $\mathbf{u}'_h \in C(\bar{S};\mathbf{H}_\mathrm{u}) \cap \mathcal{H}^\infty_\mathrm{u} \cap \mathcal{V}_\mathrm{u}$ *and Equation* (6.9).

(2) *In addition* $\mathbf{u}_h \to \mathbf{u}$ *uniformly in* $C(\bar{S};\mathbf{V}_\mathrm{u})$, $\mathbf{u}'_h \to \mathbf{u}'$ *in* $C(\bar{S};\mathbf{H}_\mathrm{u})$ *for* $h \to 0$, *where* $\mathbf{u}$ *is the solution of Problem* ($\mathbf{P}_{\mathrm{u},\sigma}$) *(the solution belongs to* $C(\bar{S};\mathbf{V}_\mathrm{u})$ *and has derivatives in* $C(\bar{S};\mathbf{H}_\mathrm{u})$, *according to Lemma 5.3.25).*

Proof. The proof consists of two parts.

(1) Using the Faedo-Galerkin method analogous to the proof of Lemma 5.3.25 and the a-priori estimates of Lemma 6.3.8 conclude the first part of the proof (cf. also [Lions and Magenes 1973a; Gajewski et al. 1974] for the existence and uniqueness result).

(2) It follows from Lemma 5.3.25 that

$$\|\mathbf{u}'_h\|_{\mathcal{H}^\infty_\mathrm{u}} \le c, \qquad \|\mathbf{u}_h\|_{\mathcal{V}^\infty_\mathrm{u}} \le c, \qquad \sqrt{h}\,\|\mathbf{u}'_h\|_{\mathcal{V}_\mathrm{u}} \le c, \tag{6.11}$$

i.e. $\mathbf{u}_h$ (resp. $\mathbf{u}'_h$) remains in a bounded set of $\mathcal{V}^\infty_\mathrm{u}$ (resp. $\mathcal{H}^\infty_\mathrm{u}$) as $h \to 0$ and that $\sqrt{h}\mathbf{u}'_h$ remains in a bounded set of $\mathcal{V}_\mathrm{u}$. Therefore, one can extract a subsequence – still denoted $\mathbf{u}_h$ – s.t. $\mathbf{u}_h \overset{*}{\rightharpoonup} \tilde{\mathbf{u}}$ in $\mathcal{V}^\infty_\mathrm{u}$ and $\mathbf{u}'_h \overset{*}{\rightharpoonup} \tilde{\mathbf{u}}'$ in $\mathcal{H}^\infty_\mathrm{u}$. Then

$$\mathbf{u}''_h = \mathbf{F} - \mathbf{A}_\mathrm{u}\mathbf{u}_h - h\mathbf{B}_\mathrm{u}\mathbf{u}'_h \rightharpoonup \mathbf{F} - \mathbf{A}_\mathrm{u}\tilde{\mathbf{u}} \qquad \text{in} \qquad \mathcal{V}^*_\mathrm{u},$$

where $\mathbf{F}$ is given as in the proof of Lemma 5.3.25. Therefore $\tilde{\mathbf{u}}'' + \mathbf{A}_\mathrm{u}\tilde{\mathbf{u}} = \mathbf{F}$ as well as $\mathbf{u}_h(0) \rightharpoonup \tilde{\mathbf{u}}(0)$ in $\mathbf{H}_\mathrm{u}$ and $\mathbf{u}'_h(0) \rightharpoonup \tilde{\mathbf{u}}'(0)$ in $\mathbf{V}^*_\mathrm{u}$, so that $\tilde{\mathbf{u}}(0) = \mathbf{u}_0$, $\tilde{\mathbf{u}}'(0) = \mathbf{u}_1$. Hence, $\tilde{\mathbf{u}} = \mathbf{u}$ and finally $\mathbf{u}_h \overset{*}{\rightharpoonup} \mathbf{u}$ in $\mathcal{V}^\infty_\mathrm{u}$ and $\mathbf{u}'_h \overset{*}{\rightharpoonup} \mathbf{u}'$ in $\mathcal{H}^\infty_\mathrm{u}$.
It remains to show the the uniform convergence. Setting $\boldsymbol{\varphi}_h := \mathbf{u}_h - \mathbf{u}$, we have

$$\boldsymbol{\varphi}''_h + \mathbf{A}_\mathrm{u}\boldsymbol{\varphi}_h = -h\mathbf{B}_\mathrm{u}\mathbf{u}'_h,$$

$\boldsymbol{\varphi}_h \in \mathcal{V}^\infty_\mathrm{u}$, $\boldsymbol{\varphi}'_h \in \mathcal{H}^\infty_\mathrm{u}$, $\boldsymbol{\varphi}_h(0) = 0$ $\boldsymbol{\varphi}'_h(0) = 0$. According to the estimate (6.10)

$$\|\boldsymbol{\varphi}'_h(t)\|^2_{\mathrm{H_u}} + \langle \mathbf{A}_\mathrm{u}\boldsymbol{\varphi}_h(t), \boldsymbol{\varphi}_h(t)\rangle_{\mathrm{V^*_uV_u}} = -2h\int_0^t \langle \mathbf{B}_\mathrm{u}\mathbf{u}'_h(s), \boldsymbol{\varphi}'_h(s)\rangle_{\mathrm{V^*_uV_u}}\,\mathrm{d}s$$

f.a.a. $t \in S$. Hence

$$\|\boldsymbol{\varphi}'_h(t)\|^2_{\mathrm{H_u}} + \|\boldsymbol{\varphi}_h(t)\|^2_{\mathrm{V_u}} \le 2h\left|\int_0^t \langle \mathbf{B}_\mathrm{u}\mathbf{u}'_h(s), \boldsymbol{\varphi}'_h(s)\rangle_{\mathrm{V^*_uV_u}}\,\mathrm{d}s\right|$$

f.a.a. $t \in S$. Using the continuity of the operator $\mathbf{B}_\mathrm{u}$, the estimate $(6.11)_3$ and Hölder's inequality we obtain

$$\|\boldsymbol{\varphi}'_h(t)\|^2_{\mathrm{H_u}} + \|\boldsymbol{\varphi}_h(t)\|^2_{\mathrm{V_u}} \le c\sqrt{h}\left(\int_0^t \|\boldsymbol{\varphi}'_h(s)\|^2_{\mathrm{V_u}}\,\mathrm{d}s\right)^{\frac{1}{2}}$$

f.a.a. $t \in S$. Gronwall's inequality yields the estimate $|\varphi_h'(t)|^2 + \|\varphi_h(t)\|^2 \leq c\sqrt{h}$. Hence, we can pass to the limit $\mathbf{u}_h \to \mathbf{u}$ in $C(\bar{S}; \mathbf{V}_u)$, $\mathbf{u}_h' \to \mathbf{u}'$ in $C(\bar{S}; \mathbf{H}_u)$ (see [Lions and Magenes 1973a] for details).

□

6.3.10 Remark
If $\mathbf{u} \in \mathcal{V}_u$ holds, then $\mathbf{A}_u \in \mathcal{V}_u^*$, so that Equation (6.9) implies $\mathbf{u}'' \in \mathcal{V}_u^*$. Then $\mathbf{u}(0)$ and $\mathbf{u}'(0)$ are well-defined, so that the initial condition has a meaning.

6.3.11 Lemma (Continuous Dependence on Parameters $\mathbf{u}_0$, $\mathbf{u}_1$, $\mathbf{f}$, θ, $\mathbf{p}$, $\boldsymbol{\varepsilon}_{trip}$ and $\boldsymbol{\varepsilon}_{cp}$)
Let Assumption 4.3.1$_{(A1)}$ be valid. Then there exists a constant $c > 0$ s.t. for two different weak solutions of Problem ($\mathbf{P}_{VE,u}$) with corresponding data holds

$$\|\mathbf{u}_1' - \mathbf{u}_2'\|_{\mathcal{H}_u^\infty \cap \mathcal{V}_u} + \|\mathbf{u}_1 - \mathbf{u}_2\|_{\mathcal{V}_u^\infty} \leq c\Big\{\|\mathbf{u}_{0,1} - \mathbf{u}_{0,2}\|_{\mathcal{V}_u} + \|\mathbf{u}_{1,1} - \mathbf{u}_{1,2}\|_{\mathcal{H}_u} + \|\mathbf{f}_1 - \mathbf{f}_2\|_{\mathcal{H}_u} + \|\theta_1 - \theta_2\|_{\mathcal{H}_\theta} + \|\mathbf{p}_1 - \mathbf{p}_2\|_{\mathcal{H}_p} + \|\boldsymbol{\varepsilon}_{trip,1} - \boldsymbol{\varepsilon}_{trip,2}\|_{\mathcal{H}_\sigma} + \|\boldsymbol{\varepsilon}_{cp,1} - \boldsymbol{\varepsilon}_{cp,2}\|_{\mathcal{H}_\sigma}\Big\}.$$

Proof. The proof is analogous to the proof of Lemma 5.3.26. □

6.3.6 Step 6: Analysis of Problem ($\mathbf{P}_{VE,FP2}$)

We continue with studying the solvability of Problem ($\mathbf{P}_{FP2}$) for fixed data $\hat{\theta}$.

6.3.12 Lemma (Existence and Uniqueness)
Let Assumption 4.3.1 be valid. Assume $\hat{\theta} \in \mathfrak{H}_\theta \cap \mathcal{V}_\theta$. Then there exists a weak solution $(\mathbf{u}, \boldsymbol{\sigma}, \boldsymbol{\varepsilon}_{cp}) \in \mathfrak{V}_u \times \mathfrak{H}_\sigma \times \mathfrak{H}_\sigma$ of Problem ($\mathbf{P}_{VE,FP2}$).
Moreover, let $(\mathbf{u}_i, \boldsymbol{\sigma}_i, \boldsymbol{\varepsilon}_{cp,i})$ be solutions w.r.t. the data $\hat{\theta}_i$, $i = 1, 2$. Then the following estimate holds:

$$\|\mathbf{u}_1 - \mathbf{u}_2\|_{\mathfrak{V}_u} + \|\boldsymbol{\sigma}_1 - \boldsymbol{\sigma}_2\|_{\mathfrak{H}_\sigma} + \|\boldsymbol{\varepsilon}_{cp,1} - \boldsymbol{\varepsilon}_{cp,2}\|_{\mathfrak{H}_\sigma} \leq c\,\|\hat{\theta}_1 - \hat{\theta}_2\|_{\mathfrak{H}_\theta \cap \mathcal{V}_\theta} \tag{6.12}$$

where the positive constant c does not depend on the data.

Proof. Again, we use a standard fixed-point argument. Before we start with the proof let us give the following results for the subproblems:
Let $(\theta, \mathbf{p}, \boldsymbol{\varepsilon}_{cp}, \boldsymbol{\varepsilon}_{trip}) \in \mathbf{H}_\theta^\infty \cap \mathfrak{H}_\theta \times \mathbf{H}_p^\infty \cap \mathfrak{H}_p \times \mathbf{H}_\sigma^\infty \cap \mathfrak{H}_\sigma \times \mathbf{H}_\sigma^\infty \cap \mathfrak{H}_\sigma$. The problem (6.1) possesses a unique solution $(\mathbf{u}, \boldsymbol{\sigma}) \in \mathcal{V}_u^\infty \cap \mathfrak{H}_u^\infty \cap \mathfrak{V}_u \times \mathfrak{H}_\sigma^\infty$ s.t.

$$\|\mathbf{u}_1 - \mathbf{u}_2\|_{\mathcal{V}_u^\infty} + \|\mathbf{u}_1' - \mathbf{u}_2'\|_{\mathcal{H}_u^\infty \cap \mathfrak{V}_u} \leq c_1\Big\{\|\theta_1 - \theta_2\|_{\mathcal{H}_\theta^\infty \cap \mathfrak{H}_\theta} + \|\mathbf{p}_1 - \mathbf{p}_2\|_{\mathcal{H}_p^\infty \cap \mathfrak{H}_p} + \|\boldsymbol{\varepsilon}_{cp,1} - \boldsymbol{\varepsilon}_{cp,2}\|_{\mathcal{H}_\sigma^\infty \cap \mathfrak{H}_\sigma} + \|\boldsymbol{\varepsilon}_{trip,1} - \boldsymbol{\varepsilon}_{trip,2}\|_{\mathcal{H}_\sigma^\infty \cap \mathfrak{H}_\sigma}\Big\},$$

and $\|\boldsymbol{\sigma}_1 - \boldsymbol{\sigma}_2\|_{\mathfrak{H}_\sigma} \leq c_2\Big\{\|\mathbf{u}_1 - \mathbf{u}_2\|_{\mathfrak{V}_u} + \|\theta_1 - \theta_2\|_{\mathfrak{H}_\theta} + \|\mathbf{p}_1 - \mathbf{p}_2\|_{\mathfrak{H}_p}$

$$+ \|\boldsymbol{\varepsilon}_{cp,1} - \boldsymbol{\varepsilon}_{cp,2}\|_{\mathfrak{H}_\sigma} + \|\boldsymbol{\varepsilon}_{trip,1} - \boldsymbol{\varepsilon}_{trip,2}\|_{\mathfrak{H}_\sigma}\Big\},$$

where c_1 and c_2 are positive constants (cf. Subsection 5.3.6 for details).
Using the definition of the stress deviator and the above estimate one gets

$$\|\boldsymbol{\sigma}_1^* - \boldsymbol{\sigma}_2^*\|_{\mathfrak{H}_\sigma} \leq c\Big\{\|\mathbf{u}_1 - \mathbf{u}_2\|_{\mathfrak{V}_u} + \|\hat{\theta}_1 - \hat{\theta}_2\|_{\mathfrak{H}_\theta} + \|\mathbf{p}_1 - \mathbf{p}_2\|_{\mathfrak{H}_p}$$
$$+ \|\boldsymbol{\varepsilon}_{trip,1} - \boldsymbol{\varepsilon}_{trip,2}\|_{\mathfrak{H}_\sigma} + \|\bar{\boldsymbol{\varepsilon}}_{cp,1} - \bar{\boldsymbol{\varepsilon}}_{cp,2}\|_{\mathfrak{H}_\sigma}\Big\}.$$

where c is a positive constant. Let $(\mathbf{u}, \boldsymbol{\sigma}, \boldsymbol{\varepsilon}_{trip}) \in \mathfrak{V}_u \times \mathcal{H}_\sigma \times \mathfrak{H}_\sigma$. The problem (6.6) possesses a unique solution $\boldsymbol{\varepsilon}_{cp} \in \mathfrak{H}_\sigma$ s.t.

$$\|\boldsymbol{\varepsilon}_{cp,1} - \boldsymbol{\varepsilon}_{cp,2}\|_{\mathfrak{H}_\sigma} \leq c\Big\{\|\mathbf{u}_1 - \mathbf{u}_2\|_{\mathfrak{V}_u} + \|\boldsymbol{\varepsilon}_{trip,1} - \boldsymbol{\varepsilon}_{trip,2}\|_{\mathfrak{H}_\sigma} + c_3\|\boldsymbol{\sigma}_1^*(t) - \boldsymbol{\sigma}_2^*(t)\|_{\mathcal{H}_\sigma}\Big\},$$

where c is a positive constant.
According to Lemma 6.3.6 we use Banach's fixed-point theorem with a (complete) weighted-norm space. Define $\mathbf{X} := \mathfrak{V}_u \times \mathfrak{H}_\sigma \times \mathfrak{H}_\sigma$. For $(\bar{\mathbf{u}}, \bar{\boldsymbol{\sigma}}, \bar{\boldsymbol{\varepsilon}}_{cp}) \in \mathbf{X}$ given, we consider the linearised Problem (6.1) for $(\mathbf{u}, \boldsymbol{\sigma}, \boldsymbol{\varepsilon}_{cp}) \in \mathbf{X}$ in which the parameter $\hat{\theta}$ is fixed and $\boldsymbol{\varepsilon}_{trip}$ is the solution of Problem $(\mathbf{P}_{VE,FP1})$ according to Lemma 6.3.6. Then there exists a unique solution $(\bar{\mathbf{u}}, \bar{\boldsymbol{\sigma}}, \bar{\boldsymbol{\varepsilon}}_{cp}) \in \mathcal{V}_u^\infty \cap \mathfrak{H}_u^\infty \cap \mathfrak{V}_u \times \mathfrak{H}_\sigma \times \mathfrak{H}_\sigma \hookrightarrow \mathbf{X}$. We define a fixed-point operator as

$$\mathbf{T}_{VE,FP2} : \mathbf{X} \to \mathbf{X}, \qquad \mathbf{T}_{VE,FP2}(\bar{\mathbf{u}}, \bar{\boldsymbol{\sigma}}, \bar{\boldsymbol{\varepsilon}}_{cp}) = (\mathbf{u}, \boldsymbol{\sigma}, \boldsymbol{\varepsilon}_{cp}) \tag{6.13}$$

and consider two different solutions $(\mathbf{u}_1, \boldsymbol{\sigma}_1, \boldsymbol{\varepsilon}_{cp,1}), (\mathbf{u}_2, \boldsymbol{\sigma}_2, \boldsymbol{\varepsilon}_{cp,2})$ corresponding to different data $(\bar{\mathbf{u}}_1, \bar{\boldsymbol{\sigma}}_1, \bar{\boldsymbol{\varepsilon}}_{cp,1}), (\bar{\mathbf{u}}_2, \bar{\boldsymbol{\sigma}}_2, \bar{\boldsymbol{\varepsilon}}_{cp,2})$. By similar arguments as for Lemma 6.3.6 we obtain f.a.a. $t \in S$ the following estimate

$$\|\mathbf{u}_1(t) - \mathbf{u}_2(t)\|_{V_u} + V\|\mathbf{u}_1'(t) - \mathbf{u}_2'(t)\|_{H_u} + h\,\|\mathbf{u}_1'(t) - \mathbf{u}_2'(t)\|_{H_u} + \|\boldsymbol{\sigma}_1(t) - \boldsymbol{\sigma}_2(t)\|_{H_\sigma}$$
$$+ \|\boldsymbol{\sigma}_1'(t) - \boldsymbol{\sigma}_2'(t)\|_{H_\sigma} + \|\boldsymbol{\varepsilon}_{cp,1}(t) - \boldsymbol{\varepsilon}_{cp,2}(t)\|_{\mathfrak{H}_\sigma} + \|\boldsymbol{\varepsilon}'_{cp,1}(t) - \boldsymbol{\varepsilon}'_{cp,2}(t)\|_{\mathfrak{H}_\sigma}$$
$$\leq c\Big\{\|\mathbf{p}_1(t) - \mathbf{p}_2(t)\|_{H_p} + \|\mathbf{p}_1'(t) - \mathbf{p}_2'(t)\|_{H_p} + \int_0^t \|\mathbf{p}_1'(s) - \mathbf{p}_2'(s)\|_{H_p}\,ds$$
$$+ \|\bar{\boldsymbol{\varepsilon}}_{cp,1}(t) - \bar{\boldsymbol{\varepsilon}}_{cp,2}(t)\|_{H_\sigma} + \int_0^t \|\bar{\boldsymbol{\varepsilon}}'_{cp,1}(s) - \bar{\boldsymbol{\varepsilon}}'_{cp,2}(s)\|_{H_\sigma}\,ds$$
$$+ \|\boldsymbol{\sigma}_1^*(t) - \boldsymbol{\sigma}_2^*(t)\|_{H_\sigma} + \|(\boldsymbol{\sigma}_1^*)'(t) - (\boldsymbol{\sigma}_2^*)'(t)\|_{H_\sigma}\Big\}$$
$$\leq c\Big\{\|\bar{\boldsymbol{\varepsilon}}_{cp,1}(t) - \bar{\boldsymbol{\varepsilon}}_{cp,2}(t)\|_{H_\sigma} + \int_0^t \|\bar{\boldsymbol{\varepsilon}}'_{cp,1}(s) - \bar{\boldsymbol{\varepsilon}}'_{cp,2}(s)\|_{H_\sigma}\,ds + \|\bar{\boldsymbol{\sigma}}_1(t) - \bar{\boldsymbol{\sigma}}_2(t)\|_{H_\sigma}$$

$$+\int_0^t \|\bar{\boldsymbol{\sigma}}_1(s)-\bar{\boldsymbol{\sigma}}_2(s)\|_{\mathrm{H}_\sigma}\,\mathrm{d}s+\int_0^t \|\bar{\mathbf{u}}_1'(s)-\bar{\mathbf{u}}_2'(s)\|_{\mathrm{V}_\mathrm{u}}\,\mathrm{d}s\Big\}$$
$$\le c\Big\{\int_0^t \|\bar{\boldsymbol{\varepsilon}}_{cp,1}'(s)-\bar{\boldsymbol{\varepsilon}}_{cp,2}'(s)\|_{\mathrm{H}_\sigma}\,\mathrm{d}s+\int_0^t \|\bar{\boldsymbol{\sigma}}_1(s)-\bar{\boldsymbol{\sigma}}_2(s)\|_{\mathrm{H}_\sigma}\,\mathrm{d}s$$
$$+\int_0^t \|\bar{\boldsymbol{\sigma}}_1'(s)-\bar{\boldsymbol{\sigma}}_2'(s)\|_{\mathrm{H}_\sigma}\,\mathrm{d}s+\int_0^t \|\bar{\mathbf{u}}_1'(s)-\bar{\mathbf{u}}_2'(s)\|_{\mathrm{V}_\mathrm{u}}\,\mathrm{d}s\Big\}$$
$$\le c\Big\{\|\bar{\mathbf{u}}_1-\bar{\mathbf{u}}_2\|_{\mathfrak{V}_\mathrm{u}}+\|\bar{\boldsymbol{\sigma}}_1-\bar{\boldsymbol{\sigma}}_2\|_{\mathcal{H}_\sigma}+\|\bar{\boldsymbol{\varepsilon}}_{cp,1}-\bar{\boldsymbol{\varepsilon}}_{cp,2}\|_{\mathfrak{H}_\sigma}\Big\}.$$

It follows (analogously to the proof of Lemma 6.3.6)

$$\|(\mathbf{u}_1,\boldsymbol{\sigma}_1,\boldsymbol{\varepsilon}_{cp,1})-(\mathbf{u}_2,\boldsymbol{\sigma}_2,\boldsymbol{\varepsilon}_{cp,2})\|_\mathrm{X}\le\sqrt{\frac{c}{\lambda}}\,\|(\bar{\mathbf{u}}_1,\bar{\boldsymbol{\sigma}}_1,\bar{\boldsymbol{\varepsilon}}_{cp,1})-(\bar{\mathbf{u}}_2,\bar{\boldsymbol{\sigma}}_2,\bar{\boldsymbol{\varepsilon}}_{cp,2})\|_\mathrm{X}$$

and hence, for λ large enough, $\mathbf{T}_{VE,FP2}$ is strictly contractive. The conditions for Banach's fixed-point theorem are fulfilled and Problem ($\mathbf{P}_{VE,FP2}$) possesses a unique solution. Moreover, the estimate (6.12) holds. □

6.3.7 Step 7: Analysis of Problem ($\mathbf{P}_{VE,\theta}$)

We continue with Problem ($\mathbf{P}_{VE,\theta}$) for given $(\mathbf{u},\boldsymbol{\sigma},\mathbf{p},\boldsymbol{\varepsilon}_{trip},\boldsymbol{\varepsilon}_{cp})\in\mathfrak{V}_\mathrm{u}\times\mathcal{H}_\sigma\times\mathfrak{X}_\mathrm{p}^\infty\times\mathcal{H}_\sigma\times\mathcal{H}_\sigma$. Consider the (linearised) parabolic problem:

Problem ($\mathbf{P}_{VE,\theta}$)
Find the temperature $\theta:\overline{\Omega}_T\to\mathbb{R}$ *s.t.*

$$\text{(6.14)}\quad \rho_0 c_e\frac{\partial\theta}{\partial t}-\operatorname{div}(\lambda_\theta\nabla\theta)+3K_\alpha\theta_0\operatorname{div}\left(\frac{\partial\mathbf{u}}{\partial t}\right)=\rho_0\sum_{i=2}^m L_i p_i'+r \quad \text{in}\quad \Omega_T$$

$$\text{(6.15)}\quad \theta(0)=\theta_0\quad\text{in}\quad\Omega\qquad\text{and}\qquad -\lambda_\theta\nabla\theta\nu=\delta(\theta-\theta_\Gamma)\quad\text{on}\quad\partial\Omega$$

6.3.13 Definition (Weak Formulation)
Under the Assumption 4.3.1$_{(A2)}$ *the function* $\theta\in\mathcal{U}_\theta$ *is called a weak solution of the Problem* ($\mathbf{P}_{VE,\theta}$), *if* $\theta(\mathbf{x},0)=\theta_0(\mathbf{x})$ *a.e. and*

$$\text{(6.16)}\quad \left\langle\rho_0\,c_e\frac{\partial}{\partial t}\theta(t),\vartheta\right\rangle_{V_\theta^*V_\theta}+\int_\Omega\lambda_\theta\,\nabla\theta(t)\,\nabla\vartheta\,\mathrm{d}\mathbf{x}+3\int_\Omega K_\alpha\,\theta_0\,\operatorname{div}\left(\frac{\partial\mathbf{u}(t)}{\partial t}\right)\vartheta\,\mathrm{d}\mathbf{x}$$
$$=\int_\Omega\rho_0\sum_{i=2}^m L_i\,p_i'(t)\,\vartheta\,\mathrm{d}\mathbf{x}+\int_\Omega r(t)\,\vartheta\,\mathrm{d}\mathbf{x}+\int_{\partial\Omega}\delta\,(\theta_\Gamma-\theta(t))\,\vartheta\,\mathrm{d}\sigma_\mathrm{x}$$

f.a. $\vartheta\in V_\theta$, *f.a.a.* $t\in S$.

6.3.14 Lemma (Existence and Uniqueness)
Under Assumption 4.3.1$_{(A2)}$ the Problem $(\mathbf{P}_{VE,\theta})$ admits a unique solution $\theta : \overline{\Omega}_T \to \mathbb{R}$, satisfying the following conditions:

$$\theta \in \mathcal{H}_\theta^\infty \cap \mathcal{V}_\theta, \qquad \theta' \in \mathcal{V}_\theta^*, \qquad \|\theta\|_{\mathcal{H}_\theta^\infty \cap \mathcal{V}_\theta} \leq c.$$

Proof. The proof follows analogue to the proof of Lemma 5.3.34 with the help of the Galerkin method using the a-priori estimates in Remark 5.4.4. □

6.3.15 Lemma (Continuous Dependence on Parameters $\mathbf{u}$, $\mathbf{p}$ and $\tilde{r}$)
Let Assumption 4.3.1$_{(A2)}$ be valid. and assume in addition that Then there exists a constant $c > 0$ s.t. for two different solutions θ_i of Problem $(\mathbf{P}_{VE,\theta})$ from Lemma 6.3.14 corresponding to the data $(\mathbf{u}_i, \boldsymbol{\sigma}_i, \mathbf{p}_i, r_i, \boldsymbol{\varepsilon}_{trip,i}, \boldsymbol{\varepsilon}_{cp,i})$, $i = 1, 2$ holds

$$\|\theta_1 - \theta_2\|_{\mathcal{H}_\theta^\infty \cap \mathcal{V}_\theta} \leq c\Big\{ \|\mathbf{u}_1' - \mathbf{u}_2'\|_{\mathcal{V}_u} + \|\mathbf{p}_1' - \mathbf{p}_2'\|_{\mathcal{H}_p} + \|\tilde{r}_1 - \tilde{r}_2\|_{\mathcal{H}_\sigma} \Big\}, \tag{6.17}$$

where $\tilde{r}_i$ is defined as in (5.32), $i = 1, 2$.

Proof. The proof is analogue to the proof of Lemma 5.3.35. □

6.3.16 Lemma (Spatial Regularity)
Let Assumptions 4.3.1$_{(A2)}$ and 4.3.2$_{(A2')}$ be valid and assume in addition zero (Dirichlet or Neumann) boundary conditions, i.e. $\theta|_{\partial\Omega} = 0$ or $\nabla\theta \cdot \nu|_{\partial\Omega} = 0$. Then any solution of Problem $(\mathbf{P}_{VE,\theta})$ satisfies $\theta \in V_\theta^\infty \cap \mathcal{W}_\theta$ and

$$\|\theta\|_{V_\theta^\infty \cap \mathcal{W}_\theta} \leq c\Big\{ \|\mathbf{u}'\|_{\mathcal{V}_u} + \|\mathbf{p}'\|_{\mathcal{H}_p} + \|\tilde{r}\|_{\mathcal{V}_\theta} \Big\},$$

where $\tilde{r}$ is defined as in (5.32).

Proof. We use the Galerkin scheme from the proof of Lemma 6.3.14, where the initial conditions are chosen s.t. $\theta_n(0) \to \theta_{n0}$ in V_θ. First we note that the Ladyzhenskaya-Sobolevski inequality $\|\theta(t)\|_{W_\theta} \leq \|\Delta\theta(t)\|_{H_\theta}$ holds (cf. [Dautray and Lions 1992; Pawlow and Zochowski 2002; Pawlow and Zajaczkowski 2005a]). Next, taking $v = -\Delta\theta_n(t)$ in the equation (6.16) and using the hypothesis of zero boundary conditions, we obtain

$$\begin{aligned}\frac{\rho_0\, c_e}{2}\frac{\mathrm{d}}{dt}\|\nabla\theta_n(t)\|_{H_\theta}^2 + \frac{\lambda_\theta}{2}\|\Delta\theta_n(t)\|_{H_\theta}^2 + \frac{\rho_0\, c_e}{2}\frac{\mathrm{d}}{dt}\|\theta(t)\|_{V_\theta}^2 \\ = \int_{\partial\Omega} \theta_n'(t)\theta_\Gamma(t)\,\mathrm{d}\sigma_x - \int_\Omega \tilde{r}(t)\Delta\theta_n(t)\,\mathrm{d}\mathbf{x}\end{aligned}$$

f.a.a. $t \in S$ with

$$\hat{r}(t) := -3K_\alpha \theta_0 \operatorname{div}(\mathbf{u}'(t)) + \rho_0 \sum_{i=1}^{N} L_i p_i'(t) + r(t), \qquad t \in S.$$

Now, integrating from 0 to t and using the Ladyzhenskaya-Sobolevski inequality as well as Young's inequality, we obtain the estimate

$$\begin{aligned} \frac{\rho_0 c_e}{2} \| \nabla \theta_n(t)\|_{H_\theta}^2 + \Big(\frac{\lambda_\theta}{2} - \varepsilon_1\Big) \int_0^t \|\theta(s)\|_{W_\theta}^2 \, ds + \Big(\frac{\rho_0 c_e}{2} - \varepsilon_2\Big) \|\theta_n(t)\|_{V_\theta}^2 \\ \leq \frac{1}{4\varepsilon_2} \|\theta_\Gamma(t)\|_{L^2(\Gamma)}^2 + \frac{1}{4\varepsilon_2} \int_0^t \|\theta_\Gamma'(s)\|_{L^2(\Gamma)}^2 \, ds + C(\varepsilon_1^{-1}) \int_0^t \|\hat{r}(s)\|_{H_\theta}^2 \, ds \end{aligned}$$

with $0 < \varepsilon_1 < \frac{\lambda_\theta}{2}$ and $0 < \varepsilon_2 < \frac{\rho_0 c_e}{2}$ f.a.a. $t \in S$. Hence, using Poincaré's and Gronwall's inequality, we get

$$(\theta_n) \quad \text{is bounded in} \quad \mathcal{V}_\theta^\infty \cap \mathcal{W}_\theta.$$

So the a-priori estimates follows, independently of n. The result follows by passing to the limit $n \to \infty$. □

6.3.8 Step 8: Analysis of Problem $(\mathbf{P}_{VE,FP3})$

Using the foregoing lemmas of this section, we are now in the position to prove the existence and uniqueness result for Problem $(\mathbf{P}_{VE,FP3})$ or rather for the full Problem $(\mathbf{P}_{VE})$.

Proof. We first observe that the solution $\theta \in \mathfrak{H}_\theta \cap \mathcal{V}_\theta$ of (6.2) satisfies

$$\begin{aligned} \|\theta_1 - \theta_2\|_{\mathcal{H}_\theta^\infty} + \|\theta_1 - \theta_2\|_{\mathcal{V}_\theta} &\leq c_1 \left\{ \|\mathbf{u}_1' - \mathbf{u}_2'\|_{\mathcal{V}_u} + \|\mathbf{p}_1 - \mathbf{p}_2\|_{\mathcal{H}_p} \right\}, \\ \|\theta_1' - \theta_2'\|_{\mathcal{H}_\theta} + \|\theta_1 - \theta_2\|_{\mathcal{V}_\theta^\infty} &\leq c_2 \left\{ \|\mathbf{u}_1' - \mathbf{u}_2'\|_{\mathcal{V}_u} + \|\mathbf{p}_1 - \mathbf{p}_2\|_{\mathcal{H}_p} \right\}, \end{aligned}$$

where c_1 and c_2 are positive constants.

We use the Lemmas 6.3.6 and 6.3.12 and Banach's fixed-point theorem with the (complete) weighted-norm space

$$\mathbf{X} := \mathfrak{H}_\theta, \qquad \|\mathbf{x}\|_{\mathrm{X}}^2 := \sup_{t \in \bar{S}} \Big\{ \exp(-\lambda t) \int_0^t \|\mathbf{x}(s)\|_{\mathrm{H}_\theta}^2 \, ds \Big\}, \qquad \lambda > 0.$$

Let $\boldsymbol{\varepsilon}_{trip}$ the unique solution to the Problem $(\mathbf{P}_{VE,FP1})$ corresponding to Lemma 6.3.6 for given θ, $\mathbf{u}$, $\boldsymbol{\sigma}$ and $\boldsymbol{\varepsilon}_{cp}$. Moreover, let $(\mathbf{u}, \boldsymbol{\sigma}, \boldsymbol{\varepsilon}_{cp})$ the unique solution to the Problem $(\mathbf{P}_{VE,FP2})$ corresponding to Lemma 6.3.12 for fixed θ and $\boldsymbol{\varepsilon}_{trip}$. For $\hat{\theta} \in \mathbf{X}$ given, we consider the linearised problem (6.2) for $\theta \in \mathbf{X}$. Finally, by Lemmas 6.3.14 and 6.3.16 there exists a unique solution $\theta \in \mathcal{V}_\theta \cap \mathfrak{H}_\theta \hookrightarrow \mathbf{X}$ for given $\mathbf{u}, \boldsymbol{\sigma}, \boldsymbol{\varepsilon}_{cp}$ and $\boldsymbol{\varepsilon}_{trip}$. We define a fixed-point operator as

$$\mathbf{T}_{VE,FP3} : \mathbf{X} \to \mathbf{X}, \qquad \mathbf{T}_{VE,FP3}(\hat{\theta}) = \theta$$

and consider solutions θ_1, θ_2 corresponding to different data $\hat{\theta}_1, \hat{\theta}_2$. By similar arguments as for Lemmas 6.3.6 and 6.3.12 (and using the estimates (6.7) and (6.12)) we obtain f.a.a. $t \in S$ the following estimate

$$\begin{aligned}
&\|\theta_1(t) - \theta_2(t)\|_{H_\theta} + \|\theta_1'(t) - \theta_2'(t)\|_{H_\theta} \\
&\quad \leq c\Big\{ \int_0^t \|\mathbf{u}_1'(s) - \mathbf{u}_2'(s)\|_{\mathrm{V_u}}^2 \,\mathrm{d}s + \int_0^t \|\mathbf{p}_1(s) - \mathbf{p}_2(s)\|_{\mathrm{H_p}}^2 \,\mathrm{d}s \Big\} \\
&\quad \leq c\Big\{ \int_0^t \|\hat{\theta}_1(t) - \hat{\theta}_2(t)\|_{H_\theta}^2 \,\mathrm{d}s + \int_0^t \|\hat{\theta}_1'(t) - \hat{\theta}_2'(t)\|_{H_\theta} \,\mathrm{d}s \Big\} \\
&\quad \leq c \int_0^t \exp(\lambda t)\,\mathrm{d}s \sup_{s\in[0,t]} \Big\{ \exp(-\lambda t)\Big(\|\hat{\theta}_1(t) - \hat{\theta}_2(t)\|_{H_\theta}^2 + \|\hat{\theta}_1'(t) - \hat{\theta}_2'(t)\|_{H_\theta}\Big)\Big\} \\
&\quad \leq c\,\frac{\exp(\lambda t)}{\lambda} \|\hat{\theta}_1 - \hat{\theta}_2\|_{\mathrm{X}}.
\end{aligned}$$

It follows

$$\|\theta_1 - \theta_2\|_{\mathrm{X}} \leq \sqrt{\frac{c}{\lambda}}\, \|\hat{\theta}_1 - \hat{\theta}_2\|_{\mathrm{X}}$$

and hence, for λ large enough, $\mathbf{T}_{VE,FP3}$ is strictly contractive. The conditions for Banach's fixed-point theorem are fulfilled and Problem $(\mathbf{P}_{VE})$ possesses a unique solution. □

6.4 Further Aspects and Remarks*

In this section some further aspects and remarks of the Problem $(\mathbf{P}_{VE})$ are given. In Subsections 6.4.1 and 6.4.2, Problem $(\mathbf{P}_{VE,\boldsymbol{\varepsilon}_{cp}})$ is investigated for time- and parameter-dependent $\mathbf{K}$. Subsection 6.4.3 provides a regularity result for Problem $(\mathbf{P}_{VE,\mathrm{u}})$.

6.4.1 Analysis of Problem $(\mathbf{P}_{VE,\boldsymbol{\varepsilon}_{cp}})$ for time-dependent K*

Similar to Subsection 5.7.2 we define $F : \mathbb{R}^{3\times 3} \times \mathbb{R} \times \mathbb{R}^3 \to \mathbb{R}$ via

$$F(\boldsymbol{\sigma}, t, \mathbf{x}) := \sqrt{\frac{3}{2}\boldsymbol{\sigma}^* : \boldsymbol{\sigma}^*} - (R_0 + R(t, \mathbf{x})),$$

$$\mathbf{K}_F(t,\mathbf{x}) := \Big\{\boldsymbol{\tau} \in \mathbb{R}^{3\times 3}_{\text{sym}}, \operatorname{tr}(\boldsymbol{\tau}) = 0 : F(\boldsymbol{\tau}, t, \mathbf{x}) \le 0\Big\},$$
$$\mathbf{K}(t) := \Big\{\boldsymbol{\tau} \in \mathbf{H}_\sigma : \boldsymbol{\tau}(\mathbf{x}) \in \mathbf{K}_F(t,\mathbf{x}) \text{ f.a.a. } \mathbf{x} \in \Omega\Big\},$$

where R is defined, e.g., as in Equation (2.66) or (2.67).

6.4.1 Lemma (Existence and Uniqueness)
Let Assumption 4.3.1$_{(A3)}$ be valid. Assume $(\mathbf{u}, \boldsymbol{\varepsilon}_{trip}) \in \mathfrak{V}_{\mathrm{u}} \times \mathfrak{H}_\sigma$. Then there exists a unique solution $\boldsymbol{\sigma}^* \in C(\bar{S}; \mathbf{H}_\sigma) \cap \mathfrak{H}_\sigma$ s.t. Problem ($\mathbf{P}_{VE,\boldsymbol{\varepsilon}_{cp}}$) is fulfilled and $\boldsymbol{\sigma}^*(0) = \boldsymbol{\sigma}^*_0$, $\boldsymbol{\sigma}^*(t) \in \mathbf{K}(t)$ f.a.a. $t \in S$, $t \to \chi_{\mathrm{K}(t)}(\boldsymbol{\sigma}^*(t))$ bounded and

$$\|\boldsymbol{\sigma}^*(t) - \boldsymbol{\sigma}^*(0)\|_{\mathrm{H}_\sigma} \le c\left\{\int_0^t \|\mathbf{u}'(s)\|_{\mathrm{V}_\mathrm{u}}\,\mathrm{d}s + \int_0^t \|\boldsymbol{\varepsilon}'_{trip}(s)\|_{\mathrm{H}_\sigma}\,\mathrm{d}s\right\} \text{ f.a.a. } t \in S,$$
$$\|(\boldsymbol{\sigma}^*)'(t)\|_{\mathrm{H}_\sigma} \le c\left\{\|\mathbf{u}'(t)\|_{\mathrm{V}_\mathrm{u}} + \|\boldsymbol{\varepsilon}'_{trip}(t)\|_{\mathrm{H}_\sigma}\right\} \text{ f.a.a. } t \in S.$$

Proof. The proof is analogous to the proof of Lemma 5.7.11. □

6.4.2 Remark (Alternative Approach)
With the help of

$$(\boldsymbol{\sigma}^*)'(t) + \partial\chi_{\mathrm{K}(t)}(\boldsymbol{\sigma}^*(t)) \ni 2\mu(\boldsymbol{\varepsilon}^*(\mathbf{u}'(t)) - \boldsymbol{\varepsilon}'_{trip}(t)), \quad \boldsymbol{\varepsilon}'_{trip}(t) = b(t)\boldsymbol{\sigma}^*(t), \quad t \in S$$

cf. Equations (3.20) *and* (2.23), *the problem can be rewritten as*

$$(\boldsymbol{\sigma}^*)'(t) + \partial\chi_{\mathrm{K}(t)}(\boldsymbol{\sigma}^*(t)) + 2\mu b(t)\boldsymbol{\sigma}^*(t) \ni 2\mu\boldsymbol{\varepsilon}^*(\mathbf{u}'(t)), \qquad t \in S$$

and treated as discussed in Remark C.2.12.

6.4.3 Lemma (Continuous Dependence on Parameters $\mathbf{u}$ and $\boldsymbol{\varepsilon}_{trip}$)
Let Assumption 4.3.1$_{(A3)}$ be valid and assume in addition $\boldsymbol{\sigma}^*(0) \in \mathbf{K}(0)$. Then two different solutions $\boldsymbol{\sigma}^*_i$ corresponding to Lemma 6.4.1 of Problem ($\mathbf{P}_{VE,\boldsymbol{\varepsilon}_{cp}}$) corresponding to the data $(\mathbf{u}_i, \boldsymbol{\varepsilon}_{trip,i})$, $i = 1, 2$ fulfill the estimate

$$\|\boldsymbol{\sigma}^*_1(t) - \boldsymbol{\sigma}^*_2(t)\|_{\mathrm{H}_\sigma} \le \|\boldsymbol{\sigma}^*_1(0) - \boldsymbol{\sigma}^*_2(0)\|_{\mathrm{H}_\sigma}$$
$$+ c\left\{\int_0^t \|\mathbf{u}'_1(s) - \mathbf{u}'_2(s)\|_{\mathrm{V}_\mathrm{u}}\,\mathrm{d}s + \int_0^t \|\boldsymbol{\varepsilon}'_{trip,1}(s) - \boldsymbol{\varepsilon}'_{trip,2}(s)\|_{\mathrm{H}_\sigma}\,\mathrm{d}s\right\} \text{ f.a.a. } t \in S.$$

Proof. Applying Remark C.2.15 gives the result. □

6.4.4 Lemma (Regularity)
Let Assumption 4.3.1$_{(A3)}$ be valid and assume in addition $(\mathbf{u}, \boldsymbol{\varepsilon}_{trip}) \in \mathfrak{W}_{\mathrm{u}} \times \mathfrak{V}_\sigma$. Define

$\mathbf{K}(t) := \left\{\boldsymbol{\tau} \in \mathbf{V}_\sigma : \boldsymbol{\tau}(\mathbf{x}) \in \mathbf{K}_F(t,\mathbf{x}) \text{ f.a.a. } \mathbf{x} \in \Omega\right\}$. *Then the solution corresponding to Lemma 6.4.1 of Problem* $(\mathbf{P}_{VE,\boldsymbol{\varepsilon}_{cp}})$ *fulfills* $\boldsymbol{\sigma}^* \in C(\bar{S};\mathbf{V}_\sigma) \cap \mathfrak{V}_\sigma$ *and*

$$\|\boldsymbol{\sigma}^*(t) - \boldsymbol{\sigma}^*(0)\|_{\mathrm{V}_\sigma} \leq c\left\{\int_0^t \|\mathbf{u}'(s)\|_{\mathrm{W}_\mathrm{u}}\,\mathrm{d}s + \int_0^t \|\boldsymbol{\varepsilon}'_{trip}(s)\|_{\mathrm{V}_\sigma}\,\mathrm{d}s\right\} \text{ f.a.a. } t \in S,$$
$$\|(\boldsymbol{\sigma}^*)'(t)\|_{\mathrm{V}_\sigma} \leq c\left\{\|\mathbf{u}'(t)\|_{\mathrm{W}_\mathrm{u}} + \|\boldsymbol{\varepsilon}'_{trip}(t)\|_{\mathrm{V}_\sigma}\right\} \text{ f.a.a. } t \in S.$$

Proof. Using Lemma 5.7.9, Lemma 5.7.10 and applying Lemma C.2.11 proves the result. □

6.4.2 Analysis of Problem $(\mathbf{P}_{VE,\boldsymbol{\varepsilon}_{cp}})$ for parameter-dependent K*

Similar to Subsection 5.7.3 we define $F : \mathbb{R}^{3\times 3} \times \mathbb{R} \to \mathbb{R}$ via

$$F(\boldsymbol{\tau},\theta) := \sqrt{\frac{3}{2}\boldsymbol{\sigma}^* : \boldsymbol{\sigma}^*} - (R_0 + R(\theta)),$$
$$\mathbf{K}_F(\theta) := \left\{\boldsymbol{\tau} \in \mathbb{R}^{3\times 3}_{\mathrm{sym}}, \mathrm{tr}(\boldsymbol{\tau}) = 0 : F(\boldsymbol{\tau},\theta) \leq 0\right\},$$
$$\mathbf{K}(\theta) := \left\{\boldsymbol{\tau} \in \mathbf{H}_\sigma : \boldsymbol{\tau}(\mathbf{x}) \in \mathbf{K}_F(\theta) \text{ f.a.a. } \mathbf{x} \in \Omega\right\},$$

where R is defined, e.g., as in Remark 2.6.2 or 2.6.3 and θ is a given parameter, e.g. the temperature. We assume in our application problem cooling or quenching processes and therefore assume a (monotonically) decreasing temperature, i.e. a growing (monotonically increasing) yield radius.

6.4.5 Lemma (Existence and Uniqueness)
Let Assumption 4.3.1$_{(A3)}$ *be valid and assume in addition* $\mathbf{u}' \in \mathcal{V}_\mathrm{u}$, $\boldsymbol{\varepsilon}'_{trip} \in \mathcal{H}_\sigma$, $\theta \in \mathfrak{H}_\theta$ *and* $\boldsymbol{\sigma}^*(0) \in \mathbf{K}(0)$. *Then, there exists a unique solution* $\boldsymbol{\sigma}^* \in C(\bar{S};\mathbf{H}_\sigma)$ *s.t. Problem* $(\mathbf{P}_{VE,\boldsymbol{\varepsilon}_{cp}})$ *is fulfilled and* $\boldsymbol{\sigma}^*(0) = \boldsymbol{\sigma}^*_0$, $\boldsymbol{\sigma}^*(t) \in \mathbf{K}(t)$ *f.a.a.* $t \in S$, $(\boldsymbol{\sigma}^*)' \in \mathcal{V}_\sigma$, $t \to \chi_{\mathrm{K}(t)}(\boldsymbol{\sigma}^*(t))$ *is bounded and*

$$\|\boldsymbol{\sigma}^*(t) - \boldsymbol{\sigma}^*(0)\|_{\mathrm{H}_\sigma} \leq c\left\{\int_0^t \|\mathbf{u}'(s)\|_{\mathrm{V}_\mathrm{u}}\,\mathrm{d}s + \int_0^t \|\boldsymbol{\varepsilon}'_{trip}(s)\|_{\mathrm{H}_\sigma}\,\mathrm{d}s\right\} \text{ f.a.a. } t \in S,$$
$$\|(\boldsymbol{\sigma}^*)'(t)\|_{\mathrm{H}_\sigma} \leq c\left\{\|\mathbf{u}'(t)\|_{\mathrm{V}_\mathrm{u}} + \|\boldsymbol{\varepsilon}'_{trip}(t)\|_{\mathrm{H}_\sigma}\right\} \text{ f.a.a. } t \in S.$$

Proof. The proof is analogous to the proof of Lemma 5.7.15. □

6.4.6 Lemma (Continuous Dependence on Parameters $\mathbf{u}$, $\boldsymbol{\varepsilon}_{trip}$ and θ)
Let Assumption 4.3.1$_{(A3)}$ *be valid and assume in addition* $\mathbf{u}' \in \mathcal{V}_\mathrm{u}$, $\boldsymbol{\varepsilon}'_{trip} \in \mathcal{H}_\sigma$, $\theta \in \mathfrak{H}_\theta$ *and*

$\boldsymbol{\sigma}^*(0) \in \mathbf{K}(0)$. *Then two different solutions $\boldsymbol{\sigma}_i^*$ of Problem ($\mathbf{P}_{VE,\boldsymbol{\varepsilon}_{cp}}$) from Lemma 6.4.5 corresponding to the data $(\mathbf{u}_i, \boldsymbol{\varepsilon}_{trip,i}, \theta_1)$, $i = 1, 2$ fulfill*

$$\|\boldsymbol{\sigma}_1^*(t) - \boldsymbol{\sigma}_2^*(t)\|_{\mathrm{H}_\sigma} \le \|\boldsymbol{\sigma}_1^*(0) - \boldsymbol{\sigma}_2^*(0)\|_{\mathrm{H}_\sigma} + c\Big\{\int_0^t \|\mathbf{u}_1'(s) - \mathbf{u}_2'(s)\|_{\mathrm{V}_\mathrm{u}}\,\mathrm{d}s + \int_0^t \|\boldsymbol{\varepsilon}_{trip,1}'(s) - \boldsymbol{\varepsilon}_{trip,2}'(s)\|_{\mathrm{H}_\sigma}\,\mathrm{d}s + \int_0^t \|\theta_1(s) - \theta_2(s)\|_{\mathrm{H}_\sigma}\,\mathrm{d}s\Big\} \text{ f.a.a. } t \in S.$$

Proof. Applying Remark C.2.21 gives the result. □

6.4.7 Lemma (Regularity)
Let Assumption 4.3.1$_{(A3)}$ hold and assume in addition $\mathbf{u}' \in \mathcal{V}_\mathrm{u}$, $\boldsymbol{\varepsilon}_{trip}' \in \mathcal{H}_\sigma$, $\theta \in \mathfrak{H}_\theta$ and $\boldsymbol{\sigma}^*(0) \in \mathbf{K}(0)$. Then for a solution of Problem ($\mathbf{P}_{VE,\boldsymbol{\varepsilon}_{cp}}$) corresponding to Lemma 6.4.5 holds

$$\|\boldsymbol{\sigma}^*(t) - \boldsymbol{\sigma}^*(0)\|_{\mathrm{V}_\sigma} \le c\Big\{\int_0^t \|\mathbf{u}'(s)\|_{\mathrm{W}_\mathrm{u}}\,\mathrm{d}s + \int_0^t \|\boldsymbol{\varepsilon}_{trip}'(s)\|_{\mathrm{V}_\sigma}\,\mathrm{d}s\Big\} \text{ f.a.a. } t \in S,$$
$$\|(\boldsymbol{\sigma}^*)'(t)\|_{\mathrm{V}_\sigma} \le c\Big\{\|\mathbf{u}'(t)\|_{\mathrm{W}_\mathrm{u}} + \|\boldsymbol{\varepsilon}_{trip}'(t)\|_{\mathrm{V}_\sigma}\Big\} \text{ f.a.a. } t \in S.$$

Proof. Using Lemma 5.7.9, Lemma 5.7.10 and applying Lemma C.2.19 gives the result. □

6.4.3 Regularity Result for Problem ($\mathbf{P}_{VE,\mathbf{u}}$)*

In this subsection we present an additional regularity result for Problem ($\mathbf{P}_{VE,\mathbf{u}}$).

6.4.8 Lemma (Regularity)
Let Assumption 4.3.1$_{(A1)}$ be valid and assume in addition either $\theta' \in \mathcal{H}_\theta$ or zero Dirichlet boundary values $\mathbf{u} = 0$ on $\partial\Omega$. Then the solution of Problem ($\mathbf{P}_{VE,\mathbf{u}}$) corresponding to Lemma 6.3.9 yields $\mathbf{u}' \in \mathcal{V}_\mathrm{u}^\infty$, $\mathbf{u}'' \in \mathcal{H}_\mathrm{u}$, $\boldsymbol{\sigma} \in \mathcal{H}_\sigma^\infty$ and

$$\|\mathbf{u}''\|_{\mathcal{H}_\mathrm{u}} + \sqrt{h}\|\mathbf{u}'\|_{\mathcal{V}_\mathrm{u}} \le c_1 + c_2\Big\{\|\mathbf{f}\|_{\mathcal{V}_\mathrm{u}} + \|\theta\|_{\mathcal{H}_\mathrm{u}^\infty} + \|\theta'\|_{\mathcal{H}_\theta} + \|\mathbf{p}\|_{\mathcal{H}_\mathrm{p}^\infty} + \|\mathbf{p}'\|_{\mathcal{H}_\mathrm{p}} + \|\boldsymbol{\varepsilon}_{trip}\|_{\mathcal{H}_\sigma^\infty} + \|\boldsymbol{\varepsilon}_{trip}'\|_{\mathcal{H}_\sigma} + \|\boldsymbol{\varepsilon}_{cp}\|_{\mathcal{H}_\sigma^\infty} + \|\boldsymbol{\varepsilon}_{cp}'\|_{\mathcal{H}_\sigma}\Big\},$$

$c_1, c_2 > 0$. *In addition we get*

$$\|\boldsymbol{\sigma}_2'\|_{\mathcal{H}_\sigma^\infty} \le c\Big\{\|\mathbf{u}'\|_{\mathcal{V}_\mathrm{u}} + \|\theta'\|_{\mathcal{H}_\theta} + \|\mathbf{p}'\|_{\mathcal{H}_\mathrm{p}} + \|\boldsymbol{\varepsilon}_{trip}'\|_{\mathcal{H}_\sigma} + \|\boldsymbol{\varepsilon}_{cp}'\|_{\mathcal{H}_\sigma}\Big\}.$$

Proof. Testing the Equation (6.9) with $\mathbf{u}''$ and using the identity

$$\int_0^t (\mathbf{A}_u(\mathbf{u}(s)), \mathbf{u}''(s))\,\mathrm{d}s = (\mathbf{A}_u(\mathbf{u}(t)), \mathbf{u}'(t)) - (\mathbf{A}_u(\mathbf{u}_0), \mathbf{u}_1) - \int_0^t (\mathbf{A}_u(\mathbf{u}'(s)), \mathbf{u}'(s))\,\mathrm{d}s$$

f.a.a. $t \in S$ concludes the proof. □

6.4.9 Remark (Regularity Result)
Differentiation (regarding time) of the Galerkin equations and testing with the second time derivative yields $\mathbf{u}'' \in \mathcal{H}_u^\infty \cap \mathcal{V}_u$ *and*

$$\|\mathbf{u}''\|_{\mathcal{H}_u^\infty} + \sqrt{h}\|\mathbf{u}''\|_{\mathcal{V}_u} + \|\mathbf{u}'\|_{\mathcal{V}_u^\infty} \le c\left\{\|\mathbf{f}'\|_{\mathcal{H}_u} + \|\theta'\|_{\mathcal{H}_\theta} + \|\mathbf{p}'\|_{\mathcal{H}_p} + \|\boldsymbol{\varepsilon}'_{trip}\|_{\mathcal{H}_\sigma} + \|\boldsymbol{\varepsilon}'_{cp}\|_{\mathcal{H}_\sigma}\right\}.$$

6.4.10 Remark (Passage to Limit of the Regularisation Parameter)
Let the assumptions of Lemma 6.4.8 be valid. Then similarly to Lemma 6.3.9 it follows $\mathbf{u}'_h \to \mathbf{u}'$ *uniformly in* $C(\bar{S}; \mathbf{V}_u)$, $\mathbf{u}''_h \to \mathbf{u}''$ *in* $C(\bar{S}; \mathbf{H}_u)$ *for* $h \to 0$, *where* $\mathbf{u}_h \in \mathcal{V}_u^\infty$ *is the solution of Problem* $(\mathbf{P}_{VE,u})$.

6.4.11 Remark (Further Regularity Results)
As in the setting of the 'Regularisation via Averaging' we obtain the spatial regularity $\mathbf{u} \in \mathcal{W}_u^\infty$ *and* $\mathbf{u}' \in \mathcal{W}_u \cap \mathcal{V}_u^\infty$ *of the solution of Problem* $(\mathbf{P}_{VE,u})$ *corresponding to Lemma 6.4.8.*
An alternative regularisation approach could be to consider

$$\rho_0 \frac{\partial^2}{\partial t^2}\mathbf{u} = \mathrm{Div}(\boldsymbol{\sigma}(\nabla \mathbf{u})) + \nu\Delta\frac{\partial}{\partial t}\mathbf{u} - \delta\Delta^2\mathbf{u} + \mathbf{F}$$

with $\nu, \delta \to 0$.

Chapter 7

Analysis of Problem ($\mathbf{P}_{QS}$)

Following the papers [Chelminski et al. 2007; Kern 2011], we study the quasi-static problem, i.e. Problem ($\mathbf{P}_{QS}$) introduced in Subsection 3.3.3 in this chapter. The modification of the fully coupled problem is only that the displacement equation for linear thermo-elasticity is governed by the quasi-static balance of momentum, i.e. the displacement $\mathbf{u}$ is given by an elliptic equation instead of an hyperbolic one.
In contrast to the literature, we consider the evolution equations in a Hilbert space setting and we deal additionally with mixed boundary values and classical plasticity.

The weak formulation of Problem ($\mathbf{P}_{QS}$) is given in Section 7.1. The existence and uniqueness result is described in Section 7.2 and the corresponding proof is given in Section 7.3.

7.1 Weak Formulation of the Problem

The weak formulation is as follows:

7.1.1 Definition (Weak Formulation of Problem ($\mathbf{P}_{QS}$))
Under the Assumption 4.3.1 a quintuple $(\mathbf{u}, \theta, \mathbf{p}, \boldsymbol{\varepsilon}_{cp}, \boldsymbol{\varepsilon}_{trip}) \in \mathfrak{V}_{\mathrm{u}} \times \mathcal{U}_\theta \times \mathfrak{X}_{\mathrm{p}}^\infty \times \mathfrak{H}_\sigma \times \mathfrak{H}_\sigma$ *is called a weak solution of the Problem* ($\mathbf{P}_{QS}$), *if*

$$
\begin{aligned}
(7.1)\quad & 2\int_\Omega \mu\,\boldsymbol{\varepsilon}(\mathbf{u}(t)) : \boldsymbol{\varepsilon}(\mathbf{v})\,\mathrm{d}\mathbf{x} + \int_\Omega \lambda\ \mathrm{div}(\mathbf{u}(t))\ \mathrm{div}(\mathbf{v})\,\mathrm{d}\mathbf{x} \\
&= 3\int_\Omega K_\alpha\ (\theta(t)-\theta_0)\ \mathrm{div}(\mathbf{v})\,\mathrm{d}\mathbf{x} + \int_\Omega K \sum_{i=1}^m \left(\frac{\rho_0}{\rho_i(\theta_0)} - 1\right) p_i(t)\ \mathrm{div}(\mathbf{v})\,\mathrm{d}\mathbf{x} \\
&\quad + 2\int_\Omega \mu\,\boldsymbol{\varepsilon}_{trip}(t) : \nabla\mathbf{v}\,\mathrm{d}\mathbf{x} + 2\int_\Omega \mu\,\boldsymbol{\varepsilon}_{cp}(t) : \nabla\mathbf{v}\,\mathrm{d}\mathbf{x} + \int_\Omega \mathbf{f}(t)\,\mathbf{v}\,\mathrm{d}\mathbf{x}
\end{aligned}
$$

f.a. $\mathbf{v} \in \mathbf{V}_u$, *f.a.a.* $t \in S$ *and* $\mathbf{u}(\mathbf{x}, 0) = \mathbf{u}_0(\mathbf{x})$ *a.e.*, $\mathbf{u}'(\mathbf{x}, 0) = \mathbf{u}_1(\mathbf{x})$ *a.e.*,

$$(7.2)\quad \left\langle \rho_0\, c_e \frac{\partial}{\partial t}\theta(t), \vartheta \right\rangle_{V_\theta^* V_\theta} + \int_\Omega \lambda_\theta \, \nabla\theta(t)\, \nabla\vartheta \,\mathrm{d}\mathbf{x} + \int_{\partial\Omega} \delta\,\theta(t)\,\vartheta\,\mathrm{d}\sigma_x$$
$$+ 3\int_\Omega K_\alpha\,\theta_0\,\operatorname{div}(\mathbf{u}'(t))\,\vartheta\,\mathrm{d}\mathbf{x} = \int_\Omega \boldsymbol{\sigma}(t) : \boldsymbol{\varepsilon}'_{trip}(t)\,\vartheta\,\mathrm{d}\mathbf{x} + \int_\Omega \boldsymbol{\sigma}(t) : \boldsymbol{\varepsilon}'_{cp}(t)\,\vartheta\,\mathrm{d}\mathbf{x}$$
$$+ \int_\Omega \rho_0 \sum_{i=2}^{m} L_i\, p_i'(t)\,\vartheta\,\mathrm{d}\mathbf{x} + \int_\Omega r(t)\,\vartheta\,\mathrm{d}\mathbf{x} + \int_{\partial\Omega} \delta\,\theta_\Gamma(t)\,\vartheta\,\mathrm{d}\sigma_x$$

f.a. $\vartheta \in V_\theta$, *f.a.a.* $t \in S$ *and* $\theta(\mathbf{x}, 0) = \theta_0(\mathbf{x})$ *a.e.*,

$$(7.3)\quad \frac{\partial \mathbf{p}}{\partial t}(\mathbf{x}, t) = \boldsymbol{\gamma}(\mathbf{p}(\mathbf{x}, t), \theta(\mathbf{x}, t), \theta'(\mathbf{x}, t), \operatorname{tr}(\boldsymbol{\sigma}(\mathbf{x}, t)), \boldsymbol{\sigma}^*(\mathbf{x}, t) : \boldsymbol{\sigma}^*(\mathbf{x}, t))$$

f.a.a. $(\mathbf{x}, t) \in \Omega_T$, $\mathbf{p}(\mathbf{x}, 0) = \mathbf{p}_0(\mathbf{x})$ *a.e.*,

$$(7.4)\quad \boldsymbol{\varepsilon}'_{trip}(\mathbf{x}, t) = \frac{3}{2}\boldsymbol{\sigma}^*(\mathbf{x}, t) \sum_{i=1}^{m} \kappa_i \frac{\partial \Phi_i}{\partial p_i}(p_i(\mathbf{x}, t)) \max\{p_i'(\mathbf{x}, t), 0\}$$

$$(7.5)\quad \boldsymbol{\varepsilon}_{cp}(\mathbf{x}, t) = \boldsymbol{\varepsilon}^*(\mathbf{u}(\mathbf{x}, t)) - \boldsymbol{\varepsilon}_{trip}(\mathbf{x}, t) - \frac{1}{2\mu}\boldsymbol{\sigma}^*(\mathbf{x}, t)$$

f.a.a. $(\mathbf{x}, t) \in \Omega_T$ *and* $\boldsymbol{\varepsilon}_{trip}(\mathbf{x}, 0) = \mathbf{0}$ *a.e.*, $\boldsymbol{\varepsilon}_{cp}(\mathbf{x}, 0) = \mathbf{0}$ *a.e.*,

$$(7.6)\quad (\boldsymbol{\sigma}^*)'(t) + \partial\chi_{\mathrm{K}}(\boldsymbol{\sigma}^*(t)) \ni 2\mu\left(\boldsymbol{\varepsilon}^*(\mathbf{u}'(t)) - \boldsymbol{\varepsilon}'_{trip}(t)\right)$$

f.a.a. $t \in S$, $\boldsymbol{\sigma}^*(0) = \boldsymbol{\sigma}_0^* \in \mathbf{K}$, $\mathbf{K} := \{\boldsymbol{\tau} \in \mathbf{H}_\sigma : \boldsymbol{\tau}(\mathbf{x}) \in \mathbf{K}_F$ *f.a.a.* $\mathbf{x} \in \Omega\}$,
$\mathbf{K}_F := \left\{\boldsymbol{\tau} \in \mathbb{R}^{3\times3}_{sym}, \operatorname{tr}(\boldsymbol{\tau}) = 0 : F(\boldsymbol{\tau}) \le 0\right\}$, $F(\boldsymbol{\sigma}) := \sqrt{\frac{3}{2}\boldsymbol{\sigma}^* : \boldsymbol{\sigma}^*} - (R_0 + R)$.

7.2 Existence and Uniqueness Results

In this section, we summarise the existence and uniqueness results for Problem ($\mathbf{P}_{QS}$) and discuss some additional remarks.

7.2.1 Theorem (Existence and Uniqueness for Problem ($\mathbf{P}_{QS}$))
Let Assumptions 4.3.1 and 4.3.2 be valid and assume in addition that the intrinsic

dissipation vanishes, i.e. $\boldsymbol{\sigma} : (\boldsymbol{\varepsilon}'_{trip} + \boldsymbol{\varepsilon}'_{cp}) = 0$. *Then the Problem* $(\mathbf{P}_{QS})$ *possesses a unique weak solution* $(\mathbf{u}, \theta, \mathbf{p}, \boldsymbol{\varepsilon}_{cp}, \boldsymbol{\varepsilon}_{trip}) \in \mathfrak{V}_{\mathrm{u}} \times \mathfrak{H}_\theta \times \mathfrak{H}_{\mathrm{p}} \times \mathfrak{H}_\sigma \times \mathfrak{H}_\sigma$.

7.2.2 Remark

We conclude this section with some remarks:

(1) *Again, we only considered the case of constant* $\mathbf{K}$ *in Theorem 7.2.1, but the proof in the situation of time- or parameter-dependent* $\mathbf{K}$ *works similarly.*

(2) *We also obtain (without going into details at this point) the continuous dependence of the solution on the parameters using suitable assumptions, i.e. it holds*

$$\begin{aligned}\|\mathbf{u}_1 - \mathbf{u}_2\|_{\mathfrak{V}_{\mathrm{u}}} + \|\theta_1 - \theta_2\|_{\mathfrak{H}_\theta} + \|\mathbf{p}_1 - \mathbf{p}_2\|_{\mathfrak{H}_{\mathrm{p}}} + \|\boldsymbol{\varepsilon}_{cp,1} - \boldsymbol{\varepsilon}_{cp,2}\|_{\mathfrak{H}_\sigma} + \|\boldsymbol{\varepsilon}_{trip,1} - \boldsymbol{\varepsilon}_{trip,2}\|_{\mathfrak{H}_\sigma} \\ \leq c\Big\{\|\mathbf{u}_{0,1} - \mathbf{u}_{0,2}\|_{\mathrm{V_u}} + \|\mathbf{u}_{1,1} - \mathbf{u}_{1,2}\|_{\mathrm{H_u}} + \|\theta_{0,1} - \theta_{0,2}\|_{V_\theta} \\ + \|\theta_{\Gamma,1} - \theta_{\Gamma,2}\|_{V_\theta} + \|\mathbf{p}_{0,1} - \mathbf{p}_{0,2}\|_{\mathrm{H_p}} + \|r_1 - r_2\|_{H_\theta} + \|\mathbf{f}_1 - \mathbf{f}_2\|_{\mathrm{H_u}}\Big\}.\end{aligned}$$

(3) *The proof of the Problem* $(\mathbf{P}_{FP3})$ *does not work with the Schauder fixed-point argument in this quasi-static setting, because of lack of compactness.*

(4) *Regularity of the solution can be shown in correspondence to the Problem* $(\mathbf{P}_A)$ *(resp. Problem* $(\mathbf{P}_{VE})$*), but as long as we do not need this explicitly in this situation, we omit the details.*

(5) *As we mentioned above, a similar situation without mixed boundary values and without classical plasticity is discussed in [Chelminski et al. 2007; Kern 2011].*

The proof of the following theorem is analogous to the proof of Theorem 5.2.3 using the analysis of the corresponding subproblems in the proof of Theorem 7.2.1.

7.2.3 Theorem (Existence for Problem $(\mathbf{P}_{QS})$)

Let Assumptions 4.3.1 and 4.3.2 be valid and assume in addition that the evolution equations for the phase transitions do not dependent on the time derivative of temperature, but on the temperature itself. Then the Problem $(\mathbf{P}_{QS})$ *possesses at least one weak solution* $(\mathbf{u}, \theta, \mathbf{p}, \boldsymbol{\varepsilon}_{cp}, \boldsymbol{\varepsilon}_{trip}) \in \mathfrak{V}_{\mathrm{u}} \times \mathfrak{H}_\theta \times \mathfrak{H}_{\mathrm{p}} \times \mathfrak{H}_\sigma \times \mathfrak{H}_\sigma$.

7.2.4 Remark (Uniqueness for Problem $(\mathbf{P}_{QS})$)

In addition to the assumptions of Theorem 7.2.3 we assume that either the intrinsic dissipation vanishes, i.e. $\boldsymbol{\sigma} : (\boldsymbol{\varepsilon}'_{trip} + \boldsymbol{\varepsilon}'_{cp}) = 0$ *(cf. the setting in Theorem 7.2.1) or that at least one solution possesses better regularity than obtained in the proof of Theorem 7.2.3, e.g.* $\boldsymbol{\sigma} \in \mathbf{X}_{\sigma}^{\infty}$ *and* $\boldsymbol{\varepsilon}_{trip}, \boldsymbol{\varepsilon}_{cp} \in \mathfrak{X}_{\sigma}^{\infty}$*. Then the weak solution* $(\mathbf{u}, \theta, \mathbf{p}, \boldsymbol{\varepsilon}_{cp}, \boldsymbol{\varepsilon}_{trip}) \in \mathfrak{V}_{\mathrm{u}} \times \mathfrak{H}_{\theta} \times \mathfrak{H}_{\mathrm{p}} \times \mathfrak{H}_{\sigma} \times \mathfrak{H}_{\sigma}$ *of Problem* ($\mathbf{P}_{QS}$) *corresponding to Theorem 7.2.3 is unique.*

7.3 Proof of Theorem 7.2.1

This section provides the proof of Theorem 7.2.1. Subsection 7.3.1 provides the outline of the proof, whereas the proof itself is analogous to the proof in Subsection 5.3.1. The main differences are discussed in Subsection 7.3.4.

7.3.1 Outline of the Proof

Again, the main idea of the proof is based on classical arguments of functional analysis concerning variational problems and fixed-point arguments, cf. Section 4.1. In order to prove the unique existence of a weak solution $(\mathbf{u}, \boldsymbol{\sigma}, \boldsymbol{\varepsilon}_{trip}, \boldsymbol{\varepsilon}_{cp}, \theta, \mathbf{p})$ of Problem ($\mathbf{P}_{QS}$), we apply the same strategy as described in Subsection 5.3.1 (cf. Figure 7.1 for a schematic representation of the proof).
The subproblems (as summarised in Tables 4.2 and 4.4) will be extended sequentially until the fully coupled problem is considered. The unique existence of a solution of the three fixed-point steps for Problem ($\mathbf{P}_{QS}$) is shown by using the Banach fixed-point theorem.

7.3.2 Step 1 – 3: Analysis of Problem ($\mathbf{P}_{QS,\boldsymbol{\varepsilon}_{cp}}$) – ($\mathbf{P}_{QS,\boldsymbol{\varepsilon}_{trip}}$)

First, we remark that the first three steps of the proof are completely analogous to the first three steps of the proof of Theorem 6.2.1 and discussed in Subsections 6.3.2 and 6.3.3.

7.3.3 Step 4: Analysis of Problem ($\mathbf{P}_{QS,FP1}$)

We continue with the solvability of Problem ($\mathbf{P}_{QS,FP1}$) for given data $\hat{\theta}$, $\bar{\mathbf{u}}$, $\bar{\boldsymbol{\sigma}}$ and $\bar{\boldsymbol{\varepsilon}}_{cp}$.

7.3.1 Lemma (Existence and Uniqueness)
Let Assumption 4.3.1 and 4.3.2 be valid. Assume that $(\hat{\theta}, \bar{\mathbf{u}}, \bar{\boldsymbol{\sigma}}, \bar{\boldsymbol{\varepsilon}}_{cp}) \in \mathfrak{H}_{\theta} \times \mathfrak{V}_{\mathrm{u}} \times \mathcal{H}_{\sigma} \times \mathcal{H}_{\sigma}$*. Then there exists a weak solution* $(\boldsymbol{\varepsilon}_{trip}, \boldsymbol{\sigma}^*, \mathbf{p}) \in \mathfrak{H}_{\sigma} \times \mathfrak{H}_{\sigma} \times \mathfrak{X}_{\mathrm{p}}^{\infty}$ *of Problem* ($\mathbf{P}_{QS,FP1}$).

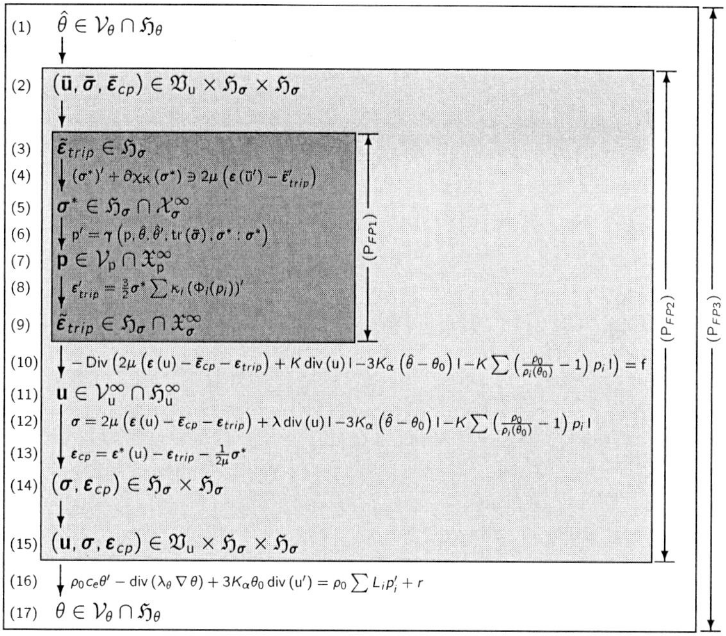

Figure 7.1: Scheme of the proof of existence and uniqueness for the quasi-static problem using the Banach fixed-point theorem for problems $(\mathbf{P}_{FPi})$, $i = 1, 2, 3$.

Moreover, let $(\boldsymbol{\varepsilon}_{trip,i}, \boldsymbol{\sigma}^*_i, \mathbf{p}_i)$ *be solutions w.r.t. the data* $(\hat{\theta}_i, \bar{\mathbf{u}}_i, \bar{\boldsymbol{\sigma}}_i, \bar{\boldsymbol{\varepsilon}}_{cp,i})$, $i = 1, 2$. *Then the following estimate holds:*

$$
\begin{aligned}
(7.7)\quad & \|\boldsymbol{\varepsilon}_{trip,1}(t) - \boldsymbol{\varepsilon}_{trip,2}(t)\|_{\mathrm{H}_\sigma} + \|\boldsymbol{\varepsilon}'_{trip,1}(t) - \boldsymbol{\varepsilon}'_{trip,2}(t)\|_{\mathrm{H}_\sigma} + \|\boldsymbol{\sigma}^*_1(t) - \boldsymbol{\sigma}^*_2(t)\|_{\mathrm{H}_\sigma} \\
& + \|\mathbf{p}_1(t) - \mathbf{p}_2(t)\|_{\mathrm{H}_{\mathrm{p}}} + \|\mathbf{p}'_1(t) - \mathbf{p}'_2(t)\|_{\mathrm{H}_{\mathrm{p}}} \le c\Big\{\|\bar{\boldsymbol{\sigma}}_1(t) - \bar{\boldsymbol{\sigma}}_2(t)\|_{\mathrm{H}_\sigma} + \\
& \int_0^t \|\bar{\boldsymbol{\sigma}}_1(s) - \bar{\boldsymbol{\sigma}}_2(s)\|_{\mathrm{H}_\sigma}\,ds + \|\hat{\theta}_1(t) - \hat{\theta}_2(t)\|_{\mathrm{H}_\sigma} + \int_0^t \|\hat{\theta}_1(s) - \hat{\theta}_2(s)\|_{\mathrm{H}_\sigma}\,ds + \\
& \|\hat{\theta}'_1(t) - \hat{\theta}'_2(t)\|_{\mathrm{H}_\sigma} + \int_0^t \|\hat{\theta}'_1(s) - \hat{\theta}'_2(s)\|_{\mathrm{H}_\sigma}\,ds + \int_0^t \|\bar{\mathbf{u}}'_1(s) - \bar{\mathbf{u}}'_2(s)\|_{\mathrm{V}_{\mathrm{u}}}\,ds\Big\}
\end{aligned}
$$

f.a.a. $t \in S$, *where the positive constant* c *does not depend on the data.*

Proof. We use a standard fixed-point argument. Before we start with the proof let us summarise the following results for the subproblems.
Let $(\mathbf{u}, \boldsymbol{\varepsilon}_{trip}) \in \mathfrak{V}_u \times \mathfrak{H}_\sigma$. The problem (7.6) possesses a unique solution $\boldsymbol{\sigma}^* \in \mathfrak{H}_\sigma \cap \mathcal{X}_\sigma^\infty$ s.t.

$$\|\boldsymbol{\sigma}_1^*(t) - \boldsymbol{\sigma}_2^*(t)\|_{\mathrm{H}_\sigma} \le c\left\{\|\boldsymbol{\varepsilon}'_{trip,1} - \boldsymbol{\varepsilon}'_{trip,2}\|_{\mathcal{H}_\sigma} + \|\mathbf{u}'_1 - \mathbf{u}'_2\|_{\mathcal{V}_u}\right\}$$

f.a.a. $t \in S$, where c is a positive constant.
Let $(\theta, \boldsymbol{\sigma}) \in \mathfrak{H}_\theta \times \mathcal{H}_\sigma$. The problem (7.3) possesses a unique solution $\mathbf{p} \in \mathfrak{H}_p \cap \mathcal{X}_p^\infty$ s.t.

$$\|\mathbf{p}_1(t) - \mathbf{p}_2(t)\|_{\mathrm{H}_p} \le c_1\left\{\|\theta_1 - \theta_2\|_{\mathfrak{H}_\theta} + \|\boldsymbol{\sigma}_1 - \boldsymbol{\sigma}_2\|_{\mathcal{H}_\sigma} + \|\boldsymbol{\sigma}_1^* + \boldsymbol{\sigma}_2^*\|_{\mathcal{H}_\sigma}\right\},$$
$$\|\mathbf{p}'_1 - \mathbf{p}'_2\|_{\mathcal{H}_p} \le c_2\left\{\|\theta_1 - \theta_2\|_{\mathfrak{H}_\theta} + \|\boldsymbol{\sigma}_1 - \boldsymbol{\sigma}_2\|_{\mathcal{H}_\sigma} + \|\boldsymbol{\sigma}_1^* + \boldsymbol{\sigma}_2^*\|_{\mathcal{H}_\sigma}\right\}$$

f.a.a. $t \in S$, where c_1, c_2 are positive constants.
Let $(\mathbf{p}, \boldsymbol{\sigma}) \in \mathfrak{H}_p \times \mathcal{H}_\sigma$. The problem (7.4) possesses a unique solution $\boldsymbol{\varepsilon}_{trip} \in \mathfrak{H}_\sigma \cap \mathcal{X}_\sigma^\infty$ s.t.

$$\|\boldsymbol{\varepsilon}_{trip,1}(t) - \boldsymbol{\varepsilon}_{trip,2}(t)\|_{\mathrm{H}_\sigma} \le c_1\left\{\|\boldsymbol{\sigma}_1^* - \boldsymbol{\sigma}_2^*\|_{\mathcal{H}_\sigma} + \|\mathbf{p}_1 - \mathbf{p}_2\|_{\mathfrak{H}_p}\right\},$$
$$\|\boldsymbol{\varepsilon}_{trip,1} - \boldsymbol{\varepsilon}_{trip,2}\|_{\mathfrak{H}_\sigma} \le c_2\left\{\|\boldsymbol{\sigma}_1^* - \boldsymbol{\sigma}_2^*\|_{\mathcal{H}_\sigma} + \|\mathbf{p}_1 - \mathbf{p}_2\|_{\mathfrak{H}_p}\right\},$$

f.a.a. $t \in S$, where c_1 and c_2 are positive constants, cf. details in Subsection 5.3.4.

We use Banach's fixed-point theorem with the (complete) weighted-norm space

$$\mathbf{X} := \mathfrak{H}_\sigma, \quad \|\mathbf{x}\|_X := \|\mathbf{x}\|_{\mathcal{H}_\sigma,\lambda} + \|\mathbf{x}'\|_{\mathcal{H}_\sigma,\lambda}, \quad \|\mathbf{x}\|^2_{\mathcal{H}_\sigma,\lambda} := \sup_{t\in\bar{S}}\left\{\exp(-\lambda t)\int_0^t \|\mathbf{x}(s)\|^2_{\mathrm{H}_\sigma}\,ds\right\}$$

with $\lambda > 0$. For $\tilde{\boldsymbol{\varepsilon}}_{trip} \in \mathbf{X}$ given, we consider the linearised problem (7.6), (7.3), (7.4) for $\boldsymbol{\varepsilon}_{trip}$ in which the parameters $\hat{\theta}$, $\bar{\mathbf{u}}$, $\bar{\boldsymbol{\sigma}}$ and $\bar{\boldsymbol{\varepsilon}}_{cp}$ are fixed. By Lemma 5.3.13 there exists a unique solution $\tilde{\boldsymbol{\varepsilon}}_{trip} \in \mathfrak{H}_\sigma \hookrightarrow \mathbf{X}$. We define a fixed-point operator as

$$\mathbf{T}_{QS,FP1} : \mathbf{X} \to \mathbf{X}, \qquad \mathbf{T}_{QS,FP1}(\tilde{\boldsymbol{\varepsilon}}_{trip}) = \boldsymbol{\varepsilon}_{trip}$$

and consider solutions $\boldsymbol{\varepsilon}_{trip,1}$, $\boldsymbol{\varepsilon}_{trip,2}$ corresponding to different data $\tilde{\boldsymbol{\varepsilon}}_{trip,1}$, $\tilde{\boldsymbol{\varepsilon}}_{trip,2}$. Using the arguments from above we obtain f.a.a. $t \in S$ the following estimate

$$\begin{aligned}&\|\boldsymbol{\varepsilon}_{trip,1}(t) - \boldsymbol{\varepsilon}_{trip,2}(t)\|_{\mathrm{H}_\sigma} + \|\boldsymbol{\varepsilon}'_{trip,1}(t) - \boldsymbol{\varepsilon}'_{trip,2}(t)\|_{\mathrm{H}_\sigma}\\ &\quad\le c\Big\{\|\boldsymbol{\sigma}_1^*(t) - \boldsymbol{\sigma}_2^*(t)\|_{\mathrm{H}_\sigma} + \|\mathbf{p}_1(t) - \mathbf{p}_2(t)\|_{\mathrm{H}_p} + \|\mathbf{p}'_1(t) - \mathbf{p}'_2(t)\|_{\mathrm{H}_p}\end{aligned}$$

$$
\begin{aligned}
&+\int_0^t \left(\|\boldsymbol{\sigma}_1^*(s)-\boldsymbol{\sigma}_2^*(s)\|_{\mathrm{H}_\sigma}+\|\mathbf{p}_1(s)-\mathbf{p}_2(s)\|_{\mathrm{H}_\mathrm{p}}+\|\mathbf{p}_1'(s)-\mathbf{p}_2'(s)\|_{\mathrm{H}_\mathrm{p}}\right) \mathrm{d}s\Big\} \\
&\le c\left\{\|\boldsymbol{\sigma}_1^*(t)-\boldsymbol{\sigma}_2^*(t)\|_{\mathrm{H}_\sigma}+\int_0^t \|\boldsymbol{\sigma}_1^*(s)-\boldsymbol{\sigma}_2^*(s)\|_{\mathrm{H}_\sigma} \mathrm{d}s\right\} \\
&\le c\int_0^t \|\tilde{\boldsymbol{\varepsilon}}'_{trip,1}(s)-\tilde{\boldsymbol{\varepsilon}}'_{trip,2}(s)\|_{\mathrm{H}_\sigma} \mathrm{d}s \\
&\le c\left\{\int_0^t \left(\|\tilde{\boldsymbol{\varepsilon}}_{trip,1}(s)-\tilde{\boldsymbol{\varepsilon}}_{trip,2}(s)\|_{\mathrm{H}_\sigma}+\|\tilde{\boldsymbol{\varepsilon}}'_{trip,1}(s)-\tilde{\boldsymbol{\varepsilon}}'_{trip,2}(s)\|_{\mathrm{H}_\sigma}\right)\mathrm{d}s\right\} \\
&\le c\int_0^t \exp(\lambda t)\,\mathrm{d}s \sup_{s\in\bar{S}}\Big\{\exp(-\lambda t)\Big(\|\tilde{\boldsymbol{\varepsilon}}_{trip,1}(s)-\tilde{\boldsymbol{\varepsilon}}_{trip,2}(s)\|^2_{\mathrm{H}_\sigma} \\
&\quad+\|\tilde{\boldsymbol{\varepsilon}}'_{trip,1}(s)-\tilde{\boldsymbol{\varepsilon}}'_{trip,2}(s)\|_{\mathrm{H}_\sigma}\Big)\Big\} \\
&\le c\,\frac{\exp(\lambda t)}{\lambda}\,\|\tilde{\boldsymbol{\varepsilon}}_{trip,1}-\tilde{\boldsymbol{\varepsilon}}_{trip,2}\|_{\mathrm{X}}.
\end{aligned}
$$

It follows

$$
\|\boldsymbol{\varepsilon}_{trip,1}-\boldsymbol{\varepsilon}_{trip,2}\|_{\mathrm{X}} \le \sqrt{\frac{c}{\lambda}}\|\tilde{\boldsymbol{\varepsilon}}_{trip,1}-\tilde{\boldsymbol{\varepsilon}}_{trip,2}\|_{\mathrm{X}}
$$

and hence, for λ large enough, $\mathbf{T}_{QS,FP1}$ is strictly contractive. The conditions for Banach's fixed-point theorem are fulfilled and Problem ($\mathbf{P}_{QS,FP1}$) possesses a unique solution. Moreover, the estimate (7.7) can be shown in the same manner. □

7.3.4 Step 5: Analysis of Problem ($\mathbf{P}_{QS,\mathbf{u}}$)

We continue with the solvability of Problem ($\mathbf{P}_{QS,\mathrm{u}}$) for given data $(\theta, \mathbf{p}, \boldsymbol{\varepsilon}_{trip}, \boldsymbol{\varepsilon}_{cp}) \in \mathcal{H}_\theta \times \mathfrak{X}_\mathrm{p} \times \mathfrak{H}_\sigma \times \mathfrak{H}_\sigma$. Consider the following problem:

Problem ($\mathbf{P}_{QS,\mathrm{u}}$)
Find the displacement field $\mathbf{u} : \overline{\Omega}_T \to \mathbb{R}^3$ *and the stress field* $\boldsymbol{\sigma} : \overline{\Omega}_T \to \mathbb{R}^{3\times 3}_{sym}$ *s.t.*

$$
\begin{aligned}
&-2\,\mathrm{Div}(\mu\boldsymbol{\varepsilon}(\mathbf{u})) - \mathrm{grad}(\lambda\,\mathrm{div}(\mathbf{u})) + 3\,\mathrm{grad}(K_\alpha(\theta-\theta_0)) \\
&+\mathrm{grad}\left(K\sum_{i=1}^m\left(\frac{\rho_0}{\rho_i(\theta_0)}-1\right)p_i\right)+2\,\mathrm{Div}(\mu\boldsymbol{\varepsilon}_{trip})+2\,\mathrm{Div}(\mu\boldsymbol{\varepsilon}_{cp})=\mathbf{f} \ \ \text{in } \Omega_T
\end{aligned}
\tag{7.8}
$$

including conditions (2.10), (2.11) *as well as initial values* $(2.24)_1$, $(2.24)_2$ *and boundary conditions*

$$
\mathbf{u}=0 \quad \text{on} \quad \Gamma_1, \qquad\qquad \boldsymbol{\sigma}\cdot\nu=\mathbf{0} \quad \text{on} \quad \Gamma_2.
$$

In accordance with [Chelminski et al. 2007; Kern 2011] we have a look at the quasi-static problem. In contrast to these references we consider the L^2-setting for the evolution equations and we take into account mixed boundary conditions.

7.3.2 Definition (Weak Formulation)
Under the Assumption 4.3.1$_{(A1)}$ a function $\mathbf{u} \in \mathcal{V}_u$ is called a weak solution of Problem ($\mathbf{P}_{QS,u}$), if

$$
\begin{aligned}
(7.9)\quad & 2\int_\Omega \mu\,\boldsymbol{\varepsilon}(\mathbf{u}(t)) : \boldsymbol{\varepsilon}(\mathbf{v})\,\mathrm{dx} + \int_\Omega \lambda \,\mathrm{div}(\mathbf{u}(t))\,\mathrm{div}(\mathbf{v})\,\mathrm{dx} \\
&= 3\int_\Omega K_\alpha\,(\theta(t)-\theta_0)\,\mathrm{div}(\mathbf{v})\,\mathrm{dx} + \int_\Omega K \sum_{i=1}^{m}(\frac{\rho_0}{\rho_i(\theta_0)} - 1)p_i(t)\,\mathrm{div}(\mathbf{v})\,\mathrm{dx} \\
&\quad + 2\int_\Omega \mu\,\boldsymbol{\varepsilon}_{trip}(t) : \nabla\mathbf{v}\,\mathrm{dx} + 2\int_\Omega \mu\,\boldsymbol{\varepsilon}_{cp}(t) : \nabla\mathbf{v}\,\mathrm{dx} + \int_\Omega \mathbf{f}(t)\,\mathbf{v}\,\mathrm{dx}
\end{aligned}
$$

f.a. $\mathbf{v} \in \mathbf{V}_u$, *f.a.a.* $t \in S$.

7.3.3 Lemma (Existence and Uniqueness)
Let Assumption 4.3.1$_{(A1)}$ be valid. Then there exists a unique weak solution $\mathbf{u} \in \mathbf{V}_u$ of the Problem ($\mathbf{P}_{QS,u}$) satisfying the estimate

$$
\begin{aligned}
\|\mathbf{u}\|_{\mathcal{V}_u} &\le c_1 + c_2\left\{\|\mathbf{f}\|_{\mathcal{V}_u} + \|\theta\|_{\mathcal{H}_\theta} + \|\mathbf{p}\|_{\mathcal{H}_p} + \|\boldsymbol{\varepsilon}_{trip}\|_{\mathcal{H}_\sigma} + \|\boldsymbol{\varepsilon}_{cp}\|_{\mathcal{H}_\sigma}\right\}, \\
\|\boldsymbol{\sigma}\|_{\mathcal{H}_\sigma} &\le c_1 + c_2\left\{\|\mathbf{u}\|_{\mathcal{V}_u} + \|\theta\|_{\mathcal{H}_\theta} + \|\mathbf{p}\|_{\mathcal{H}_p} + \|\boldsymbol{\varepsilon}_{trip}\|_{\mathcal{H}_\sigma} + \|\boldsymbol{\varepsilon}_{cp}\|_{\mathcal{H}_\sigma}\right\},
\end{aligned}
$$

where $c_1, c_2 > 0$ do not depend on $\mathbf{u}$, $\boldsymbol{\sigma}$, $\mathbf{f}$, θ, $\mathbf{p}$, $\boldsymbol{\varepsilon}_{trip}$ and $\boldsymbol{\varepsilon}_{cp}$.

Proof. Testing the weak formulation (7.9) with $\mathbf{u}$ and applying Lemma B.3.1 shows the result, cf. [Chelminski et al. 2007; Kern 2011]. □

7.3.4 Lemma (Continuous Dependence on Parameters $\mathbf{f}$, θ, $\mathbf{p}$, $\boldsymbol{\varepsilon}_{trip}$ and $\boldsymbol{\varepsilon}_{cp}$)
Let Assumption 4.3.1$_{(A1)}$ be valid and let $(\mathbf{u}_1, \boldsymbol{\sigma}_1)$, $(\mathbf{u}_2, \boldsymbol{\sigma}_2)$ be two different solutions of Problem ($\mathbf{P}_{QS,u,\sigma}$) with corresponding data. Then

$$
\begin{aligned}
\|\mathbf{u}_1 - \mathbf{u}_2\|_{\mathcal{V}_u} \le c\Big\{&\|\mathbf{f}_1 - \mathbf{f}_2\|_{\mathcal{V}_u} + \|\theta_1 - \theta_2\|_{\mathcal{H}_\theta} \\
&+ \|\mathbf{p}_1 - \mathbf{p}_2\|_{\mathcal{H}_p} + \|\boldsymbol{\varepsilon}_{trip,1} - \boldsymbol{\varepsilon}_{trip,2}\|_{\mathcal{H}_\sigma} + \|\boldsymbol{\varepsilon}_{cp,1} - \boldsymbol{\varepsilon}_{cp,2}\|_{\mathcal{H}_\sigma}\Big\}
\end{aligned}
$$

and

$$
\begin{aligned}
\|\boldsymbol{\sigma}_1 - \boldsymbol{\sigma}_2\|_{\mathcal{H}_\sigma} \le c\Big\{&\|\mathbf{u}_1 - \mathbf{u}_2\|_{\mathcal{V}_u} + \|\theta_1 - \theta_2\|_{\mathcal{H}_\theta} \\
&+ \|\mathbf{p}_1 - \mathbf{p}_2\|_{\mathcal{H}_p} + \|\boldsymbol{\varepsilon}_{trip,1} - \boldsymbol{\varepsilon}_{trip,2}\|_{\mathcal{H}_\sigma} + \|\boldsymbol{\varepsilon}_{cp,1} - \boldsymbol{\varepsilon}_{cp,2}\|_{\mathcal{H}_\sigma}\Big\}.
\end{aligned}
$$

Proof. The proof is similar to the proof of Lemmas 5.3.26 and 6.3.11. □

7.3.5 Lemma (Time Regularity)
Let Assumption 4.3.1$_{(A1)}$ be valid. Denote by $\mathbf{u}$ *the weak solution of Problem* $(\mathbf{P}_{QS,\mathrm{u}})$*, then* $\mathbf{u}$ *satisfies the estimate*

$$\|\mathbf{u}'\|_{\mathcal{V}_\mathrm{u}} \leq c_1 + c_2 \left\{ \|\mathbf{f}'\|_{\mathcal{V}_\mathrm{u}} + \|\theta'\|_{\mathcal{H}_\theta} + \|\mathbf{p}'\|_{\mathcal{H}_\mathrm{p}} + \|\boldsymbol{\varepsilon}'_{trip}\|_{\mathcal{H}_\sigma} + \|\boldsymbol{\varepsilon}'_{cp}\|_{\mathcal{H}_\sigma} \right\},$$
$$\|\boldsymbol{\sigma}'\|_{\mathcal{H}_\sigma} \leq c_1 + c_2 \left\{ \|\mathbf{u}'\|_{\mathcal{V}_\mathrm{u}} + \|\theta'\|_{\mathcal{H}_\theta} + \|\mathbf{p}'\|_{\mathcal{H}_\mathrm{p}} + \|\boldsymbol{\varepsilon}'_{trip}\|_{\mathcal{H}_\sigma} + \|\boldsymbol{\varepsilon}'_{cp}\|_{\mathcal{H}_\sigma} \right\},$$

where $c_1, c_2 > 0$ *do not depend on* $\mathbf{u}'$, $\boldsymbol{\sigma}'$, $\mathbf{f}'$, θ', $\mathbf{p}'$, $\boldsymbol{\varepsilon}'_{trip}$ *and* $\boldsymbol{\varepsilon}'_{cp}$*.*

Proof. Differentiation of the weak formulation (7.9) (w.r.t. time), testing with $\mathbf{u}'$ and applying Lemma B.3.1 shows the estimate. Details can be found in [Chelminski et al. 2007; Kern 2011].
This does not show the existence of $\mathbf{u}'$. We only sketch the idea how to do this and refer to the literature (we follow the argumentation in [Wloka 1987], where the time regularity is proven for parabolic and hyperbolic equations, e.g.). Using the given assumptions, Lemma 7.3.3 implies $\mathbf{u} \in \mathcal{V}_\mathrm{u}$. We formally differentiate the equation $\mathbf{A}_\mathrm{u}\mathbf{u} = \mathbf{F}$. Again, we do not know a-priori whether the corresponding derivative exists and lies in the correct space and obtain

$$\mathbf{A}_\mathrm{u}\mathbf{u}' = \mathbf{F}', \qquad \mathbf{u}(0) = \mathbf{u}_0, \qquad \mathbf{u}'(0) = \mathbf{u}_1.$$

Therefore we consider $\mathbf{A}_\mathrm{u}\mathbf{v} = \mathbf{F}'$ and show that $\mathbf{v} = \mathbf{u}'$. Lemma 7.3.3 implies the solution $\mathbf{v} \in \mathcal{V}_\mathrm{u}$ of this problem. We form the auxiliary function $\mathbf{w}(t) := \mathbf{u}(0) + \int_0^t \mathbf{v}(s)\,\mathrm{d}s$ which is absolutely continuous since $\mathbf{v} \in \mathcal{V}_\mathrm{u}$ and fulfills $\mathbf{w} \in \mathcal{V}_\mathrm{u}$ with $\mathbf{w}(0) = \mathbf{u}(0)$. Integration by parts gives

$$\mathbf{A}_\mathrm{u}\mathbf{w}(t) - \mathbf{A}_\mathrm{u}\mathbf{u}(0) = \int_0^t \mathbf{A}_\mathrm{u}\mathbf{w}'(s)\,\mathrm{d}s = \int_0^t \mathbf{A}_\mathrm{u}\mathbf{v}(s)\,\mathrm{d}s = \mathbf{F}(t) - \mathbf{F}(0), \qquad t \in S.$$

Subtracting the original problem and applying the uniqueness result for the elliptic problem yields $\mathbf{w} = \mathbf{u}$ in S.
With the help of an induction argument it is possible to obtain even more regularity (depending on the regularity of the right-hand side). □

7.3.6 Lemma (Continuous Dependence on Parameters $\mathbf{f}$, θ, $\mathbf{p}$, $\boldsymbol{\varepsilon}_{trip}$ and $\boldsymbol{\varepsilon}_{cp}$)
Let Assumption 4.3.1$_{(A1)}$ be valid and let $(\mathbf{u}_1, \boldsymbol{\sigma}_1)$, $(\mathbf{u}_2, \boldsymbol{\sigma}_2)$ *be two different solutions of Problem* $(\mathbf{P}_{QS,\mathrm{u}})$ *with corresponding data. Then*

$$\|\mathbf{u}'_1 - \mathbf{u}'_2\|_{\mathcal{V}_\mathrm{u}} \leq c \Big\{ \|\mathbf{f}'_1 - \mathbf{f}'_2\|_{\mathcal{V}_\mathrm{u}} + \|\theta'_1 - \theta'_2\|_{\mathcal{H}_\theta}$$

$$+ \|\mathbf{p}_1' - \mathbf{p}_2'\|_{\mathcal{H}_p} + \|\boldsymbol{\varepsilon}_{trip,1}' - \boldsymbol{\varepsilon}_{trip,2}'\|_{\mathcal{H}_\sigma} + \|\boldsymbol{\varepsilon}_{cp,1}' - \boldsymbol{\varepsilon}_{cp,2}'\|_{\mathcal{H}_\sigma}\Big\}$$

and

$$\|\boldsymbol{\sigma}_1' - \boldsymbol{\sigma}_2'\|_{\mathcal{H}_\sigma} \le c\Big\{\|\mathbf{u}_1' - \mathbf{u}_2'\|_{\mathcal{V}_u} + \|\theta_1' - \theta_2'\|_{\mathcal{H}_\theta} + \|\mathbf{p}_1' - \mathbf{p}_2'\|_{\mathcal{H}_p} + \|\boldsymbol{\varepsilon}_{trip,1}' - \boldsymbol{\varepsilon}_{trip,2}'\|_{\mathcal{H}_\sigma} + \|\boldsymbol{\varepsilon}_{cp,1}' - \boldsymbol{\varepsilon}_{cp,2}'\|_{\mathcal{H}_\sigma}\Big\}.$$

7.3.7 Remark (Further Regularity Results)
Referring to Section 4.5, we collect some regularity results for mixed boundary value problems (MBVP) for second order elliptic differential equations (DE) from the literature. $W^{1,p}$-estimates for solutions to MBVP for second order elliptic DE are given in [Gröger 1989; Gallout and Monier 1999]. The regularisation of mixed second order elliptic problems is discussed in [Ito 1990b; Shamir 1968]. Strongly coupled elliptic systems are treated in [Bensoussan and Frehse 2002; Ebmeyer and Frehse 1997; Ebmeyer 1999]. In [Griepentrog 1999; Rehberg 2009] linear elliptic BVP with non-smooth data are investigated. Regularity results for linear MBVP of second order can be found in [Kaßmann and Madych 2004; Mazja et al. 1991; Maz'ya et al. 1991; Savaré 1997], and especially regularity results in elasticity with mixed boundary conditions are given in [Herzog et al. 2011a,b; Brown and Mitrea 2009]. A brief mathematical review of linearised elasticity is discussed in [Ciarlet and Ciarlet Jr. 2004; Grisvard 1985; Nečas 1983; Valent 1988]. Problems of elasto-plasticity and in particular global spatial regularity in elasto-plasticity are intensively treated in [Alber 1994, 1998; Knees 2005, 2006, 2008, 2009; Nesenenko 2009; Hömberg and Khludnev 2006a,b].

7.3.5 Step 6: Analysis of Problem ($\mathbf{P}_{FP2}$)

We continue with studying the solvability of Problem ($\mathbf{P}_{QS,FP2}$) for fixed data $\hat{\theta}$.

7.3.8 Lemma (Existence and Uniqueness)
Let Assumptions 4.3.1 and 4.3.2 be valid. Assume $\hat{\theta} \in \mathfrak{H}_\theta$. Then there exists a weak solution $(\mathbf{u}, \boldsymbol{\sigma}, \boldsymbol{\varepsilon}_{cp}) \in \mathfrak{V}_u \times \mathfrak{H}_\sigma \times \mathfrak{H}_\sigma$ of Problem ($\mathbf{P}_{QS,FP2}$).
Moreover, let $(\mathbf{u}_i, \boldsymbol{\sigma}_i, \boldsymbol{\varepsilon}_{cp,i})$ be solutions w.r.t. the data $\hat{\theta}_i$, $i = 1, 2$. Then the following estimate holds:

$$\|\mathbf{u}_1 - \mathbf{u}_2\|_{\mathfrak{V}_u} + \|\boldsymbol{\sigma}_1 - \boldsymbol{\sigma}_2\|_{\mathfrak{H}_\sigma} + \|\boldsymbol{\varepsilon}_{cp,1} - \boldsymbol{\varepsilon}_{cp,2}\|_{\mathfrak{H}_\sigma} \le c\,\|\hat{\theta}_1 - \hat{\theta}_2\|_{\mathfrak{H}_\theta}, \tag{7.10}$$

where the positive constant c does not depend on the data.

Proof. We use the same standard fixed-point argument as in the proof of the preceding lemma. Before we start with the proof let us collect the following results for the involved subproblems.

Let $(\theta, \mathbf{p}, \boldsymbol{\varepsilon}_{cp}, \boldsymbol{\varepsilon}_{trip}) \in \mathfrak{H}_\theta \times \mathfrak{H}_\mathrm{p} \times \mathfrak{H}_\sigma \times \mathfrak{H}_\sigma$. The problem (7.1) possesses a unique solution $(\mathbf{u}, \boldsymbol{\sigma}) \in \mathfrak{H}^\infty_\mathrm{u} \cap \mathcal{V}^\infty_\mathrm{u} \times \mathcal{H}^\infty_\sigma$ s.t.

$$\|\mathbf{u}_1 - \mathbf{u}_2\|_{\mathfrak{V}^\infty_\mathrm{u}} \leq c_1\Big\{\|\theta_1 - \theta_2\|_{\mathfrak{H}_\theta} + \|\mathbf{p}_1 - \mathbf{p}_2\|_{\mathfrak{H}_\mathrm{p}} + \|\boldsymbol{\varepsilon}_{cp,1} - \boldsymbol{\varepsilon}_{cp,2}\|_{\mathfrak{H}_\sigma} + \|\boldsymbol{\varepsilon}_{trip,1} - \boldsymbol{\varepsilon}_{trip,2}\|_{\mathfrak{H}_\sigma}\Big\}$$

$$\text{and } \|\boldsymbol{\sigma}_1 - \boldsymbol{\sigma}_2\|_{\mathfrak{H}_\sigma} \leq c_2\Big\{\|\mathbf{u}_1 - \mathbf{u}_2\|_{\mathfrak{V}_\mathrm{u}} + \|\theta_1 - \theta_2\|_{\mathfrak{H}_\theta} + \|\mathbf{p}_1 - \mathbf{p}_2\|_{\mathfrak{H}_\mathrm{p}}$$
$$\|\boldsymbol{\varepsilon}_{cp,1} - \boldsymbol{\varepsilon}_{cp,2}\|_{\mathfrak{H}_\sigma} + \|\boldsymbol{\varepsilon}_{trip,1} - \boldsymbol{\varepsilon}_{trip,2}\|_{\mathfrak{H}_\sigma}\Big\},$$

where c_1 and c_2 are positive constants.
Using the definition of the stress deviator and the preceding estimates one gets

$$\|\boldsymbol{\sigma}^*_1 - \boldsymbol{\sigma}^*_2\|_{\mathfrak{H}_\sigma} \leq c\Big\{\|\mathbf{u}_1 - \mathbf{u}_2\|_{\mathfrak{V}_\mathrm{u}} + \|\theta_1 - \theta_2\|_{\mathfrak{H}_\theta} + \|\mathbf{p}_1 - \mathbf{p}_2\|_{\mathfrak{H}_\mathrm{p}}$$
$$+ \|\boldsymbol{\varepsilon}_{trip,1} - \boldsymbol{\varepsilon}_{trip,2}\|_{\mathfrak{H}_\sigma} + \|\bar{\boldsymbol{\varepsilon}}_{cp,1} - \bar{\boldsymbol{\varepsilon}}_{cp,2}\|_{\mathfrak{H}_\sigma}\Big\}.$$

Let $(\mathbf{u}, \boldsymbol{\sigma}, \boldsymbol{\varepsilon}_{trip}) \in \mathfrak{V}_\mathrm{u} \times \mathfrak{H}_\sigma \times \mathfrak{H}_\sigma$. The problem (7.6) possesses a unique solution $\boldsymbol{\varepsilon}_{cp} \in \mathfrak{H}_\sigma$ s.t.

$$\|\boldsymbol{\varepsilon}_{cp,1} - \boldsymbol{\varepsilon}_{cp,2}\|_{\mathfrak{H}_\sigma} \leq c\Big\{\|\mathbf{u}_1 - \mathbf{u}_2\|_{\mathfrak{V}_\mathrm{u}} + \|\boldsymbol{\varepsilon}_{trip,1} - \boldsymbol{\varepsilon}_{trip,2}\|_{\mathfrak{H}_\sigma} + \|\boldsymbol{\sigma}^*_1(t) - \boldsymbol{\sigma}^*_2(t)\|_{\mathfrak{H}_\sigma}\Big\},$$

where c is a positive constant.
According to Lemma 7.3.1 we use Banach's fixed-point theorem with a weighted-norm space. Define $\mathbf{X} := \mathfrak{V}_\mathrm{u} \times \mathfrak{H}_\sigma \times \mathfrak{H}_\sigma$. For $(\bar{\mathbf{u}}, \bar{\boldsymbol{\sigma}}, \bar{\boldsymbol{\varepsilon}}_{cp}) \in \mathbf{X}$, we consider the linearised problem (7.1) for $(\mathbf{u}, \boldsymbol{\sigma}, \boldsymbol{\varepsilon}_{cp}) \in \mathbf{X}$ in which the parameter $\hat{\theta}$ is fixed and $\boldsymbol{\varepsilon}_{trip}$ is the solution of Problem $(\mathbf{P}_{QS,FP1})$ according to Lemma 7.3.1. Then there exists a unique solution $(\bar{\mathbf{u}}, \bar{\boldsymbol{\sigma}}, \bar{\varepsilon}_{cp}) \in \mathfrak{V}_\mathrm{u} \times \mathfrak{H}_\sigma \times \mathfrak{H}_\sigma \hookrightarrow \mathbf{X}$. We define a fixed-point operator as

$$\mathbf{T}_{QS,FP2} : \mathbf{X} \to \mathbf{X}, \qquad \mathbf{T}_{QS,FP2}(\bar{\mathbf{u}}, \bar{\boldsymbol{\sigma}}, \bar{\varepsilon}_{cp}) = (\mathbf{u}, \boldsymbol{\sigma}, \boldsymbol{\varepsilon}_{cp})$$

and consider two different solutions $(\mathbf{u}_1, \boldsymbol{\sigma}_1, \boldsymbol{\varepsilon}_{cp,1}), (\mathbf{u}_2, \boldsymbol{\sigma}_2, \boldsymbol{\varepsilon}_{cp,2})$ corresponding to different data $(\bar{\mathbf{u}}_1, \bar{\boldsymbol{\sigma}}_1, \bar{\boldsymbol{\varepsilon}}_{cp,1}), (\bar{\mathbf{u}}_2, \bar{\boldsymbol{\sigma}}_2, \bar{\boldsymbol{\varepsilon}}_{cp,2})$. By similar arguments as for Lemma 7.3.1 (using the preceding estimates) we obtain f.a.a. $t \in S$ the following estimate

$$\|\mathbf{u}_1(t) - \mathbf{u}_2(t)\|_{\mathrm{V}_\mathrm{u}} + \|\mathbf{u}'_1(t) - \mathbf{u}'_2(t)\|_{\mathrm{V}_\mathrm{u}} + \|\boldsymbol{\sigma}_1(t) - \boldsymbol{\sigma}_2(t)\|_{\mathrm{H}_\sigma}$$
$$+ \|\boldsymbol{\sigma}'_1(t) - \boldsymbol{\sigma}'_2(t)\|_{\mathrm{H}_\sigma} + \|\boldsymbol{\varepsilon}_{cp,1}(t) - \boldsymbol{\varepsilon}_{cp,2}(t)\|_{\mathrm{H}_\sigma} + \|\boldsymbol{\varepsilon}'_{cp,1}(t) - \boldsymbol{\varepsilon}'_{cp,2}(t)\|_{\mathrm{H}_\sigma}$$
$$\leq c\Big\{\|\mathbf{p}_1(t) - \mathbf{p}_2(t)\|_{\mathrm{H}_\mathrm{p}} + \|\mathbf{p}'_1(t) - \mathbf{p}'_2(t)\|_{\mathrm{H}_\mathrm{p}} + \int_0^t \|\mathbf{p}'_1(s) - \mathbf{p}'_2(s)\|_{\mathrm{H}_\mathrm{p}}\, ds$$
$$+ \|\bar{\boldsymbol{\varepsilon}}_{cp,1}(t) - \bar{\boldsymbol{\varepsilon}}_{cp,2}(t)\|_{\mathrm{H}_\sigma} + \int_0^t \|\bar{\boldsymbol{\varepsilon}}'_{cp,1}(s) - \bar{\boldsymbol{\varepsilon}}'_{cp,2}(s)\|_{\mathrm{H}_\sigma}\, ds$$

$$
\begin{aligned}
&\quad + \|\boldsymbol{\sigma}_1^*(t) - \boldsymbol{\sigma}_2^*(t)\|_{\mathsf{H}_\sigma} + \|(\boldsymbol{\sigma}_1^*)'(t) - (\boldsymbol{\sigma}_2^*)'(t)\|_{\mathsf{H}_\sigma}\Big\} \\
&\le c\Big\{\|\bar{\boldsymbol{\varepsilon}}_{cp,1}(t) - \bar{\boldsymbol{\varepsilon}}_{cp,2}(t)\|_{\mathsf{H}_\sigma} + \int_0^t \|\bar{\boldsymbol{\varepsilon}}'_{cp,1}(s) - \bar{\boldsymbol{\varepsilon}}'_{cp,2}(s)\|_{\mathsf{H}_\sigma}\,\mathrm{d}s + \|\bar{\boldsymbol{\sigma}}_1(t) - \bar{\boldsymbol{\sigma}}_2(t)\|_{\mathsf{H}_\sigma} \\
&\quad + \int_0^t \|\bar{\boldsymbol{\sigma}}_1(s) - \bar{\boldsymbol{\sigma}}_2(s)\|_{\mathsf{H}_\sigma}\,\mathrm{d}s + \int_0^t \|\bar{\mathbf{u}}'_1(s) - \bar{\mathbf{u}}'_2(s)\|_{\mathsf{V}_u}\,\mathrm{d}s\Big\} \\
&\le c\Big\{\int_0^t \|\bar{\boldsymbol{\varepsilon}}'_{cp,1}(s) - \bar{\boldsymbol{\varepsilon}}'_{cp,2}(s)\|_{\mathsf{H}_\sigma}\,\mathrm{d}s + \int_0^t \|\bar{\boldsymbol{\sigma}}_1(s) - \bar{\boldsymbol{\sigma}}_2(s)\|_{\mathsf{H}_\sigma}\,\mathrm{d}s \\
&\quad + \int_0^t \|\bar{\boldsymbol{\sigma}}'_1(s) - \bar{\boldsymbol{\sigma}}'_2(s)\|_{\mathsf{H}_\sigma}\,\mathrm{d}s + \int_0^t \|\bar{\mathbf{u}}'_1(s) - \bar{\mathbf{u}}'_2(s)\|_{\mathsf{V}_u}\,\mathrm{d}s\Big\} \\
&\le c\Big\{\|\bar{\mathbf{u}}_1 - \bar{\mathbf{u}}_2\|_{\mathfrak{V}_u} + \|\bar{\boldsymbol{\sigma}}_1 - \bar{\boldsymbol{\sigma}}_2\|_{\mathcal{H}_\sigma} + \|\bar{\boldsymbol{\varepsilon}}_{cp,1} - \bar{\boldsymbol{\varepsilon}}_{cp,2}\|_{\mathfrak{H}_\sigma}\Big\}.
\end{aligned}
$$

It follows (analogously to the proof of Lemma 7.3.1)

$$
\|(\mathbf{u}_1, \boldsymbol{\sigma}_1, \boldsymbol{\varepsilon}_{cp,1}) - (\mathbf{u}_2, \boldsymbol{\sigma}_2, \boldsymbol{\varepsilon}_{cp,2})\|_X \le \sqrt{\frac{c}{\lambda}}\|(\bar{\mathbf{u}}_1, \bar{\boldsymbol{\sigma}}_1, \bar{\boldsymbol{\varepsilon}}_{cp,1}) - (\bar{\mathbf{u}}_2, \bar{\boldsymbol{\sigma}}_2, \bar{\boldsymbol{\varepsilon}}_{cp,2})\|_X
$$

and hence, for λ large enough, $\mathbf{T}_{QS,FP2}$ is strictly contractive. The conditions for Banach's fixed-point theorem are fulfilled and Problem ($\mathbf{P}_{QS,FP2}$) possesses a unique solution. Moreover, the estimate (7.10) holds. □

7.3.6 Step 7: Analysis of Problem ($\mathbf{P}_{QS,\theta}$)

Next, we remark that the analysis of Problem ($\mathbf{P}_{QS,\theta}$) is completely analogous to the analysis of Problem ($\mathbf{P}_{VE,\theta}$) and investigated in detail in Subsection 6.3.7.

7.3.7 Step 8: Analysis of Problem ($\mathbf{P}_{QS,FP3}$)

Using the foregoing lemmas of this section, we are now able to prove the existence and uniqueness result for Problem ($\mathbf{P}_{QS,FP3}$) or rather for the full Problem ($\mathbf{P}_{QS}$).

Proof. We first observe that the solution $\theta \in \mathfrak{H}_\theta \cap \mathcal{V}_\theta$ of (7.2) satisfies

$$
\begin{aligned}
\|\theta_1 - \theta_2\|_{\mathcal{H}_\theta^\infty} + \|\theta_1 - \theta_2\|_{\mathcal{V}_\theta} &\le c_1\left\{\|\mathbf{u}'_1 - \mathbf{u}'_2\|_{\mathcal{V}_u} + \|\mathbf{p}_1 - \mathbf{p}_2\|_{\mathcal{H}_p}\right\}, \\
\|\theta'_1 - \theta'_2\|_{\mathcal{H}_\theta} + \|\theta_1 - \theta_2\|_{\mathcal{V}_\theta^\infty} &\le c_2\left\{\|\mathbf{u}'_1 - \mathbf{u}'_2\|_{\mathcal{V}_u} + \|\mathbf{p}_1 - \mathbf{p}_2\|_{\mathcal{H}_p}\right\},
\end{aligned}
$$

where c_1, c_2 are positive constants.
We use the Lemmas 7.3.1 and 7.3.8 and Banach's fixed-point theorem with the (complete) weighted-norm space

$$
\mathbf{X} := \mathfrak{H}_\theta, \qquad \|\mathbf{x}\|_X^2 := \sup_{t\in\bar{S}}\left\{\exp(-\lambda t)\int_0^t \|\mathbf{x}(s)\|_{\mathsf{H}_\theta}^2\,\mathrm{d}s\right\}, \qquad \lambda > 0.
$$

Let $\boldsymbol{\varepsilon}_{trip}$ the unique solution to Problem ($\mathbf{P}_{QS,FP1}$) corresponding to Lemma 7.3.1 for given θ, $\mathbf{u}$, $\boldsymbol{\sigma}$ and $\boldsymbol{\varepsilon}_{cp}$. Moreover, let $(\mathbf{u}, \boldsymbol{\sigma}, \boldsymbol{\varepsilon}_{cp})$ the unique solution to Problem ($\mathbf{P}_{QS,FP2}$) corresponding to Lemma 7.3.8 for fixed θ and $\boldsymbol{\varepsilon}_{trip}$. For $\hat{\theta} \in \mathbf{X}$ given, we consider the linearised problem (7.2) for $\theta \in \mathbf{X}$. Finally, by Lemmas 6.3.14 and 5.3.36 there exists a unique solution $\theta \in \mathcal{V}_\theta \cap \mathfrak{H}_\theta \hookrightarrow \mathbf{X}$ for given $\mathbf{u}$,$\boldsymbol{\sigma}$,$\boldsymbol{\varepsilon}_{cp}$ and $\boldsymbol{\varepsilon}_{trip}$. We define a fixed-point operator as

$$\mathbf{T}_{QS,FP3} : \mathbf{X} \to \mathbf{X}, \qquad \mathbf{T}_{QS,FP3}(\hat{\theta}) = \theta$$

and consider solutions θ_1, θ_2 corresponding to different data $\hat{\theta}_1, \hat{\theta}_2$. By similar arguments as for Lemmas 7.3.1 and 7.3.8 (and using the estimates (7.7) and (7.10)) we obtain f.a.a. $t \in S$ the following estimate

$$\begin{aligned}
&\|\theta_1(t) - \theta_2(t)\|_{H_\theta} + \|\theta_1'(t) - \theta_2'(t)\|_{H_\theta} \\
&\qquad \le c\left\{\int_0^t \|\mathbf{u}_1'(s) - \mathbf{u}_2'(s)\|^2_{\mathrm{V_u}}\,\mathrm{d}s + \int_0^t \|\mathbf{p}_1(s) - \mathbf{p}_2(s)\|^2_{\mathrm{H_p}}\,\mathrm{d}s\right\} \\
&\qquad \le c\left\{\int_0^t \|\hat{\theta}_1(t) - \hat{\theta}_2(t)\|^2_{H_\theta}\,\mathrm{d}s + \int_0^t \|\hat{\theta}_1'(t) - \hat{\theta}_2'(t)\|_{H_\theta}\,\mathrm{d}s\right\} \\
&\qquad \le c\int_0^t \exp(\lambda t)\,\mathrm{d}s \sup_{s\in[0,t]}\left\{\exp(-\lambda t)\Big(\|\hat{\theta}_1(t) - \hat{\theta}_2(t)\|^2_{H_\theta} + \|\hat{\theta}_1'(t) - \hat{\theta}_2'(t)\|_{H_\theta}\Big)\right\} \\
&\qquad \le c\,\frac{\exp(\lambda t)}{\lambda}\|\hat{\theta}_1 - \hat{\theta}_2\|_{\mathrm{X}}.
\end{aligned}$$

It follows

$$\|\theta_1 - \theta_2\|_{\mathrm{X}} \le \sqrt{\frac{c}{\lambda}}\,\|\hat{\theta}_1 - \hat{\theta}_2\|_{\mathrm{X}}$$

and hence, for λ large enough, $\mathbf{T}_{QS,FP3}$ is strictly contractive. The conditions for Banach's fixed-point theorem are fulfilled and Problem ($\mathbf{P}_{QS}$) possesses a unique solution. □

Chapter 8

Numerical Simulation

In this chapter we present some numerical simulations with realistic material data to demonstrate the distortion effect of metallurgical phase transitions. We implement the Problem (**P**), which has been introduced in Section 3.1, in COMSOL Multiphysics® and simulate the Jominy-End-Quench-Test in two dimensions (due to rotation symmetry) for steel 100Cr6. Although we try to present a complete numerical simulation, several simplifications have been done to the general model in Chapter 3. However, an independent contribution to the numerical analysis is not aspired and a full numerical implementation would go beyond the scope of this work – the simulation is only a means to an end.
In Section 8.1 we state the modified PDE system based on Problem (**P**) and discuss the Jominy-End-Quench-Test. The numerical implementation is explained in Section 8.2. In Section 8.3 we discuss the simulation results. References and related works are given in Section 8.4.

8.1 Formulation of the Problem

In order to show phase induced thermo-mechanical effects, we simulate the Jominy-End-Quench-Test in COMSOL Multiphysics®.
We assume that the steel specimen has already been heated up to the high temperature phase austenite and consider only the cooling stage of a heat treatment process. The heat treatment process for most steels involves heating the alloy until austenite forms, then cooling it so rapidly that the transformation into martensite occurs almost immediately. The presence of two phases leads to plastic deformation due to the volume differences in the phases. For a sufficiently high cooling rate, i.e. close to the quenched boundary parts where the temperature descends very quickly, austenite is transformed into the hard phase martensite, whereas a slower temperature change, e.g. in the interior of a big workpiece, causes usually softer phases like ferrite, pearlite or bainite to grow. A correlation of the temperature change and the phase transformation can be obtained

from the time-temperature-transformation diagram (see Figure 1.6(b)). Due to the fact that the product phases (e.g. pearlite and martensite) have different thermal expansion (different (lower) densities in comparison with austenite), their formation influences the deformation of the workpiece (cf. Section 1.2).
In order to avoid technicalities, we consider only the cooling of a steel specimen from high temperature phase austenite with fraction p_A to one product phase (martensite) with phase fraction p_M. This applies for instance in martensitic hardening of eutectoid carbon steel.

8.1.1 The Jominy-End-Quench-Test

The Jominy-End-Quench-Test is a standard test to determine the hardenability of steels. Moreover, it is useful for simulation testing, too. In this work, it is used to test the thermo-mechanical model with phase transitions and TRIP in Problem ($\mathbf{P}_{NUM}$) and to illustrate the results (cf. e.g. [Hunkel et al. 2004; Hömberg 1996, 1997; Narazaki et al. 2005] for more information).

(a) Photo from [Liedtke 2005b].

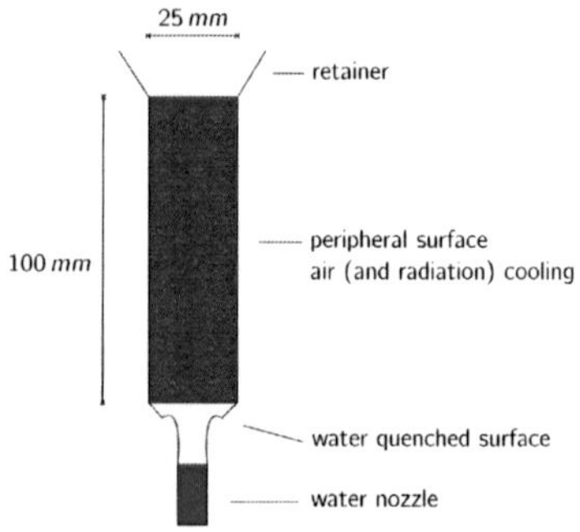

(b) Scheme from [Hunkel et al. 2004].

Figure 8.1: Jominy-End-Quech-Test.

8.1.1 Remark (Hardenability)
Hardenability is the ability of a steel to partially or completely transform from austenite to some fraction of martensite at a given depth below the surface, when cooled under a given condition. For example, a steel of a high hardenability can transform to a high

fraction of martensite to depths of several millimetres under relatively slow cooling, whereas a steel of low hardenability may only form a high fraction of martensite to a depth of less than a millimetre, even under rapid cooling.
Steels with high hardenability are needed for large high strength components or for small high precision components, e.g. in mining or aircraft industry, whereas steels with low hardenability may be used for smaller components, such as tools, or for surface hardened components such as gears. High hardenability allows slower quenches to be used (e.g. an oil quench), which reduces the distortion and residual stress from thermal gradients. More about hardening and hardenability can be found in the literature, cf. [Liedtke 2005b, 2009] for an overview.

The Jominy-End-Quench-Test involves heating a test piece from the particular steel (cylinder with 25mm diameter and 100mm long) to an austenitising temperature and quenching from one end with a controlled and standardised jet of water. After quenching, the hardness profile is measured at intervals from the quenched end after the surface has been ground to remove any effects of decarburisation (less than a millimetre is removed from the surface).
The hardness variation along the test surface is a result of micro-structural variation which arises since the cooling rate decreases with distance from the quenched end. The cooling rate along the test specimen varies from a few hundred $\frac{K}{s}$ up to one to two $\frac{K}{s}$. The accuracy of the quenching simulations of steel components depends very much on the accuracy of the values of material properties of the steel material used for the simulation. Therefore, we need a validation method for material property data. Generally, the simulation of the quench test of a steel specimen with simple shape is useful for the validation of the accuracy of these data. In addition, the accurate heat transfer coefficients on the surfaces of specimen during the quenching must be known, cf. for instance [Hunkel et al. 2004].

8.1.2 The PDE Problem

We aim at simulating the quenching process of a steel specimen. We consider a two-dimensional (due to axially symmetric reasons) cross-section of a steel cylinder made of the eutectoid carbon steel 100Cr6[1] of dimension 100mm × 25mm and infer specific cooling to obtain corresponding phase distribution. We assume the specimen to be the closure of a Lipschitz domain $\Omega_1 = \Omega$. As mentioned above, we consider only the cooling and not the preliminary heating process. Therefore we assume a sufficiently

[1] The bearing steel 100Cr6 (SAE 52100) is the subject of intense research within the SFB 570. The chemical composition is defined in EN ISO 683-17.

high, homogeneous initial temperature and complete austenitisation. Moreover, we distinguish between the reference and the initial geometry. The dark grey rectangle in Figure 8.2(b) gives the reference geometry of dimension 100mm × 25mm at room temperature and the light grey rectangle represents the thermally expanded specimen at the initial (austenitisation) temperature. Using the initial value $\mathbf{u}_0(\mathbf{x}) = \alpha_A(\theta_0 - \theta_{\text{air}})\,\mathbf{x}$ for the displacement, we take this thermal expansion into account[2].
In the model the workpiece is quenched from the bottom and it is assumed to be fixed in only one single point to avoid additional stress caused by thermoelastic effects. The fixed point is the centre of the top edge (point P_3 in Figure 8.2(a)), i.e. the displacement in this point is set to be zero. Furthermore, we suppose that the whole upper boundary part may only shrink horizontally. Moreover, it is assumed to be thermally insulated. The quenching of the workpiece corresponds to nonzero Robin boundary conditions at

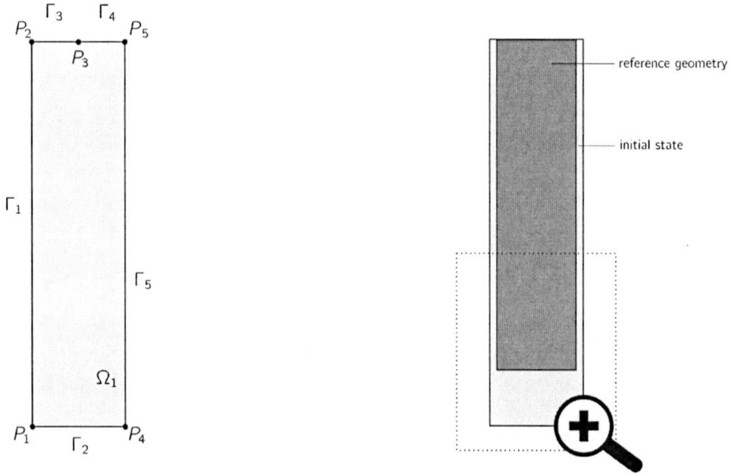

(a) Labelling of the boundaries. (b) Initial state.

Figure 8.2: Introduction of notation. The detail enlargement refers to the simulation results in Section 8.3.

the bottom. The left and the right edges fulfill Robin boundary conditions as well, but

[2]The reason for this approach is, that we do not simulate the heating phase of the experiment.

the cooling rate is considerably lower. The boundary conditions are given in Equations (8.8) – (8.9). See Figure 8.2(a) for the labelling of the boundary parts.
The following Problem ($\mathbf{P}_{NUM}$) corresponds to the simplification of the fully coupled Problem (**P**) in Section 3.1.

Problem ($\mathbf{P}_{NUM}$)
Find the displacement field $\mathbf{u} : \overline{\Omega}_T \to \mathbb{R}^3$*, the stress field* $\boldsymbol{\sigma} : \overline{\Omega}_T \to \mathbb{R}^{3\times3}_{sym}$*, the temperature field* $\theta : \overline{\Omega}_T \to \mathbb{R}$ *and the phase fractions* $\mathbf{p} : \overline{\Omega}_T \to \mathbb{R}^m$ *s.t.*

(8.1)
$$\rho_0 \frac{\partial^2 \mathbf{u}}{\partial t^2} - 2\,\mathrm{Div}\,(\mu\boldsymbol{\varepsilon}(\mathbf{u})) - \mathrm{grad}\,(\lambda\,\mathrm{div}(\mathbf{u})) + 3\,\mathrm{grad}\,(K_\alpha(\theta-\theta_0))$$
$$+\,\mathrm{grad}\left(K\sum_{i=1}^{2}\left(\frac{\rho_0}{\rho_i(\theta_0)}-1\right)p_i\right) + 2\,\mathrm{Div}\,(\mu\boldsymbol{\varepsilon}_{trip}) + 2\,\mathrm{Div}\,(\mu\boldsymbol{\varepsilon}_{cp}) = \mathbf{f} \quad \text{in } \Omega_T$$

(8.2)
$$\rho_0 c_e \frac{\partial\theta}{\partial t} - \mathrm{div}\,(\lambda_\theta \nabla\theta) + 3K_\alpha\theta\,\mathrm{div}\left(\frac{\partial\mathbf{u}}{\partial t}\right) = \boldsymbol{\sigma} : \left(\boldsymbol{\varepsilon}'_{trip} + \boldsymbol{\varepsilon}'_{cp}\right) + \rho_0 L_2 p'_2 + r \quad \text{in } \Omega_T$$

(8.3)
$$\frac{\partial\mathbf{p}}{\partial t} = \boldsymbol{\gamma}(\mathbf{p},\theta) \quad \text{in } \Omega_T$$

(8.4)
$$\boldsymbol{\varepsilon}'_{cp} = \Lambda\boldsymbol{\sigma}^*, \quad \Lambda \geq 0 \text{ and } \Lambda = 0 \text{ for } F(\boldsymbol{\sigma}^*, R_0) < 0 \quad \text{in } \Omega_T$$

(8.5)
$$\boldsymbol{\varepsilon}'_{trip} = \frac{3}{2}\kappa\boldsymbol{\sigma}^*\sum_{i=1}^{2}\max\{p'_i, 0\} \quad \text{in } \Omega_T$$

including the following initial conditions f.a. $\mathbf{x}$ *in* Ω*:*

(8.6) $\mathbf{u}(\mathbf{x},0) = \mathbf{u}_0(\mathbf{x})$, $\quad \mathbf{u}'(\mathbf{x},0) = \mathbf{0}$, $\quad \theta(\mathbf{x},0) = \theta_0$,
(8.7) $\mathbf{p}(\mathbf{x},0) = \mathbf{p}_0(\mathbf{x})$, $\quad \boldsymbol{\varepsilon}_{trip}(\mathbf{x},0) = \mathbf{0}$, $\quad \boldsymbol{\varepsilon}_{cp}(\mathbf{x},0) = \mathbf{0}$

and the mixed boundary conditions for $\mathbf{u}$ *and for* θ

(8.8) $\boldsymbol{\sigma}\cdot\boldsymbol{\nu}_{\Gamma_1} = \mathbf{0}$ *and* $-\lambda_\theta \nabla\theta\cdot\boldsymbol{\nu}_{\Gamma_1} = \delta_{\mathrm{air}}\left(\theta - \theta_{\mathrm{air}}\right)$ *on* Γ_1
(8.9) $\boldsymbol{\sigma}\cdot\boldsymbol{\nu}_{\Gamma_2} = \mathbf{0}$ *and* $-\lambda_\theta \nabla\theta\cdot\boldsymbol{\nu}_{\Gamma_2} = \delta_{\mathrm{water}}\left(\theta - \theta_{\mathrm{water}}\right)$ *on* Γ_2
(8.10) $\mathbf{u} = \mathbf{0}$ *and* $-\lambda_\theta \nabla\theta\cdot\boldsymbol{\nu}_{\Gamma_3} = 0$ *on* Γ_3
(8.11) $\mathbf{u} = \mathbf{0}$ *and* $-\lambda_\theta \nabla\theta\cdot\boldsymbol{\nu}_{\Gamma_4} = 0$ *on* Γ_4
(8.12) $\boldsymbol{\sigma}\cdot\boldsymbol{\nu}_{\Gamma_5} = \mathbf{0}$ *and* $-\lambda_\theta \nabla\theta\cdot\boldsymbol{\nu}_{\Gamma_5} = \delta_{\mathrm{air}}\left(\theta - \theta_{\mathrm{air}}\right)$ *on* Γ_5

We consider only two phases, a parent phase, austenite and a forming phase, martensite (cf. [Burtchen et al. 2006; Denis et al. 1983; Grostabussiat et al. 2001; Gautier et al. 2000] for martensite quenching). A prototypical model (cf. Example 2.4.2 for this simple law and Example 5.7.7 for some analytical results) to describe this behaviour is given by the so-called Leblond-Devaux model:

$$\begin{aligned} p_A'(\mathbf{x}, t) &= -p_M'(\mathbf{x}, t), & p_A(\mathbf{x}, 0) &= 1, \\ p_M'(\mathbf{x}, t) &= \mu\left(\bar{p}_M(\theta(\mathbf{x}, t)) - p_M(\mathbf{x}, t)\right), & p_M(\mathbf{x}, 0) &= 0 \end{aligned}$$

for $\mathbf{x} \in \Omega$, where $p_1 = p_A$ – austinite phase fraction, $p_2 = p_M$ – martensite phase fraction, $\mathbf{p}_0 = (p_A(0), p_M(0))^T$, θ – given temperature and $\bar{p}_M$ represents the maximal martensitic phase fraction that can be attained. We use the following Koistinen-Marburger ansatz (cf. Example 2.4.4):

$$\bar{p}_M(\theta) = 1 - \exp\left(\frac{\theta - \theta_{ms}}{\theta_{m0}}\right) \quad \text{for} \quad \theta \in [\theta_{mf}, \theta_{ms}] \quad \text{and} \quad \bar{p}_M(\theta) = 0 \quad \text{for} \quad \theta \in [\theta_{ms}, \theta_0].$$

Moreover, we assume $\mathbf{f} = (0, \rho_0 g)^T$ and $r = 0$. Furthermore, we assume a linear mixture rule between both phases f.a. material parameters, i.e. the value of a material parameter is calculated by $\pi = \pi_A p_A + \pi_M p_M$, where p_A, p_M are the austenitic and martensitic phase fraction and π_A, π_M are the corresponding material parameters, which are taken from data table 8.1 for eutectoid carbon steel 100Cr6.
In order to study TRIP and stress-dependent phase transformation in steel one usually performs tests on small specimens in special devices like dilatometers or Gleeble™ machines, which can measure temperature, length, diameter and applied stress as functions of times (cf. [Hömberg et al. 2009; Wolff et al. 2005a, 2004, 2007d]). A dilatometer is an instrument used for magnifying and measuring expansion and contraction of a solid during heat and subsequent cooling (cf. [Dalgic and Löwisch 2004, 2006; Dalgic et al. 2006]) and is often used in the determination of the TRIP-strain and phase transitions, especially the evolution of the forming phases, occurring during heat treatment of steels. The parameter μ is calculated via the measured data from such experiments provided by the Institute for Materials Science within the framework of the SFB 570.

8.2 Numerical Implementation

We solve this problem by the Finite Element Method (FEM) using the Structural Mechanics Module in COMSOL Multiphysics®. COMSOL Multiphysics® (formerly FEMLAB™) is a commercial FEM solver and simulation software package for various

	Property	Unit	Phase	
			Austenite	*Martensite*
λ	Thermal Conductivity	$\frac{W}{mm°C}$	0.02012	0.04273
c_e	Specific Heat Capacity	$\frac{J}{kg°C}$	560.7	501.3
ρ	Density	$\frac{kg}{m^3}$	7798	7741
E	Elastic Modulus	MPa	170665	204671
ν	Poisson Ratio	–	0.3318	0.3556
R_0	Initial Yield Strength	MPa	173.3	1278.4
α	Thermal Expansion Coefficient	$\frac{10^{-6}}{°C}$	23.8	10.9
ρ_0	Initial Density	$\frac{kg}{m^3}$	8041.4	
ΔH_{a-m}	Transformation Enthalpy	$\frac{J}{kg}$	78520	
θ_0	Initial Temperature	°C	850	
θ_{air}	Air Temperature	°C	22	
θ_{water}	Water Temperature	°C	18	
θ_{ms}	Martensite Start Temperature	°C	211	
θ_{mf}	Martensite Finish Temperature	°C	−174	
θ_{m0}	Koistinen-Marburger Temperature	°C	93.4	
δ_{air}	Heat Transfer Coefficient (Air)	$\frac{W}{m^2 K}$	150	
δ_{water}	Heat Transfer Coefficient (Water)	$\frac{W}{m^2 K}$	3200	
κ	Greenwood-Johnson Parameter	$\frac{1}{MPa}$	$7 \cdot 10^{-5}$	
μ	Leblond-Devaux Parameter	–	5.95	
g	Gravitational Acceleration	$\frac{m}{s^2}$	9.81	

Table 8.1: Average material properties and their units and dimensions, cf. [Acht et al. 2008a,b; Şimşir et al. 2009; Suhr 2010].

Number of Vertex Elements	5
Number of Boundary Elements	38
Number of Elements	156

Minimum Element Quality	0.9414
Minimum Element Size	0.03mm
Maximum Element Size	6.7mm

Table 8.2: Overview of the mesh quality.

physics and engineering applications, especially coupled phenomena, or multiphysics (cf. e.g. [Zimmermann 2006]).
The reference geometry is discretised by a free triangular mesh with 156 elements using the boundary layer meshing feature (cf. Figure 8.3 and Table 8.2). Adaptive mesh refinement or remeshing is not applied. The dependent variables are discretised by quadratic Lagrange shape functions.

The entire problem is solved fully coupled by using semi-discretisation via vertical line method. The initialisation of the mass matrix is done by the Backward Euler method, for the time-stepping the time-dependent implicit linear multistep solver BDF (Backward Differentiation Formulas for the numerical integration of the ODEs with maximal order 2 and minimal order 1) is used and the direct solver MUMPS (MUltifrontal Massively Parallel sparse direct Solver) is employed for the system of algebraic equations.

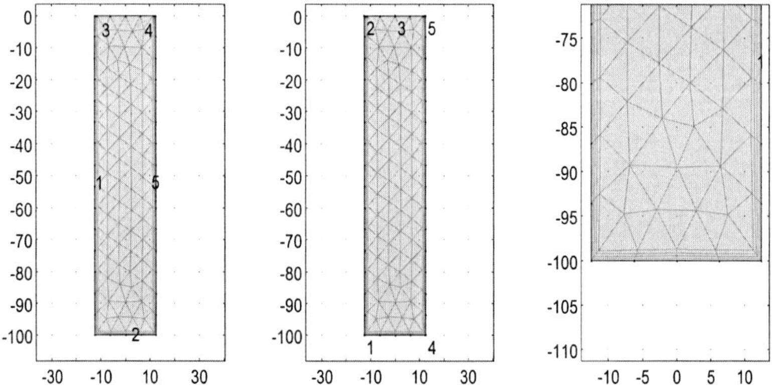

(a) Labelling of the edges. (b) Labelling of the vertexes. (c) Detail enlargement.

Figure 8.3: Discretisation of the reference geometry.

In the following, we give some hints for the implementation of the Problem ($\mathbf{P}_{NUM}$) in COMSOL Multiphysics®. We used a time-dependent study for the 'Solid Mechanics Physics' with an elasto-plastic material model (with von Mises yield function for perfect plasticity) and thermal expansion in order to implement the Equations (8.1) and (8.4). The 'flow rule' for plasticity is time-independent and already implemented in the elasto-plastic material model. Also, we added a 'Heat Transfer in Solids Physics' to take the Equation (8.2) into account. The dissipation term $\boldsymbol{\sigma} : \left(\boldsymbol{\varepsilon}'_{trip} + \boldsymbol{\varepsilon}'_{cp}\right)$ was added as a heat source together with another heat source for the term $\rho_0 L_2 \gamma_2$, describing the influence of the latent heat. Moreover, the contribution from the volumetric strain rate, i.e. the term $3K_\alpha \theta \operatorname{div}(\mathbf{u}')$ was added. The main difficulty was the implementation of the Equations (8.3) and (8.5). We solved this problem by manually adding an extra 'Mathematics Physics – PDE Interface' for the evolution of the phase transitions. In the same way, we added the evolution for the transformation-induced plastic tensor $\boldsymbol{\varepsilon}_{trip}$. The additive decomposition of strains is realised by adding initial stresses and strains for $\boldsymbol{\varepsilon}_{trip}$. The plastic strain tensor $\boldsymbol{\varepsilon}_{cp}$ is added internally by the elasto-plastic material model.

8.3 Simulation Results and Conclusion

In this section we present the simulation results. The running time of about one and a half hours in Table 8.3 refers to an Intel®Core™ 2 Duo processor with 2.1GHz and 3GB RAM. The Figures 8.4 – 8.13 demonstrate the process at a sample of time steps.

Number of Degrees of Freedom	9933
Solution Time	$\approx$ 95min

Table 8.3: Typical running time.

The workpiece is shown at times $t = 0$s, 0.2s, 2s, 5s, 10s, 20s, 30s, 40s, 50s and 100s. Colours indicate the temperature in the left figures and the martensite fraction in the right figures, where blue corresponds to low values and red to high values of temperature or phase fraction. The background rectangle gives the initial geometry. We only look at a detail enlargement, cf. Figures 8.2(b) and 8.3(c). The deformation of the workpiece is magnified by a factor of 10.
At time $t = 0$s, the initially rectangular specimen is completely heated and stress-free, cf. Figure 8.4. A strong cooling is applied to the bottom, the left and right side is cooled moderately and there is no heat transfer at the top of the sample. When quenching sets in, we can observe the shrinking of the quenched area in Figure 8.5. Because of the strong cooling, the sample contracts more at the left and right bottom vertexes in vertical direction. After a while, thermoelastic effects may be observed first, i.e. the shrinking due to classical thermo-elasticity can be observed in Figure 8.6. After some time, the austenite-martensite-transformation sets in (cf. Figure 8.7), beginning at the left and right bottom vertexes, where the temperature has already dropped below the martensite start temperature, i.e. the rapid cooling effected the growth of martensite in the lower part of the specimen. Martensite has a lower density and thereby a larger volume and higher expansion than austenite. Therefore the sample starts to expand in this area. Thus, the transformation pushes the material outward, which causes a huge bulge in the specimen (cf. Figures 8.8 – 8.11) and remains until the heat treatment is completed, cf. Figures 8.12 and 8.13.
This corresponds to the observed behaviour in experiments, cf. [Hunkel et al. 2004; Narazaki et al. 2005] for instance.

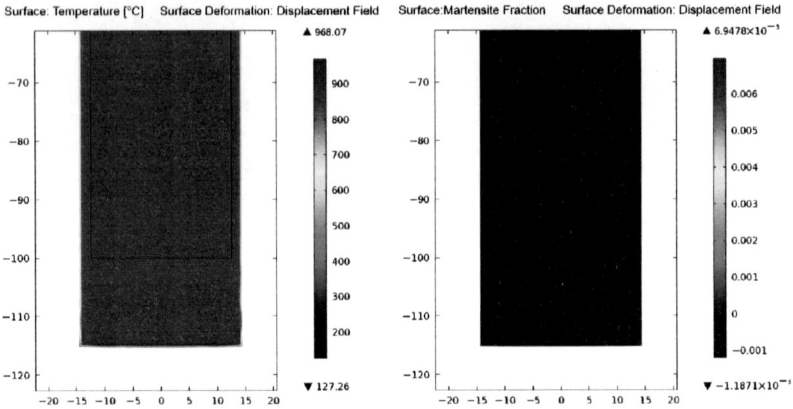

(a) Temperature

(b) Martensite Fraction

Figure 8.4: Workpiece at time $t = 0$s with deformation scale factor of 10.

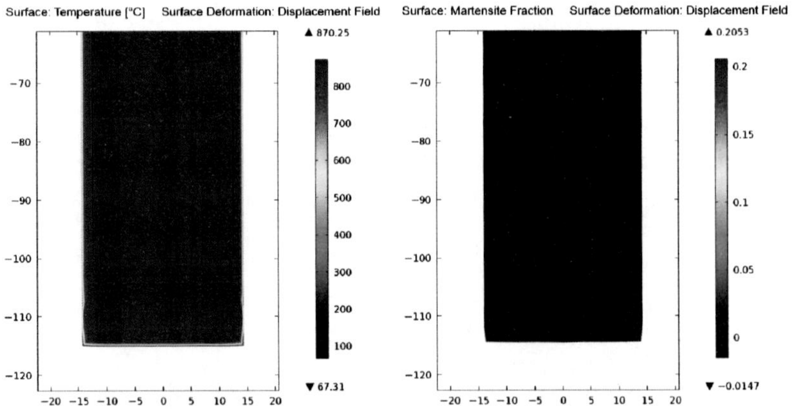

(a) Temperature

(b) Martensite Fraction

Figure 8.5: Workpiece at time $t = 0.2$s with deformation scale factor of 10.

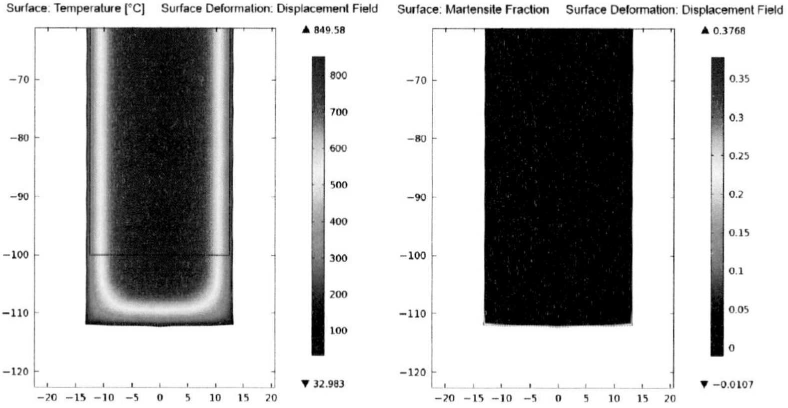

(a) Temperature (b) Martensite Fraction

Figure 8.6: Workpiece at time $t = 2$s with deformation scale factor of 10.

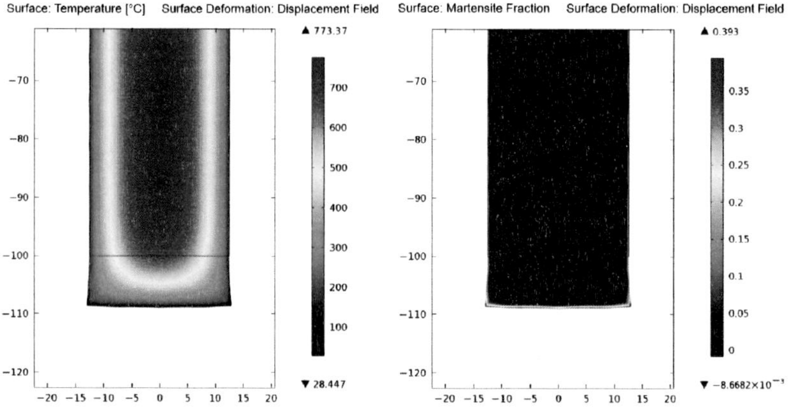

(a) Temperature (b) Martensite Fraction

Figure 8.7: Workpiece at time $t = 5$s with deformation scale factor of 10.

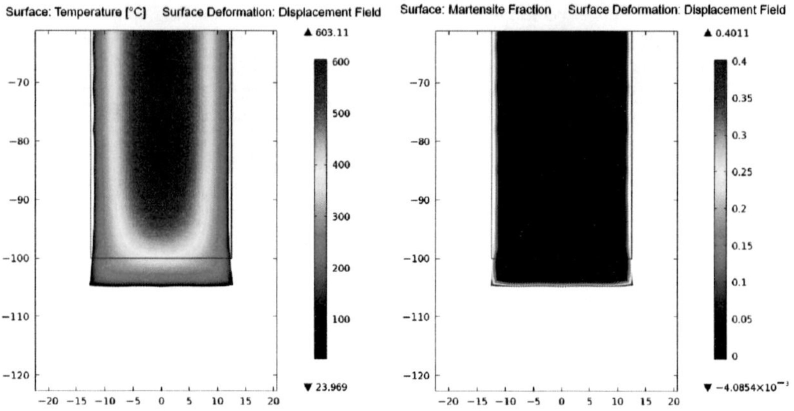

(a) Temperature (b) Martensite Fraction

Figure 8.8: Workpiece at time $t = 10$s with deformation scale factor of 10.

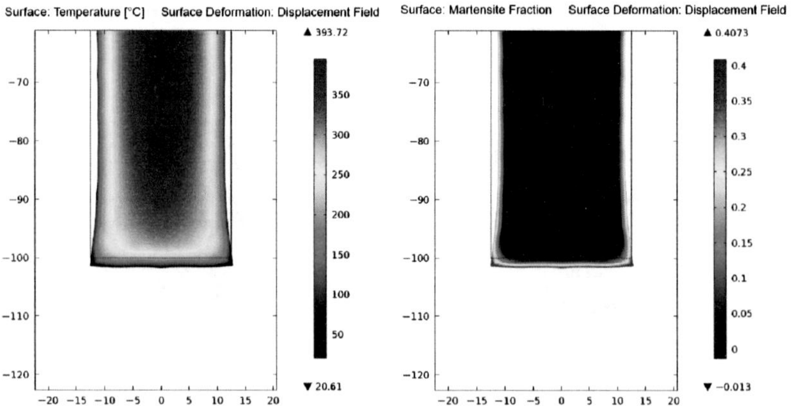

(a) Temperature (b) Martensite Fraction

Figure 8.9: Workpiece at time $t = 20$s with deformation scale factor of 10.

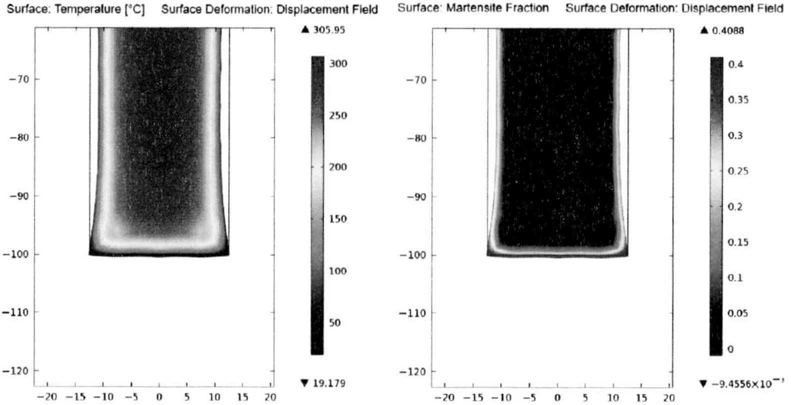

(a) Temperature (b) Martensite Fraction

Figure 8.10: Workpiece at time $t = 30$s with deformation scale factor of 10.

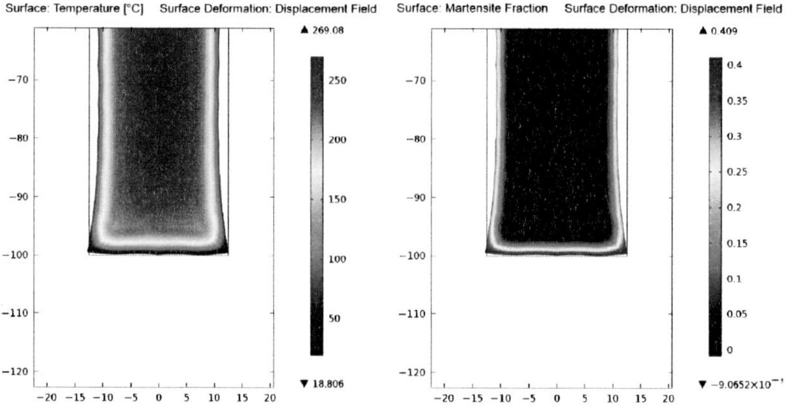

(a) Temperature (b) Martensite Fraction

Figure 8.11: Workpiece at time $t = 40$s with deformation scale factor of 10.

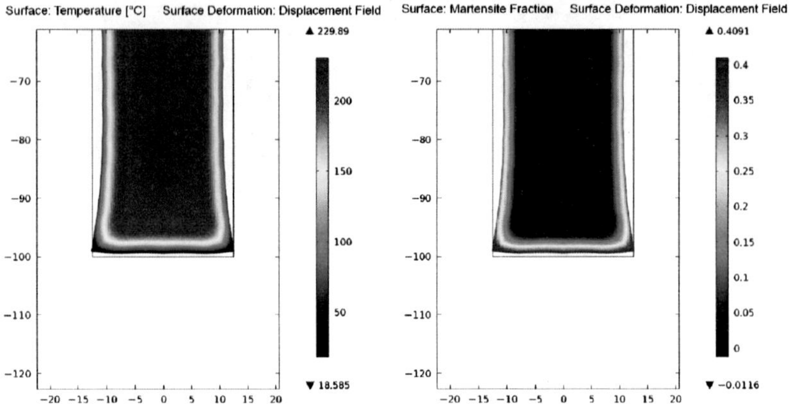

(a) Temperature (b) Martensite Fraction

Figure 8.12: Workpiece at time $t = 50$s with deformation scale factor of 10.

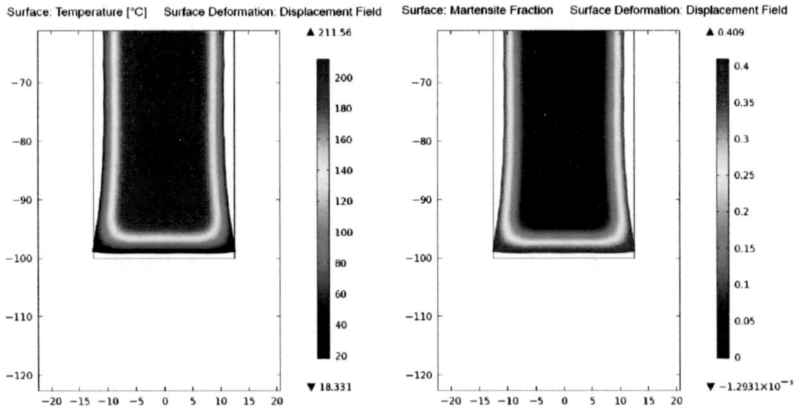

(a) Temperature (b) Martensite Fraction

Figure 8.13: Workpiece at time $t = 100$s with deformation scale factor of 10.

8.4 References and Comments

In this section we give some references for numerical analysis and computations of problems related to the fully coupled Problem (**P**) and the corresponding regularised Problems ($\mathbf{P}_A$), ($\mathbf{P}_{VE}$) and ($\mathbf{P}_{QS}$), cf. Sections 3.1 and 3.3. Of course, this list is not exhaustive.

- Some general (introductory) references for finite element methods are for instance [Atkinson and Han 2001; Grossmann et al. 2005; Knabner and Angermann 2000]. Especially applications of the finite element method in solid mechanics are treated in [Braess 1997]. Whereas [Braess 1997] discusses problems in elasticity, [Dunne and Petrinic 2005; Simo and Hughes 1998; de Souza Neto et al. 2008] are dealing with computational plasticity. Numerical analysis for problems of elasto-plasticity can be found e.g. in [Alberty and Carstensen 2000; Hartmann 1993; Han et al. 1997; Mahnken 1999; Valdman 2001; Wieners 2000].
- Numerical treatment of phase transitions can be found in [Inoue and Wang 1982] in combination with thermo-inelasticity. For the combination with carbon diffusion we refer to [Bergheau et al. 1999], with surface hardening to [Fuhrmann and Hömberg 1999] and the combination with quenching processes is treated in [Réti et al. 2001; Schmidt et al. 2003, 2006; Wolff et al. 2000].
- The numerical finite element simulation using the adaptive hierarchical finite element toolbox ALBERTA for a simplified model, consisting of the coupled system of heat conduction and elasticity equations can be found in the thesis [Suhr 2005]. Adaptive finite element methods with ALBERTA are investigated in the work of [Suhr 2010] for numerical simulations, concerning analysis of distortion for workpieces with and without phase changes (cf. [Schmidt and Siebert 2005] for an introduction to the open-source FEM-toolbox ALBERTA). Throughout the thesis a mathematical model for steel quenching is introduced, a numerical scheme is developed and in simulations for conical rings it is shown that the model is in accordance with the reality. Besides a systematical comparison of different models for phase transition and TRIP which is a continuation of [Suhr 2005], the major contribution is in the field of numerical treatment of the mechanics of the problem. For the classical plasticity, two hardening models are introduced for combined isotropic and kinematic hardening, which both include a coupling with phase transitions and TRIP. Numerical treatment of such models cannot be found in the literature so far. A semi-implicit numerical scheme is derived for both models. In order to identify the model parameters for the more complex hardening model an optimisation procedure is developed. This work was also part of the SFB 570.

- The papers [Chelminski et al. 2007; Hömberg and Kern 2009] and the thesis [Kern 2011] are concerned with thermo-mechanical modelling and numerical treatment of metallurgical phase transitions in steel during quenching. Their model is similar to Problem ($\mathbf{P}_{QS}$) without classical plasticity and mixed boundary conditions, but it still captures the effect of TRIP. The implementation is done within the finite element framework provided by the WIAS-toolbox p∂elib (cf. [Fuhrmann et al. 1998]) using a semi-implicit approach for time-stepping. The resulting code is applied to an experimental setup within the SFB 570 in order to investigate the effect of inhomogeneous quenching strategies of roller bearing rings. Moreover, a strategy for distortion compensation by means of a gradient method obtained from optimal control theory is introduced.
- A semi-implicit algorithm (predictor-corrector approach) for incorporating the interaction between (classical) plasticity and TRIP in steel is developed in [Wolff et al. 2011b]. Contrary to the usual elasto-plasticity, the underlying model of material behaviour of steel is far more complex, comparable with Problem (**P**).

Chapter 9

Discussion and Outlook

In this thesis we introduced and investigated a mathematical model for steel quenching. In Chapters 2 – 8 modelling, analysis and simulation were presented for a model of linear thermo-elasto-plasticity with phase transitions and TRIP. The main results are the proofs of the unique existence via fixed-point argumentation of a (global-in-time) weak solution of the regularised IBVPs ($\mathbf{P}_A$),($\mathbf{P}_{VE}$) and ($\mathbf{P}_{QS}$) (of the corresponding fully coupled problem ($\mathbf{P}$), cf. Sections 3.1 and 3.3) under suitable conditions taking into account mixed BCs for different settings. In this work the following issues were covered:

Steklov Regularisation The proof of the unique existence of a weak solution for the full problem of regularised linear thermo-elasto-plasticity coupled with TRIP and phase transitions was given in Chapter 5. In the full setting the regularisation replaces the first time derivative of the displacement vector by the difference quotient in the variational inequality (differential inclusion) for the plastic flow law as well as in the dissipation term in the heat equation. Moreover, the temperature is replaced by the Steklov average in the law of thermo-elasticity in the balance equation of momentum.
The unique existence of a weak solution for the subproblem of phase transitions and the subproblem of TRIP were proven in Subsections 5.3.4 and 5.3.3. The ODEs for the evolution of the phase fractions were solved by applying the Banach fixed-point theorem. Besides the existence of a unique global solution of the phase fractions, continuous dependence on a parameter, regularity in space, a balance condition and a non-negativity condition were shown. A weak solution of the parabolic equation for the temperature and the coupled system of hyperbolic equations for the displacement was obtained by using the Galerkin approximation (cf. Subsections 5.3.8 and 5.3.6). Furthermore, continuity w.r.t. the data was shown for both equations. In case of the heat equation one obtained in addition the positivity, asymptotic behaviour and additional spatial regularity of the solution (cf. Subsections 5.7.4 and 5.6.8). The improved regularity results for the weak solution of the displacement vector were achieved by using special bases (using

fractional powers of the Lamé operator as test functions, cf. Subsection 5.6.6) and therefore improved regularity for the fully coupled problem followed while using the Banach fixed-point theorem for zero Dirichlet boundary conditions (cf. Section 5.5). Tracking other approaches (cf. e.g. [Mielke 2007]) might also work in special cases and need further attention. Finally, the proof of the weak solvability of the coupled problem for displacement, temperature and phase fractions followed by using Banach's and Schauder's fixed-point theorem.

Visco-elastic Regularisation The proof of the unique existence of a weak solution for the visco-elastic problem of regularised linear thermo-elasto-plasticity coupled with TRIP and phase transitions was provided in Chapter 6. Now the regularised problem means that an additional term including the first time derivative of the displacement vector with a small prefactor appears in the law of thermo-elasticity in the balance equation of momentum. In the literature this approach is called 'parabolic regularisation', 'visco-elastic regularisation' or 'visco-plastic regularisation' if an additional regularisation is used in the plastic flow law. A weak solution of the system of hyperbolic equations for the displacement and the parabolic equation for the temperature was obtained by using the Galerkin approximation. Again, the proof of the weak solvability of the coupled problem for displacement, temperature and phase fractions followed by using Banach's and Schauder's fixed-point theorem.

Quasi-static Problem The proof of the unique existence of a weak solution for the quasi-static problem of linear thermo-elasto-plasticity coupled with TRIP and phase transitions was given in Chapter 7. Now the quasi-static problem means that the second time derivative of the displacement is neglected in the equation of momentum. A weak solution of the system of elliptic equations for the displacement and the parabolic equation for the temperature was obtained by using the Galerkin approximation. Again, the proof of the weak solvability of the coupled problem followed by using Banach's and Schauder's fixed-point theorem.

The main objective of this thesis is the formulation and the analysis of the mathematical problem of TRIP and their interaction with the classical plasticity. However, some of the results are not completely satisfactory and some questions remain to be answered in a possible continuation of this work. Some open problems are discussed below.

Modelling A further continuation at the level of mathematical modelling would be the connection of this problem with micro- and meso-investigations (e.g. micro-models

for TRIP are presented in [Fischer 1990; Diani et al. 1995]), in particular of phase transformations, transformation-induced plastic behaviour and the interaction with classical plasticity, which leads to the method of mathematical homogenisation (cf. [Oleinik et al. 1992; Visintin 2006, 2008; Schweizer 2009; Schweizer and Veneroni 2010] for homogenisation approaches of models in visco-elasticity or elasto-plasticity).
Moreover, dimensional analysis could be helpful to reduce the number of key parameters in order to accomplish qualitative investigations or to prepare approximate calculations (cf. Section 2.8 for an ansatz).
In order to investigate special heat treatment processes, like carburising (or carbonisation, cf. [Liedtke 2008]), the description of the carbon diffusion would be interesting. In this case, the model has to be extended by adding an equation describing the carbon concentration. The following ansatz can be found in [Hüßler 2007; Wolff et al. 2006d]:

$$\begin{aligned} \frac{\partial}{\partial t}(p_1 c_1) - \operatorname{div}(d_{c1} p_1 \nabla c_1) &= -\sum_{i=2}^{m} \frac{\partial}{\partial t}(p_i c_i) && \text{in } \Omega_T, \\ -d_{c1} p_1 \nabla c_1 \cdot \nu_\Gamma &= \delta_{c1} p_1 \mu_c \left(c_1 - \frac{c_\Gamma}{\rho} \right) && \text{in } \Gamma_T, \\ c_1(\mathbf{x}, 0) &= c_1^0(\mathbf{x}) && \text{in } \Omega, \end{aligned}$$

where p_i – phase fraction of the ith phase, c_i – carbon concentration of the ith phase, c_Γ – carbon concentration of the surrounding medium and $\rho, d_{c1}, \delta_{c1}, \mu_c, c_1^0$ appropriate parameters (constants). PDEs of this type are sometimes called degenerate parabolic (or pseudo-parabolic) equations and treated in [Meier 2008] or in [Glitzky and Hünlich 2006; Haller-Dintelmann and Rehberg 2008; Hieber and Rehberg 2008] in a general context.
More complex phenomena, like additional inelastic dissipation, isotropic and kinematic hardening, which leads to additional back-stresses and accumulated strains, stress-dependent transformation behaviour and parameters depending on the temperature and the phase fractions, could be taken into account. The investigation of more general saturation functions (cf. Example 2.5.1) or more general evolution equations for the phase fractions p_i, like

$$p_i'(t, \mathbf{x}) = \gamma_i(t, \mathbf{x}, \mathbf{p}, \theta, \theta', \boldsymbol{\sigma}_{vM}, \boldsymbol{\sigma}_m, \boldsymbol{\xi})$$

could also be discussed.
Finally, in the context of heat treatment, creep (cf. [Altenbach 2010; Naumenko and Altenbach 2007; Bökenheide et al. 2011]) and damage (e.g. fatigue, ratcheting, friction (cf. [Abdel-Malek and Meyer 2007; Amassad et al. 1999b; Chau 2006; Ellyin 1997; Han et al. 2001; Lemaitre and Desmorat 2005])) may be relevant and need to be investigated (cf. [Wolff et al. 2011a]). To describe phase transformations during the complete heating and cooling process cf. e.g. the ansatz in [Surm et al. 2008].

Analysis An important objective of one possible continuation of this work is to obtain better regularity results for the solution of the full problem for mixed boundary conditions in order to get rid of the various regularisations used in the different settings. A restriction for better regularity results might be simplified boundary conditions or local solvability of the full problem.
In the mathematical and analytical part of this work we used basically energy methods (Galerkin approximation) and Banach and/or Schauder fixed-point argumentation in the L^2-setting for the investigation of the differential equations. The application of the Rothe method, semi-group methods or various fixed-point principles in a different L^p-setting, $p \in]1, \infty[$, might provide new insights.

Simulation The numerical simulation should be extended in order to incorporate additional forming phases like ferrite and bainite. Moreover, the simulation of other geometries (especially those that are relevant in SFB 570 such as conical rings, shafts and gearwheels) should be taken into account.
An complex implementation (3D calculations in ALBERTA or MATLAB™ with real data and comparison with commercial software packages like ABAQUS™, ANSYS™, COMSOL™ or SYSWELD™ and the comparison of the simulation results with experimental data provided by the Institute for Materials Science in Bremen) could follow up this work as a larger project (cf. [Kern 2011; Suhr 2010] as an ansatz).

Application In addition, special heat treatment processes such as nitriding or nitrocarburising (cf. e.g. [Hoffmann et al. 1997, 2001; Hoffmann 1981; Liedtke 2005a]) could be investigated in order to continue the works [Hüßler 2007; Tijani 2008] to a certain extent.
Based on the experimental setup in the SFB 570, the effect of inhomogeneous quenching strategies could be investigated and the numerical computations should be compared with the experimental results in order to follow up the work [Chelminski et al. 2007; Hömberg and Kern 2009; Kern 2011] to some extent. Therefore, a strategy for distortion compensation by means of a optimisation method obtained from optimal control theory has to be introduced.
Possible extensions are also testing and evaluating models for phase transformations, TRIP and stress-dependent transformation behaviour for under-eutectoid steels with different carbon content as well as the evaluation of relevant creep models based on measured data (cf. [Bökenheide et al. 2011; Wolff et al. 2011a]).

In summary, the results give a theoretical basis for further mathematical investigation or the efficient implementation of numerical algorithms suitable for real-world applications.

Appendix A

Function Spaces

We recall the (standard) definitions of function spaces used in this thesis and collect a few important results for the proofs in Chapters 5 – 7 from [Renardy and Rogers 1996]. Properties of Lebesgue spaces are discussed in [Kolmogorov and Fomin 1970; Natanson 1975; Siddiqi 2004]. For details on Sobolev spaces we refer to [Adams and Fournier 2003; Amann 2000; Maz'ya 1985; Naumann 2005b; Simon 1986; Tartar 2000]. Further information on Banach space-valued functions in can be found in [Naumann 2005a; Wloka 1967, 1987; Zeidler 1990a].

A.1 Spaces of Continuously Differentiable Functions

Let $\Omega \subset \mathbb{R}^n$ ($n \leq 3$ for most applications) be a bounded domain. We denote by $C(\Omega)$ the space of all real-valued functions that are continuous on Ω. We denote further by $C(\overline{\Omega})$ the space of functions that are bounded and uniformly continuous on Ω. The space $C(\overline{\Omega})$ is a Banach space with the norm

$$\|u\|_{C(\overline{\Omega})} = \sup\{|u(\mathbf{x})| : \mathbf{x} \in \Omega\}.$$

For any non-negative integer m, $C^m(\Omega)$ is defined to be the space of functions that together with their derivatives of order less than or equal to m are continuous, i.e.

$$C^m(\Omega) = \{u \in C(\Omega) : \mathrm{D}^\alpha u \in C(\Omega) \text{ for } |\alpha| \leq m\}.$$

Analogously

$$C^m(\overline{\Omega}) = \{u \in C(\overline{\Omega}) : \mathrm{D}^\alpha u \in C(\overline{\Omega}) \text{ for } |\alpha| \leq m\}.$$

We write $C(\Omega)$ and $C(\overline{\Omega})$ instead of $C^0(\Omega)$ and $C^0(\overline{\Omega})$. The space is a Banach space when endowed with the norm

$$\|u\|_{C(\overline{\Omega})} = \sum_{|\alpha| \leq m} \|\mathrm{D}^\alpha u\|_{C(\overline{\Omega})}.$$

Finally,

$$C^\infty(\Omega) = \{u \in C(\Omega) : \; u \in C^m(\Omega) \; \forall \, m \in \mathbb{N}\}$$

and

$$C^\infty(\overline{\Omega}) = \{u \in C(\overline{\Omega}) : u \in C^m(\overline{\Omega}) \; \forall \, m \in \mathbb{N}\}.$$

These are spaces of infinitely differentiable functions.

A.2 Lebesgue Spaces

For any number $p \in [1, \infty[$, we denote by $L^p(\Omega)$ the space of (equivalence classes of) Lebesgue-measurable functions u for which

$$\int_\Omega |u(\mathbf{x})|^p \, \mathrm{d}\mathbf{x} < \infty,$$

where integration is to be understood in the sense of Lebesgue. The space $L^p(\Omega)$ is a Banach space when endowed with the norm

$$\|u\|_{L^p(\Omega)} = \left(\int_\Omega |u(\mathbf{x})|^p \, \mathrm{d}\mathbf{x} \right)^{\frac{1}{p}}.$$

The definition of the spaces $L^p(\Omega)$ can be extended to include the case $p = \infty$. We define the essential supremum of any Lebesgue-measurable function by

$$\operatorname*{ess\,sup}_\Omega u = \inf \{M \in]-\infty, \infty] : u(\mathbf{x}) \leq M \, \text{a.e. in } \Omega\}.$$

Then u is said to be essentially bounded above if $\operatorname{ess\,sup}_\Omega u < \infty$. A similar definition of essential infimum leads to the notion of a function that is essentially bounded below. We say that u is essentially bounded if both infima are finite. Then we may define

$$L^\infty(\Omega) = \{u : u \text{ is essentially bounded on } \Omega\}.$$

This space is a Banach space when endowed with the norm

$$\|u\|_{L^\infty(\Omega)} = \operatorname*{ess\,sup}_\Omega |u|.$$

Since all continuous functions on a bounded closed set are bounded, we have

$$C(\overline{\Omega}) \hookrightarrow L^\infty(\Omega).$$

Let u be a function defined on Ω. We say that $u \in L^p_{loc}(\Omega)$ if for any proper subset $\Omega' \subset\subset \Omega$, $u|_{\Omega'} \in L^p(\Omega')$.

A.3 Sobolev Spaces

For any non-negative integer m and real number $p \geq 1$ or $p = \infty$ we define

$$W^{m,p}(\Omega) = \left\{ u \in L^p(\Omega) : \mathrm{D}^\alpha u \in L^p(\Omega) \text{ for any } \alpha \in \mathbb{Z}_+^n \text{ with } |\alpha| \leq m \right\},$$

where derivatives are taken in distributional sense. Norms in the spaces $W^{m,p}(\Omega)$ are defined by

$$\text{(A.1)} \qquad \|u\|_{W^{m,p}(\Omega)} = \left(\sum_{|\alpha| \leq m} \|\mathrm{D}^\alpha u\|_{L^p(\Omega)}^p \right)^{\frac{1}{p}}, \qquad p \in [1, \infty[$$

$$\text{(A.2)} \qquad \|u\|_{W^{m,\infty}(\Omega)} = \max_{|\alpha| \leq m} \|\mathrm{D}^\alpha u\|_{L^\infty(\Omega)}.$$

With the norms defined above, the space $W^{m,p}(\Omega)$ is a Banach space. The space $W^{m,p}(\Omega)$ is reflexive if and only if $p \in]1, \infty[$. We note that $W^{0,p}(\Omega) = L^p(\Omega)$.

A.4 Banach Space-Valued Functions[1]

When dealing with initial-boundary value problems, it makes some sense to treat functions of space and time as maps from a time interval into a Banach space such as those that have been discussed before. To begin, let X, Y be two Banach spaces, T a positive number and $S :=]0, T[$ a finite time interval; then the space $C^m(\bar{S}; X)$ ($m \in \mathbb{N}$) consists of all continuous functions $u : \bar{S} \to X$ that have derivatives of order less than or equal to m. This is a Banach space when endowed with the norm

$$\|u\|_{C^m(\bar{S};X)} = \sum_{k=0}^{m} \max_{t \in \bar{S}} \|u^{(k)}(t)\|_X,$$

where $u^{(k)}(t)$ denotes the kth time derivative of u. We write $C(\bar{S}; X)$ for the case $m = 0$.

Turning to Lebesgue spaces, for $p \in [1, \infty[$ the space $L^p(S; X)$ consists of all Bochner-measurable functions[2] $u : \bar{S} \to X$ for which

$$\|u\|_{L^p(S;X)} := \left(\int_0^T \|u(t)\|_X^p \, \mathrm{d}t \right)^{\frac{1}{p}} < \infty.$$

[1] In some references the corresponding function space is also called 'Lebesgue Space of Vector-Valued Functions' or simply 'Space of Abstract Functions'.

[2] A Bochner-measurable function taking values in a Banach space is a function that equals a.e. the limit of a sequence of step functions.

This is a Banach space, provided that the members are understood to represent equivalence classes of functions that are equal a.e. on S. The extension of this definition to include the case $p = \infty$ is carried out in the usual way: The space $L^\infty(S;X)$ consists of all Bochner-measurable functions $u : \bar{S} \to X$ that are essentially bounded. This is a Banach space with the norm

$$\|u\|_{L^\infty(S;X)} := \operatorname*{ess\,sup}_{t\in\bar{S}} \|u(t)\|_X.$$

If X is a Hilbert space with inner product $(\cdot,\cdot)_X$, then $L^2(S;X)$ is a Hilbert space with the inner product

$$(u,v)_{L^2(S;X)} = \int_0^T (u(t),v(t))_X \,dt.$$

The following theorem summarizes some properties of these spaces.

A.4.1 Theorem

Let $p \in [1,\infty[$. Then

(1) $C(\bar{S};X)$ is dense in $L^p(S;X)$ and the embedding is continuous.

(2) If $X \hookrightarrow Y$, then $L^p(S;X) \hookrightarrow L^q(S;Y)$ for $1 \le q \le p \le \infty$.

Proof. See [Zeidler 1990a]. □

Let X^* be the topological dual of a separable normed space X. Then for $p \in]1,\infty[$, the dual space of $L^p(S;X)$ is given by

$$[L^p(S;X)]^* = L^q(S;X^*) \qquad \text{with} \qquad \frac{1}{p} + \frac{1}{q} = 1.$$

Furthermore, if X is reflexive, then so is $L^p(S;X)$ for $p \in]1,\infty[$.

It is necessary to define in an appropriate way derivatives w.r.t. the time variable for functions that lie in the spaces $L^p(S;X)$. The approach is similar to that taken in the case of generalised derivatives of distributions; that is, we take as a starting point the elementary integration by parts formula

$$\int_0^T \Phi^{(m)}(t)u(t)\,dt = (-1)^m \int_0^T \Phi(t)u^m(t)\,dt,$$

which holds f.a. functions $\Phi \in C_0^\infty(S)$ and $u \in C^m(\bar{S};X)$. A function $u \in L^1_{loc}(S;X)$ is then said to possess an mth generalised derivative $u^{(m)} \in L^1_{loc}(S;Y)$ if there exists a function $v \in L^1_{loc}(S;Y)$ s.t.

$$\text{(A.3)} \qquad \int_0^T \Phi^{(m)}(t)u(t)\,dt = (-1)^m \int_0^T \Phi(t)v(t)\,dt, \qquad \forall\, \Phi \in C_0^\infty(S),$$

where X and Y are appropriate Banach spaces. When (A.3) hold, we simply write $v = u^{(m)}$. The following lemma gives an important property of generalised derivatives.

A.4.2 Lemma

Let V be a reflexive Banach space and let H be a Hilbert space with the property that $V \hookrightarrow H \hookrightarrow V^*$, the continuous embedding $V \hookrightarrow H$ being dense.[3] *Let $p, q \in [1, \infty]$ be conjugate exponents. Then any function $u \in L^p(S; V)$ possesses a unique generalised derivative $u^{(m)} \in L^q(S; V^*)$ if and only if there is a function $v \in L^q(S; V^*)$ s.t.*

$$\int_0^T (u(t), w)_H \, \Phi^{(m)}(t) \, \mathrm{d}t = (-1)^m \int_0^T \Phi(t) \, \langle v(t), w \rangle_{V^*V} \, \mathrm{d}t$$

f.a. $w \in V$, $\Phi \in C_0^\infty(S)$. Then $u^{(m)} = v$ and f.a.a. $t \in S$

$$\frac{\mathrm{d}^m}{\mathrm{d}t^m} (u(t), w)_H = \langle v(t), w \rangle_{V^*V} \qquad \text{f.a. } w \in V.$$

Proof. See [Zeidler 1990a]. □

For an integer $m \geq 0$ and a real number $p \geq 1$ we define by $W^{m,p}(S; X)$ the space of functions $f \in L^p(S; X)$ s.t. $f^{(i)} \in L^p(S; X)$, $i \leq m$. This is a Banach space with the norm

$$\|f\|_{W^{m,p}(S;X)} = \left(\sum_{i=0}^{m} \|f^{(i)}\|^p_{L^p(S;X)} \right)^{\frac{1}{p}}.$$

If X is a Hilbert space, $W^{m,2}(S; X)$ is also a Hilbert space with the inner product

$$(f, g)_{W^{m,p}(S;X)} = \int_0^T \sum_{i=0}^{m} \left(f^{(i)}(t), g^{(i)}(t) \right)_X \mathrm{d}t.$$

We note the fundamental inequality

$$\|f(t) - f(s)\|_X \leq \int_s^t \|f'(\tau)\|_X \, \mathrm{d}\tau,$$

which holds for $s < t$ and $f \in W^{1,p}(S; X)$, $p \geq 1$. On several occasions we will also need the continuous embedding property

$$W^{1,2}(S; X) \hookrightarrow C(\bar{S}; X).$$

[3] This relation is usually called evolution triple or Gelfand triple, cf. e.g. [Zeidler 1990a] for details.

In particular, there exists a constant $c > 0$ s.t.

$$\|u\|_{C(\bar{S};X)} \leq c\, \|u\|_{W^{1,2}(S;X)} \qquad \forall\, u \in W^{1,2}(S;X).$$

We will also need the property

$$C^\infty(\bar{S};X) \quad \text{is dense in} \quad W^{1,2}(S;X).$$

Moreover, we define

$$W^{1,p}(S;X,Y) := \Big\{u \in L^p(S;X) : u' \in L^q(S;Y)\Big\}, \quad \text{for } 1 < p < \infty,\ \tfrac{1}{p} + \tfrac{1}{p} = 1.$$

A.4.3 Lemma
Let $V \hookrightarrow H \hookrightarrow V^*$ be an evolution triple and let $p, q \in]1,\infty[$ be conjugate exponents. Then the following hold:

(1) The space $W^{1,p}(S;V,V^*)$, i.e. the set of all $u \in L^p(S;V)$ that have generalised derivatives $u' \in L^q(S;V^*)$, is a Banach space with the norm

$$\|u\|_{W^{1,p}(S;V,V^*)} := \|u\|_{L^p(S;V)} + \|u'\|_{L^q(S;V^*)}.$$

(2) The embedding

$$W^{1,p}(S;V,V^*) \hookrightarrow C(\bar{S};H)$$

is continuous. Further, there exists a constant $c > 0$ s.t.

$$\|u\|_{C(\bar{S};H)} \leq c\, \|u\|_{W^{1,p}(S;V,V^*)} \qquad \forall\, u \in W^{1,p}(S;V,V^*).$$

(3) For all $u, v \in W^{1,p}(S;V,V^*)$ and arbitrary $s, t \in S$, $0 \leq s \leq t \leq T$, the following generalised integration by parts formula holds:

$$(u(t),v(t))_H - (u(s),v(s))_H = \int_s^t \langle u'(\tau), v(\tau)\rangle_{V^*V} + \langle v'(\tau), u(\tau)\rangle_{V^*V}\, \mathrm{d}\tau.$$

Proof. See [Zeidler 1990a]. □

A.5 Inequalities

A.5.1 Lemma (Generalised Gronwall Inequality)
Let the functions $v(t)$, $w(t)$, $v(t)u(t)$ and $w(t)u^p(t)$ be locally integrable non-negative functions on S. If $u_0 > 0$ and $p \geq 0$, $p \neq 1$, then the inequality

$$u(t) \leq u_0 + \int_0^t v(s)u(s)\,\mathrm{d}s + \int_0^t w(s)u^p(s)\,\mathrm{d}s$$

implies that

$$u(t) \le \exp\left(\int_0^t v(s)\,\mathrm{d}s\right)\left\{u_0^{1-p} + (1-p)\int_0^t w(s)\exp\left((p-1)\int_0^s v(r)\,\mathrm{d}r\right)\mathrm{d}s\right\}^{\frac{1}{1-p}}$$

for $t \in S$.

Proof. See [Willett and Wong 1965]. □

A.5.2 Lemma (Generalised Discrete Gronwall Inequality)
Suppose that v_n, w_n *and* u_{n+1} ($n \in \mathbb{N}$) *are non-negative sequences of numbers with* $v_0 = w_0 = 0$ *and that* u_0 *and* p *are constants with* $u_0 > 0$ *and* $p \ge 0$, $p \ne 1$. *Then the inequality*

$$u_{n+1} \le u_0 + \sum_{i=0}^{n} v_i u_i + \sum_{i=0}^{n} w_i u_i^p \tag{A.4}$$

implies that

$$u_{n+1} \le \prod_{i=0}^{n}(1+v_i)\left\{u_0^p + (1-p)\sum_{j=0}^{n} w_j \prod_{k=0}^{j}(1+v_k)^{p-1}\right\}^{\frac{1}{1-p}} \tag{A.5}$$

for $n \in \mathbb{N}$.

Proof. See [Willett and Wong 1965]. □

A.5.3 Remark (Gronwall-Bihari-Bellmann Type)
The (standard) Gronwall's inequality can be found in [Emmrich 2004; Renardy and Rogers 1996]. For various variants and generalisations, sometimes also called 'integral inequalities of Gronwall-Bihari-Bellmann type', see e.g. [Böhm 1992; Babolian and Shaerlar 2011; Dhongade and Deo 1973, 1976; Lakshmikantham 1973; Pachpatte 1973, 1975a,b] and the references therein.

A.5.4 Lemma (Minkowski's Integral Inequality)
Let $p \in [1, \infty[$. *If* $u \in L^1(S; L^p(\Omega))$, *then*

$$\left(\int_\Omega \left|\int_S u(t,x)\,\mathrm{d}t\right|^p \mathrm{d}x\right)^{\frac{1}{p}} \le \int_S \left(\int_\Omega |u(t,x)|^p\,\mathrm{d}x\right)^{\frac{1}{p}} \mathrm{d}t.$$

Proof. See [Adams and Fournier 2003]. □

A.5.5 Lemma (Interpolation Inequality)
Let $\Omega \subset \mathbb{R}^n$ be a bounded $\mathcal{C}^k$-smooth domain. Then for indices $p, p_1, p_2 \in]1, +\infty[$, $k, k_1, k_2 \geq 0$, $\lambda \in]0, 1[$ that satisfy $k = \lambda k_1 + (1-\lambda)k_2$, $\frac{1}{p} = \frac{\lambda}{p_1} + \frac{1-\lambda}{p_2}$ and any function $u \in W^{k_1,p_1}(\Omega) \cap W^{k_2,p_2}(\Omega)$ there holds $u \in W^{k,p}(\Omega)$, and the interpolation estimate

$$\|u\|_{W^{k,p}(\Omega)} \leq c\,\|u\|^{\lambda}_{W^{k_1,p_1}(\Omega)}\,\|u\|^{1-\lambda}_{W^{k_2,p_2}(\Omega)}$$

is valid with a constant c independent of u.

Proof. This Lemma can be concluded from Theorem 6.4.5 in [Bergh and Löfström 1976]. More details on embedding and interpolation theorems can be found in [Amann 2000, 1993; Böhm 1992; Canavati and Galaz-Fontes 1990; Eck 2004; Griepentrog et al. 2002; Maligranda and Persson 1993; Mazja and Shaposhnikova 2009]. □

A.5.6 Lemma (Multiplicator Theorem)
Assume that Ω has the cone property (e.g. Ω is a bounded Lipschitz domain), and that $p \geq r$, $q \geq r$ and

$$\frac{m}{n} > \frac{1}{p} + \frac{1}{q} - \frac{1}{r}.$$

If the volume of Ω is infinite, assume further that $mp \leq n$ when $q \neq r$, that $mq \leq n$ when $p \neq r$ and that

$$\frac{m-1}{n} \leq \frac{1}{p} + \frac{1}{q} - \frac{1}{r} \qquad \text{when} \qquad p \neq r, \qquad q \neq r.$$

Then, if $u \in W^{m,p}(\Omega)$ and $v \in W^{m,q}(\Omega)$, we have $uv \in W^{m,r}(\Omega)$, and there exists a positive number c independent of u and v s.t.

$$\|uv\|_{W^{m,r}(\Omega)} \leq c\,\|u\|_{W^{m,p}(\Omega)}\,\|v\|_{W^{m,q}(\Omega)}.$$

Proof. See [Valent 1985]. □

Appendix B

Elliptic Systems

Let $\Omega \subset \mathbb{R}^n$ be a smooth, bounded domain and let $\Gamma_1 \subset \partial\Omega$ with $\text{meas}(\Gamma_1) > 0$, $\Gamma_2 := \partial\Omega \setminus \Gamma_1$ with outward normal ν be given. Consider the elliptic mixed boundary-value problem (MVBP), cf. [Bensoussan and Frehse 2002] :

(B.1) $\quad -\mathrm{D}_i(a^i_j(x, \mathrm{D}u)) = f^j \quad$ in $\quad \Omega, \quad j = 1, \ldots, N,$

(B.2) $\quad u^j = b^j_1 \quad$ on $\quad \Gamma_1,$

(B.3) $\quad -a^i_j(x, \nabla u)\, \nu^j(x) = b^j_2 \quad$ on $\quad \Gamma_2,$

where $u = (u^1, \cdots, u^N)$.

B.0.1 Remark (Lamé-Navier Equations)
Such systems are motivated in particular by the theory of nonlinear elasticity. Consider the Lamé-Navier equations

$$-\mu\Delta u - (\lambda + \mu)\,\nabla(\text{div}(u)) = f,$$

which describe the displacement $u(x)$ of the point x of an elastic body Ω under a volume force f. This system is an example of (B.1), *with*

$$a^i_j(x, \mathrm{D}u) := \mu \mathrm{D}_i u^j + (\lambda + \mu)\mathrm{D}^j u^i.$$

The properties of the difference quotient and the Steklov average are discussed in [Naumann 2005a; Meier 2008; Ladyženskaya et al. 1968; Wolf 2002]. For convenience, we summarise the results which are needed in Chapter 5 in the next two sections. Finally, we deal with existence and uniqueness of the second order elliptic MBVP (B.1)–(B.3).

B.1 Difference Quotient

Let X be a Banach space with dual X^* and $T > h > 0$.

B.1.1 Definition (Difference Quotients)
For functions $u : [0, T+h] \to X$ and $v : [-h, T] \to X$ we denote the right- and left-hand-side difference quotients by

$$\mathcal{D}_h u(t) := \frac{1}{h}\Big(u(t+h) - u(t)\Big), \qquad t \in S \qquad \text{and}$$
$$\mathcal{D}_{-h} v(t) := \frac{1}{h}\Big(v(t) - v(t-h)\Big), \qquad t \in S.$$

B.1.2 Lemma (Properties of the Difference Quotient)
Let $p \geq 1$.

(1) For $u \in L^2(-h, T; X^*)$ and $v \in L^2(-h, T; X)$ it holds

$$\int_0^T \langle u(t), \mathcal{D}_{-h} v(t)\rangle \,\mathrm{d}t = -\int_{-h}^{T-h} \langle \mathcal{D}_h u(t), v(t)\rangle \,\mathrm{d}t + \langle u, v\rangle\big|_{-h}^{T-h}.$$

(2) Let $u \in W^{1,p}(S; X)$. For each $\delta > 0$ we have

$$\mathcal{D}_h u \to \frac{\partial}{\partial t} u \quad \text{in} \quad L^p(0, T-\delta; X), \qquad \text{for} \quad h \to 0.$$

Proof. See [Meier 2008]. □

B.2 Steklov Average

B.2.1 Definition (Steklov Average)
For $u \in L^1(0, T+h; X)$ we define the (right-hand-side) Steklov average of u via

$$\mathcal{S}_h u(t) := \frac{1}{h}\int_t^{t+h} u(s)\,\mathrm{d}s, \qquad t \in S.$$

For $v \in L^1(-h, T; X)$ the (left-hand-side) Steklov average of v is defined via

$$\mathcal{S}_{-h} v(t) := \frac{1}{h}\int_{t-h}^{t} v(s)\,\mathrm{d}s, \qquad t \in S.$$

B.2.2 Lemma (Properties of the Steklov Average)
Let $p \geq 1$.

(1) If $u \in L^p(0, T+h; X)$, then $\mathcal{S}_h u \in W^{1,p}(S; X)$ and $\frac{\partial}{\partial t}\Big(\mathcal{S}_h u\Big) = \mathcal{D}_h u$.

(2) For $u \in L^2(-h, T; X^*)$ and $v \in L^2(-h, T; X)$ it holds

$$\int_0^T \langle u(t), \mathcal{S}_{-h} v(t)\rangle \,\mathrm{d}t = \int_{-h}^{T-h} \langle \mathcal{S}_h u(t), v(t)\rangle \,\mathrm{d}t + \frac{1}{h}\int_{T-h}^{T}\int_s^T \langle u(t), v(s)\rangle \,\mathrm{d}t\,\mathrm{d}s.$$

(3) Let $u \in L^p(S;X)$. For each $\delta > 0$ we have

$$\mathcal{S}_h u \to u \quad \text{in} \quad L^p(0, T-\delta; X), \qquad h \to 0.$$

(4) If $u \in C(S;X)$, then for each $\delta > 0$ we have

$$\mathcal{S}_h u \to u \quad \text{in} \quad C([0, T-\delta]; X), \qquad h \to 0.$$

Proof. See [Meier 2008]. □

B.3 Existence and Uniqueness

Using the theory of monotone operators, we can show that the following lemma is valid (cf. [Agmon et al. 1964; Bensoussan and Frehse 2002; Chen et al. 1998; Wloka et al. 1995]).

B.3.1 Lemma (Existence and Uniqueness)
Let Ω be a bounded Lipschitz domain, $\Gamma_1 \subset \partial\Omega$, $\Gamma_2 := \partial\Omega \setminus \Gamma_1$ as above, $A = (a_i^j(\cdot,\cdot))_{i,j}$ be symmetric, elliptic and bounded, $f \in [W^{-1,2}(\Omega)]^N$, $b_1 \in [W^{\frac{1}{2},2}(\Gamma_1)]^N$ and $b_2 \in [W^{-\frac{1}{2},2}(\Gamma_2)]^N$. Then the unique solution u of problem (B.1) – (B.3) is an element of $[W^{1,2}(\Omega)]^N$ and the a-priori estimate

$$\|u\|_{[W^{1,2}(\Omega)]^N} \leq c\Big(\|f\|_{[W^{-1,2}(\Omega)]^N} + \|b_1\|_{[W^{\frac{1}{2},2}(\Gamma_1)]^N} + \|b_2\|_{[W^{-\frac{1}{2},2}(\Gamma_2)]^N}\Big)$$

holds with a constant c that depends on Ω, the lower and upper bounds of A and the modulus of continuity of A.

Proof. See [Boettcher 2007] for the special situation of Remark B.0.1. The result follows as an application of the Lax-Milgram theorem (cf. [Drivaliaries and Yannakakis 2007] for a generalised Lax-Milgram theorem). □

B.3.2 Remark (Regularity)
There exists a vast amount of literature on existence, uniqueness and regularity for second order elliptic equations *(cf. [Amann 1993; Ebmeyer and Frehse 1997; Ladyženskaya et al. 1968; Giaquinta 1978; Gilbarg and Trudinger 2001; Renardy and Rogers 1996; Roubíček 2005; Savaré 1997; Shamir 1968; Skrypnik 1986; Wloka 1987; Wolf 2002; Wloka et al. 1995; Zeidler 1990a,b]), but there are only a few references dealing with existence and uniqueness results for* second order elliptic systems *(cf. [Clark et al. 1998; Chen et al. 1998; Ebmeyer 1999; Escher 1992; Giaquinta and Modica 1979; Grisvard 1985; Mazja*

et al. 1991; Maz'ya et al. 1991; Wloka et al. 1995]). Especially for regularity results of second order elliptic systems exists very few literature (cf. [Bensoussan and Frehse 2002; Kaßmann and Madych 2004]) and there are hardly any references for regularity for second order elliptic mixed boundary value systems (cf. [Bensoussan and Frehse 2002; Kaßmann and Madych 2004]). In [Bensoussan and Frehse 2002] the problem (B.1) – (B.3) is considered on a cube and under appropriate assumptions it is shown that the corresponding solution satisfies $u \in [W^{1+\frac{1-\delta}{2},2}(\Omega)]^N$, $\delta \in]0,1[$ *for* $f \in [L^2(\Omega)]^N$ *and* $b_1 = b_2 = 0$.

Further information about elliptic boundary value problems can be found e.g. in [Agmon et al. 1964; Agmon 1959; Arkeryd 1966; Ebmeyer and Frehse 1997; Gallout and Monier 1999; Gröger 1989, 1994; Haller-Dintelmann et al. 2007; Haller-Dintelmann and Rehberg 2008; Koshelev and Chelkak 1985; Nečas 1983]. Moreover, elliptic-parabolic boundary value problems are investigated in [Griepentrog 1999; Haller-Dintelmann 2008; Kenmochi and Pawlow 1986; Wolf 2002].

Appendix C

Variational Inequalities and Differential Inclusions of Parabolic Type

Here we summarise some results based on the Yosida regularisation for proving existence, uniqueness and regularity of differential inclusions of the form

$$\frac{\partial}{\partial t}u(t) + Au(t) \ni f(t), \qquad\qquad t > 0,$$

where A is a maximal monotone operator.

Let V be a real reflexive Banach space and V^* its dual space. Let $S :=]0, T[$ be a finite time interval. We assume that there exists a real Hilbert space H identified with its dual, s.t. $V \subset H \equiv H^* \subset V^*$, where $V \subset H$ and $H^* \subset V^*$ are both densely and continuously embedded. Hence $\langle u, v\rangle_{V^*V} = (u, v)_H$ hold f.a. $u \in H$ and $v \in V$.

Basics of the theory of monotone operators can be found e.g. in [Aubin and Ekeland 1984; Aubin and Frankowska 1990; Deimling 1980; Roubíček 2005; Showalter 1997; Zeidler 1990b]. An extensive summary for variational analysis is presented in [Rockafellar and Wets 2004].

C.1 Preliminaries on Set-Valued Analysis

A *multivalued map* $A : V \to \mathcal{P}(V^*)$ is called *monotone* if, f.a. $f_1 \in A(u_1)$, $f_2 \in A(u_2)$, it holds that $\langle f_1 - f_2, u_1 - u_2\rangle \geq 0$. We admit $A(u) = \emptyset$ and denote the *effective domain* of A by $D(A) := \{u \in V : A(u) \neq \emptyset\}$. The set $G(A) := \{(u, f) \in V \times V^* : u \in D(A), f \in A(u)\}$ is called the *graph* of A. $A : V \to \mathcal{P}(V^*)$ is called *maximal monotone*, if the monotone multivalued map A is maximal, i.e. if there is no other monotone set-valued map whose graph strictly contains the graph of A. Besides, we call $A : V \to \mathcal{P}(V^*)$

coercive if

$$\lim_{\|u\|\to\infty} \inf_{f\in A(u)} \frac{\langle f, u\rangle}{\|u\|} = +\infty. \tag{C.1}$$

Moreover, $\Phi : V \to \mathbb{R} \cup \{+\infty\}$ is called *proper* if it is not identically equal to $+\infty$ and does not take the value $-\infty$. In the case that Φ is convex, $A : V \to \mathcal{P}(V^*)$ will be the *subdifferential* of Φ, i.e. $A := \partial\Phi$, defined by

$$\partial\Phi(u) := \{f \in V^* : \forall v \in V : \Phi(u) - \Phi(v) \leq \langle f, u - v\rangle\}, \qquad u \in V.$$

C.1.1 Remark (Subdifferential Calculus)
The concepts of subderivative, subgradient and subdifferential arise in convex analysis as a generalisation of the (usual) derivative, gradient and differential. Good references for subdifferential calculus are e.g. [Aubin and Ekeland 1984; Aubin and Frankowska 1990; Barbu and Precupanu 1978; Deimling 1980; Ekeland and Temam 1976; Roubíček 2005; Růžička 2004; Showalter 1997; Zeidler 1990b].

C.1.2 Example (Normal-Cone Mapping)
If K is a closed convex subset of V and $\Phi(u) = \chi_K(u)$ is the indicator function of K, i.e. $\chi_K(u) = 0$ if $u \in K$ and $\chi_K(u) = +\infty$ if $u \notin K$, then

$$\partial\chi_K(u) = N_K(u) := \begin{cases} \{f \in V^* : \forall v \in K : \langle f, u - v\rangle \geq 0\}, & u \in K \\ \emptyset, & u \notin K \end{cases},$$

where $N_K(u)$ is the normal cone to K at u (cf. [Han and Reddy 1999b; Roubíček 2005]).

C.1.3 Remark (Hilbert space Setting)
Let $\mathbb{P}_K : H \to K$ be the (orthogonal) projection. Then

$$\begin{aligned}\partial\chi_K(u) &= \begin{cases} \{f \in H : \forall v \in K : (f, u - v)_H \geq 0\}, & u \in K \\ \emptyset, & u \notin K \end{cases} \\ &= \begin{cases} \{v - \mathbb{P}_K(v) \in K : \mathbb{P}_K(v) = \mathbb{P}_K(u)\}, & u \in K \\ \emptyset, & u \notin K \end{cases}\end{aligned}$$

We cite the following statements (cf. [Roubíček 2005]).

C.1.4 Lemma (Convex Case)
Let $A : V \to \mathcal{P}(V^*)$ have a proper convex potential $\Phi : V \to \mathbb{R} \cup \{+\infty\}$, i.e. $A = \partial\Phi$. Then:

(1) A is closed-valued, convex-valued and monotone.

(2) (Rockafellar) If Φ is lower semi-continuous, then A is maximal monotone.

(3) If Φ is also coercive, then A is surjective, i.e. $\forall f \in V^* \; \exists u : A(u) \ni f$.

C.1.5 Lemma (Special Non-Convex Case)
Let $\Phi = \Phi_1 + \Phi_2 : V \to \mathbb{R} \cup \{+\infty\}$ be coercive, let Φ_1 be a proper convex lower semi-continuous functional and let Φ_2 be a weakly lower semi-continuous and Gâteaux differentiable functional. Then, for any $f \in V^*$, there exists an $u \in V$ solving the inclusion:

$$\partial\Phi_1(u) + \Phi_2'(u) \ni f. \tag{C.2}$$

C.1.6 Remark (Special Case)
If $\Phi_1 = \chi_K$ with $K \subset V$ convex, then the inclusion (C.2) turns into the variational inequality:

$$\text{Find } u \in K\text{: } \forall\, v \in K\text{:} \quad \langle \Phi_2'(u), v - u\rangle \geq \langle f, v - u\rangle, \tag{C.3}$$

cf. e.g. [Roubíček 2005].

C.1.7 Remark (Differential Variational Inequalities)
Problems of this form are known as 'differential variational inequalities'. They arise in theoretical mechanics in the study of elasto-plastic and in mathematical economics in the study of planning processes, cf. [Cottle et al. 1980; Kinderlehrer and Stampacchia 1980; Hlavácek et al. 1988]. In the situation $H = \mathbb{R}^N$ those problems are also known as 'projected differential inclusion', cf. [Aubin and Cellina 1984; Aubin and Ekeland 1984; Aubin and Frankowska 1990].

C.2 Existence and Uniqueness Results

In the sequel, we collect some important existence and uniqueness results for the proofs in Chapters 5 – 7. We consider the three different situations: constant, time-dependent and parameter-dependent convex set K.

C.2.1 The Case of constant K

C.2.1 Lemma (Existence and Uniqueness)
Let $\Phi : H \to]-\infty, +\infty]$ be a lower semi-continuous, proper convex function on a real

Hilbert space H. Then, f.a. $u_0 \in D(\Phi)$ and $f \in L^2(S;H)$, there exists a unique strong solution $u \in C(\bar{S};H)$ of

$$\text{(C.4)} \qquad \frac{\mathrm{d}u}{\mathrm{d}t}(t) + \partial\Phi(u(t)) \ni f(t), \qquad t \in S, \qquad u(0) = u_0$$

satisfying:

- $u(t) \in D(\partial\Phi)$ f.a.a. $t \in S$,
- $u \in W^{1,2}(S;H)$,
- $t \mapsto \varphi(u(t))$ *is absolutely continuous on* $\bar{S}$.

C.2.2 Remark (Proof of Lemma C.2.1)
The proof can be carried out with the help of the Yosida approximation, which is a common tool used in the context of semi-groups. The multi-valued maximal monotone operator $\partial\Phi$ is approximated by a single-valued Lipschitz continuous operator $(\partial\Phi)_\lambda$. Therefore we can use a generalised version of the theorem of Picard-Lindelöf for vector-valued spaces in order to prove the unique existence of a classical solution $u_\lambda \in C^1(\bar{S};H)$ f.a. $\lambda > 0$, $f_\lambda \in C(\bar{S};H)$ of the approximative problem

$$\frac{\mathrm{d}u_\lambda}{\mathrm{d}t}(t) + (\partial\Phi)_\lambda(u_\lambda(t)) \ni f_\lambda(t), \qquad t \in S, \qquad u_\lambda(0) = u_0.$$

Using adequate a-priori estimates, we can show that the sequence of approximative solutions u_λ converges to a limit function u for $\lambda \to 0$, which solves the original problem, cf. [Barbu 1976; Brézis 1973; Showalter 1997; Zeidler 1985].

An alternative approach is the following. Since the time derivative is a linear, maximal monotone operator on $\left\{u \in L^2(S;V) : \frac{\mathrm{d}u}{\mathrm{d}t} \in L^2(S;V^*), u(0) = 0\right\}$ we can use the sum rule for maximal monotone operators and apply the theorem of Browder to prove the result, cf. [Růžička 2004; Zeidler 1990b].

An important special case of the above result is $\Phi = \chi_K$ (cf. [Barbu 1976; Růžička 2004]), which is used in this thesis.

C.2.3 Remark (Evolution Triple Setting)
In [Akagi and Ôtani 2004, 2005; Kenmochi 1977b; Ruess 2009; Savaré 1993, 1996] the evolution triple setting $V \subset H \subset V^*$ is considered and the unique existence of a weak solution of the problem

$$u'(t) + \partial\Phi(u(t)) \ni f(t) \quad \text{in} \quad V^*, \qquad u(0) = u_0 \in H$$

in $L^2(S;V) \cap C(\bar{S};H) \cap W^{1,2}(S;V^*)$ has been proved.

C.2.4 Remark (Continuous Dependence on Right-hand Side)
Let u_1 be the strong solution of (C.4) with a right-hand side $f_1 \in L^2(S;H)$. If u_2 is a strong solution of (C.4) with a right-hand side $f_2 \in L^2(S;H)$, it is easy to check via the monotonicity of $\partial\Phi(t,\cdot)$ that

$$|u_1(t) - u_2(t)| \le |u_1(s) - u_2(s)| + \int_s^t |f_1(\tau) - f_2(\tau)|\, d\tau$$

holds f.a. $0 \le s \le t \le T$.

C.2.2 The Case of time-dependent K

We shall consider the existence and uniqueness of the solution of the equation

$$\frac{du}{dt}(t) + \partial\Phi(t, u(t)) \ni f(t), \qquad t \in S \tag{C.5}$$

for a given initial condition u_0.

C.2.5 Remark (Yosida Approximation of the Subdifferential of a Characteristic Function)
For each $\lambda > 0$ and $t \in \bar{S}$ define the Yosida approximation Φ_λ of the convex potential Φ as

$$\Phi_\lambda(t,v) := \inf\left\{\frac{\|v-z\|^2}{2\lambda} + \Phi(t,z) : z \in H\right\} \quad \text{for} \quad v \in H.$$

Let $\Phi(t,u) = \chi_{K(t)}(u)$. Then,

$$\Phi_\lambda(t,u) = \frac{1}{2\lambda}\operatorname{dist}(u, K(t))^2 = \frac{1}{2\lambda}\|v - \mathbb{P}_{K(t)}(u)\|^2$$

is a convex and Gateaux differentiable, cf. e.g. [Han and Reddy 1999b], and

$$\partial\Phi_\lambda(t,u) = \frac{1}{\lambda}\|u - \mathbb{P}_{K(t)}(u)\|$$

is maximal monotone.

C.2.6 Assumption
Suppose that

(1) *there are positive constants α and β s.t. $\Phi(t,z) + \alpha\|z\| + \beta \ge 0$ for any $t \in \bar{S}$ and $z \in H$,*

(2) *for each $\lambda > 0$ and $z \in H$ there is a non-negative function $\rho \in L^1(S)$ s.t. $\Phi_\lambda(t,z) - \Phi_\lambda(s,z) \le \int_s^t \rho(\tau)\, d\tau$ for $s,t \in \bar{S}$ with $s \le t$,*

(3) *(i)* *for each* $r \geq 0$, *there exists a number* $a_r \in [0, 1[$ *and functions* $b_r, c_r \in L^1(S)$ *s.t.* $\frac{d}{dt}\Phi_\lambda(t, z) \leq a_r\|\partial\Phi_\lambda(t, z)\|^2 + b_r(t)|\Phi_\lambda(t, z)| + c_r(t)$ *a.e. on* $\bar{S}$ *for* $z \in H$ *with* $\|z\| \leq r$ *and* $\lambda \in [0, 1[$ *and*

(ii) *there exists an H-valued function h on* $\bar{S}$ *and a partition* $\{0 = t_0 < t_1 < \cdots < t_N = T\}$ *of* $\bar{S}$ *s.t.* $\Phi(t, h(t)) \in L^1(S)$ *and the restriction of h to* $]t_{k-1}, t_k[$ *belongs to* $W^{1,1}(t_{k-1}, t_k; H)$ *for* $k = 1, 2, \ldots, N$.

C.2.7 Remark

When (3) in Assumption C.2.6 is replaced by

(3') *there exists a number* $a \in [0, 1[$ *and functions* $b, c \in L^1(S)$ *s.t.*

$$\frac{d}{dt}\Phi_\lambda(t, z) \leq a\|\partial\Phi_\lambda(t, z)\|^2 + b(t)|\Phi_\lambda(t, z)| + (1 + \|z\|^2)c_r(t), \qquad t \in \bar{S}$$

f.a. $z \in H$ *and* $\lambda \in]0, 1]$

the same conclusion remains valid.

C.2.8 Assumption

T denotes a positive number and $\{\Phi(t, \cdot)\}_{t\in\bar{S}}$ *satisfies*

(1) *For each* $t \in \bar{S}$, $\Phi(t, \cdot)$ *is a lower semi-continuous convex function from H into* $]-\infty, +\infty]$ *with the nonempty effective domain* $D(\Phi(t, \cdot))$.

(2) *Let* K_r *be a non-negative constant, let* $g_r \in W^{1,2}(S)$ *and let* $h_r \in W^{1,1}(S)$ *for any* $r > 0$. *Then, for* $t_0 \in \bar{S}$ *and for each* $x_0 \in D(\Phi(t_0, \cdot))$ *s.t.* $\|x_0\| \leq r$, *there exists a function* $x : \bar{S} \to H$ *s.t.*

$$\|x(t) - x(t_0)\| \leq |g_r(t) - g_r(t_0)|(\Phi(t_0, x_0) + K_r)^{\frac{1}{2}}, \qquad t \in \bar{S}$$

$$\Phi(t, x(t)) \leq \Phi(t_0, x_0) + |h_r(t) - h_r(t_0)|(\Phi(t_0, x_0) + K_r), \qquad t \in \bar{S}.$$

C.2.9 Proposition

Assumption C.2.6 follows from the second part of Assumption C.2.8.

The proof can be found in [Kenmochi 1977a].

C.2.10 Definition (Strong Solution)

Let $u : \bar{S} \to H$. *Then u is called a strong solution of (C.5) on* $\bar{S}$ *if*

(1) $u \in C(\bar{S}; H)$

(2) *u is strongly absolutely continuous on any compact subset of S and*

(3) $u(t)$ *is in* $D(\Phi(t, \cdot))$ f.a.a. $t \in \bar{S}$ *and satisfies (C.5)* f.a.a. $t \in \bar{S}$.

C.2.11 Lemma (Existence and Uniqueness)
Let $f \in L^2(S; H)$ and let $\{\varphi(t,\cdot)\}_{t\in\bar{S}}$ satisfy Assumption C.2.8. Then, for each $u_0 \in \overline{D(\varphi(0,\cdot))}$, the equation (C.5) has a unique strong solution u on $\bar{S}$ with $u(0) = u_0$. In particular, if $u_0 \in D(\varphi(0,\cdot))$, then u satisfies

(1) For all $t \in \bar{S}$, $u(t)$ is in $D(\varphi(t,\cdot))$ and $\varphi(t, u(t))$ is absolutely continuous on $\bar{S}$.

(2) u is strongly absolutely continuous on $\bar{S}$ and satisfies $\frac{du}{dt} \in L^2(S; H)$.

C.2.12 Remark (Proof of Lemma C.2.11)
The proof via the Yosida approximation technique can be found e.g. [Attouch and Damlamian 1972; Guillaume and Syam 2005; Hu and Papageorgiou 1998; Kenmochi 1975; Kenmochi and Pawlow 1986; Papageorgiou 1992a; Tolstonogov 2009; Yamada 1976]. Furthermore, the solution mapping $p : L^2(S; H) \to C(\bar{S}; H)$, $f \mapsto u$ is compact, cf. [Hu and Papageorgiou 1998] .

See [Akagi and Ôtani 2004; Kenmochi 1977a,b; Kenmochi and Toshitaka 1975] for the evolution triple setting mentioned in Remark C.2.3, i.e. $u'(t) + \partial\Phi(t, u(t)) \ni f(t)$ in V^*. Equations of the type $u'(t) + \partial\varphi(t, u(t)) + g(t, u(t)) \ni f(t)$ with effective domain D independent of t are treated in [Watanabe 1973]. A generalisation (for a perturbed problem) is presented in [Ahmed 1992].

See e.g. [Ruess 2009] for partial differential delay equations of the type

$$u'(t) + Bu(t) \ni F(u'(t)), \qquad t \geq 0, \qquad u_{|I} = \varphi \in E$$

with B a (generally) nonlinear and multivalued differential expression in a Banach space X, $I = [-r, 0]$, $r > 0$ and space E of functions from I to X.

Now let

$$\text{(C.6)} \qquad \Phi(t,x) = \chi_{K(t)}(x) \quad \text{with} \quad \chi_{K(t)}(x) = \begin{cases} 0 & \text{if } x \in K(t) \\ +\infty & \text{otherwise} \end{cases} .$$

Then the problem (C.5) becomes

$$-u'(t) \in N_{K(t)}(u(t)) + f(t), \qquad t \in S, \qquad u(0) = u_0 \in K(0).$$

We remark that equations of the type $u'(t) \in F(t, u(t)) - N_K(u(t))$ are e.g. treated in [Aubin and Cellina 1984]. We make the following hypothesis concerning the multifunction $t \to K(t)$.

C.2.13 Assumption
Let $X := \{A : A \subset H \text{ nonempty, closed (and convex)}\}$ *and let*

$$\mathrm{h}(A, C) := \max\{\sup[\mathrm{d}(a, C) : a \in A], \sup[\mathrm{d}(c, A) : c \in C]\}$$

be the Hausdorff distance. There exists a $v \in L^2(S)$ *and* $K : \bar{S} \to X$ *is a multifunction s.t.*

$$\mathrm{h}(K(s), K(t)) \leq \int_t^s v(\tau)\,\mathrm{d}\tau.$$

C.2.14 Remark
In the situation of (C.6), it is easy to check that Assumption C.2.8 holds with $g_r(t) = \int_0^t v(s)\,\mathrm{d}s$, $g_r'(t) = v(t)$, $\beta = 2$, $\alpha = 0$ *and* $K_r = 1$, *cf. e.g. [Kubo and Yamazaki 2007; Papageorgiou 1992b, 1995; Papageorgiou and Yannakakis 2002; Yamazaki 2005].*

C.2.15 Remark (Continuous Dependence on Right-hand Side)
Let u_1 *be a strong solution of (C.5) with (C.6) with a right-hand side* $f_1 \in L^2(S; H)$. *If* u_2 *is a strong solution of (C.5) with a right-hand side* $f_2 \in L^2(S; H)$, *it is easy to check via the monotonicity of* $\partial\Phi(t, \cdot)$ *that*

$$|u_1(t) - u_2(t)| \leq |u_1(s) - u_2(s)| + \int_s^t |f_1(\tau) - f_2(\tau)|\,\mathrm{d}\tau$$

holds f.a. $0 \leq s \leq t \leq T$.

Proof. Note that the differential inclusion

$$u'(t) + \partial\chi_{K(t)}(u(t)) \ni f(t), \qquad t \in S \tag{C.7}$$

is equivalent to the variational inequality

$$\forall\, u(t) \in K(t): \quad \big(u'(t), v - u(t)\big) + \chi_{K(t)}(v) - \chi_{K(t)}(u(t)) \geq \big(f(t), v - u(t)\big) \tag{C.8}$$

for all $v \in K(t)$. Let u_1, u_2 be two different strong solutions of (C.7) with corresponding right-hand sides f_1, f_2. Putting $v = u_2(t)$ in (C.8) (resp. $v = u_1(t)$ in (C.8)), we get

$$\big(u_1'(t), u_2(t) - u_1(t)\big) - \chi_{K(t)}(u_2(t)) + \chi_{K(t)}(u_1(t)) \geq \big(f_1(t), u_2(t) - u_1(t)\big),$$
$$\big(u_2'(t), u_1(t) - u_2(t)\big) - \chi_{K(t)}(u_1(t)) + \chi_{K(t)}(u_2(t)) \geq \big(f_2(t), u_1(t) - u_2(t)\big).$$

Adding both relations gives

$$\big(u_1'(t) - u_2'(t), u_1(t) - u_2(t)\big) \leq \big(f_1(t) - f_2(t), u_1(t) - u_2(t)\big).$$

Integration and using Gronwall's inequality completes the proof. □

C.2.3 The Case of parameter-dependent K

Let E be a complete metric space (the parameter space) and consider the following problem:

$$\text{(C.9)} \qquad u'(t) + \partial\Phi(t, u(t), \lambda) \ni f(t), \qquad t \in S,$$
$$\text{(C.10)} \qquad u(0) = u_0(\lambda), \qquad \lambda \in E.$$

We make the following hypotheses in order to show the continuity regarding the parameter λ.

C.2.16 Assumption

Let $\Phi : S \times H \times E \to \mathbb{R}^+ \cup \{\infty\}$ be a function s.t.

(1) $\Phi(t, \cdot, \lambda) : H \to \mathbb{R}^+ \cup \{\infty\}$ is proper, convex and lower semi-continuous f.a. $(t, \lambda) \in S \times E$,

(2) for any compact $B \subset E$ and $r > 0$, there exist $K_r^B > 0$, $g_r^B \in W^{1,\infty}(S)$ and $h_r^B \in W^{1,1}(S)$ s.t. if $\lambda \in B$, $t \in S$, $u \in D(\Phi(t, \cdot, \lambda))$ with $\|u\| \leq r$ and $s \in [t, b]$ then there exists $\hat{u} \in D(\Phi(s, \cdot, \lambda))$ s.t.

$$\|\hat{u} - u\| \leq |g_r^B(s) - g_r^B(t)|(\Phi(t, u, \lambda) + K_r^B) \qquad \text{and}$$
$$\Phi(s, \hat{u}, \lambda) \leq \Phi(s, \hat{u}, \lambda) + |h_r^B(s) - h_r^B(t)|(\Phi(t, u, \lambda) + K_r^B),$$

(3) $\overline{\bigcup_{\lambda \in B}\{z \in H : \|z\|^2 + \Phi(t, z, \lambda) \leq 0\}}$ is compact in H f.a. $t \in S$, $B \subset E$ compact and all $\lambda \in \mathbb{R}^+$,

(4) $\Phi(t\cdot, \lambda_n) \overset{M}{\to} \Phi(t\cdot, \lambda)$ if $\lambda_n \to \lambda$ in E,

(5) $u_0 : E \to H$ is continuous, $u_0(\lambda) \in D(\Phi(0, \cdot, \lambda))$ f.a. $\lambda \in E$ and for any compact $B \subset E$: $\sup_{\lambda \in B} \Phi(0, u_0, \lambda) < \infty$.

Moreover, there exists a $v \in L^2(S)$ and $K : S \times E \to \mathcal{P}(H)$ is a multi-function, s.t.

$$\mathrm{h}(K(t, \lambda), K(s, \lambda)) \leq \int_s^t v(s)\,\mathrm{ds} \quad \text{for} \quad 0 \leq s \leq t \leq b, \qquad \lambda \in E$$

and $K(t, \lambda_n) \overset{M}{\to} K(t, \lambda)$ for $\lambda_n \to \lambda$.

C.2.17 Remark

In the situation of (C.6), Assumption C.2.16 is fulfilled with $g_r(t) = \int_0^t v(s)\,\mathrm{ds}$, $t \in S$, $g_r' \in L^2(S)$, $\beta = 2$, $\alpha = 0$, $K_r = 1$ for all $r > 0$, cf. Remark C.2.14. Moreover, $K(t, \lambda_n) \overset{M}{\to} K(t, \lambda) \implies \Phi(t, \cdot, \lambda_n) \overset{M}{\to} \Phi(t, \cdot, \lambda)$, cf. e.g. [Tolstonogov 2009].

C.2.18 Remark (Mosco Convergence)
Let $\Phi : V \to \mathbb{R} \cup \{\infty\}$ be convex, lower semi-continuous, proper and possess, for any $n > 0$, a convex, Gâteaux differentiable regularisation $\Phi_n : V \to \mathbb{R}$. Then Φ converges in the sense of Mosco, i.e. $\Phi_n \xrightarrow{M} \Phi$ if

$$\begin{aligned} v_n \to v &\implies \liminf_{n\to 0} \Phi_n(v_n) \geq \Phi(v) \qquad \text{and} \\ \forall\, v \in V\, \exists\, v_n \to v &\implies \limsup_{n\to 0} \Phi_n(v_n) \leq \Phi(v), \end{aligned}$$

cf. [Roubíček 2005].

C.2.19 Lemma (Existence and Uniqueness)
Let $f \in L^2(S;H)$ and let $\{\Phi(t,\cdot,\lambda)\}_{t\in S}$ satisfy Assumption C.2.16. Then for each $u_0 \in \overline{D(\Phi(0,\cdot))}$, the problem (C.9),(C.10) has a unique strong solution u on S with $u(0) = u_0$. In particular, if $u_0 \in D(\Phi(0,\cdot))$, then u satisfies

(1) For all $t \in S$, $u(t) \in D(\Phi(t,\cdot))$ and $\Phi(t,u(t))$ is absolutely continuous on S.

(2) u is strongly absolutely continuous on S and satisfies $\frac{\mathrm{d}u}{\mathrm{d}t} \in L^2(S;H)$.

C.2.20 Lemma (Continuity w.r.t. the Parameter)
If Assumption C.2.16 hold, $f_n \rightharpoonup f$ in $L^2(S;H)$ and $\lambda_n \to \lambda$ in E, then for the solution mapping it follows $p(f_n,\lambda_n) \to p(f,\lambda)$ in $C(S;H)$.

Proof. See [Hu and Papageorgiou 1998; Papageorgiou 1992b, 1995]. □

C.2.21 Remark (Continuous Dependence on the Right-hand Side and the Parameter)
Put $\Phi(t,x,\lambda) := \chi_{K(t,\lambda)}(x)$ in (C.9) and let u_1, u_2 be two different strong solutions of (C.9),(C.10) with corresponding right-hand sides f_1, f_2 and parameters λ_1, λ_2. Moreover, let $\mathrm{h}(K_1(t,\lambda_1), K_2(t,\lambda_2)) \leq c\,|\lambda_2 - \lambda_1|$. Then

$$\begin{aligned} \|u_1(t) - u_2(t)\|_X^2 \leq c \Big\{ \|u_1(s) - u_2(s)\|_X^2 &+ \int_s^t \|f_1(\tau) - f_2(\tau)\|_X^2 \,\mathrm{d}\tau \\ &+ \int_s^t \|\lambda_1(\tau) - \lambda_2(\tau)\|_X^2 \,\mathrm{d}\tau \Big\} \end{aligned}$$

f.a. $0 \leq s \leq t \leq T$, analogous to Remark C.2.15.

Proof. Let $u_1(t) \in K_1(t,\lambda_1)$, $u_2(t) \in K_2(t,\lambda_2)$ f.a.a. $t \in S$. Then $\mathbb{P}_{K_1(t,\lambda_1)} u_2(t) \in K_1(t,\lambda_1)$, $\mathbb{P}_{K_2(t,\lambda_2)} u_1(t) \in K_2(t,\lambda_2)$ f.a.a. $t \in S$. Consider

$$\begin{aligned} \langle u_1'(t), \mathbb{P}_1 u_2(t) - u_1(t)\rangle + \chi_1(\mathbb{P}_1 u_2(t)) - \chi_1(u_1(t)) &\geq \langle f_1(t), \mathbb{P}_1 u_2(t) - u_1(t)\rangle, \\ \langle u_2'(t), \mathbb{P}_2 u_1(t) - u_2(t)\rangle + \chi_2(\mathbb{P}_2 u_1(t)) - \chi_2(u_2(t)) &\geq \langle f_2(t), \mathbb{P}_2 u_1(t) - u_2(t)\rangle, \end{aligned}$$

where $\mathbb{P}_i := \mathbb{P}_{K_i(t,\lambda_i)}$ and $\chi_i := \chi_{K_i(t,\lambda_i)}$, $i = 1, 2$. Adding both equations and using the fact that

$$\chi_1(\mathbb{P}_1 u_2(t)) = \chi_1(u_1(t)) = \chi_1(\mathbb{P}_2 u_1(t)) = \chi_2(u_2(t)) = 0$$

gives for $t \in S$

$$\begin{aligned}&\langle u_1'(t) - u_2'(t), u_1(t) - u_2(t)\rangle + \langle u_1'(t), u_2(t) - \mathbb{P}_1 u_2(t)\rangle + \langle u_2'(t), u_1(t) - \mathbb{P}_2 u_1(t)\rangle \\ &\le \langle f_1(t) - f_2(t), u_1(t) - u_2(t)\rangle + \langle f_1(t), u_2(t) - \mathbb{P}_1 u_2(t)\rangle + \langle f_2(t), u_1(t) - \mathbb{P}_2 u_1(t)\rangle.\end{aligned}$$

Hence, for $t \in S$,

$$\begin{aligned}\frac{1}{2}\frac{\mathrm{d}}{\mathrm{d}t}\|u_1(t)-u_2(t)\|^2 &\le \frac{1}{2}\|f_1(t)-f_2(t)\|^2+\frac{1}{2}\|u_1(t)-u_2(t)\|^2+\|f_1(t)\|\,\|u_2(t)-\mathbb{P}_1 u_2(t)\| \\ &+ \|f_2(t)\|\,\|u_1(t) - \mathbb{P}_2 u_1(t)\| + \|u_1'(t)\|\|u_2(t) - \mathbb{P}_1 u_2(t)\| + \|u_2'(t)\|\,\|u_1(t) - \mathbb{P}_2 u_1(t)\|.\end{aligned}$$

Moreover, for $t \in S$, $\|u_i'(t)\| \le c$, $\|f_i(t)\| \le c$ and

$$\begin{aligned}\left\|u_1(t) - \mathbb{P}_2 u_1(t)\right\| &= \mathrm{dist}(u_1(t), K_2(t,\lambda_2)) \le \mathrm{h}(K_1(t,\lambda_1), K_2(t,\lambda_2)) \le c\,|\lambda_2 - \lambda_1|, \\ \left\|u_2(t) - \mathbb{P}_1 u_2(t)\right\| &= \mathrm{dist}(u_2(t), K_1(t,\lambda_1)) \le \mathrm{h}(K_1(t,\lambda_1), K_2(t,\lambda_2)) \le c\,|\lambda_2 - \lambda_1|.\end{aligned}$$

Finally, integration and using Gronwall's inequality completes the proof. □

C.2.4 References

There exists a huge amount of literature about (parabolic) variational inequalities and differential inclusions. We collect the references for variational inequalities [Attouch and Damlamian 1978; Arseni-Benou et al. 1999; Attouch and Damlamian 1972; Barbu 1976; Brézis 1973; Browder 1968, 1969; Cottle et al. 1980; Jeong et al. 2000; Kano 2009; Kenmochi 1975, 1977a,b; Kinderlehrer and Stampacchia 1980; Kano et al. 2009; Kunze and Monteiro Marques 1998; Kenmochi and Niezgódka 1994; Kenmochi and Toshitaka 1975; Kubo 2010, 1989; Kubo and Yamazaki 2007; Maruo 1975; Naumann 1984; Ôtani 1982, 1993/94; Savaré 1993, 1996, 1997; Watanabe 1973] and differential inclusions [Aubin and Cellina 1984; Akagi and Ôtani 2004, 2005; Guillaume and Syam 2005; Hu and Papageorgiou 1998; Migórski 1996; Papageorgiou 1989, 1990, 1992a,b, 1995; Papageorgiou and Yannakakis 2002; Yamazaki 2005; Yamada 1976; Yotsutani 1978; Zeidler 1985].
Variational inequalities with applications in solid mechanics are treated in [Hlavácek et al. 1988; Han and Sofonea 2001].

Bibliography

S. Abdel-Malek and L. W. Meyer. Modellierung des Fließverhaltens und Beschreibung des duktilen Versagens in einem weiten Bereich von Dehngeschwindigkeiten. *Mat.-wiss. u. Werkstofftech.*, 38(2):101–107, 2007.

C. Acht, M. Dalgic, F. Frerichs, M. Hunkel, A. Irretier, T. Lübben, and H. Surm. Ermittlung der Materialdaten zur Simulation des Durchhärtens von Komponenten aus 100Cr6 – Teil 1. *HTM*, 63(5):234–244, 2008a.

C. Acht, M. Dalgic, F. Frerichs, M. Hunkel, A. Irretier, T. Lübben, and H. Surm. Ermittlung der Materialdaten zur Simulation des Durchhärtens von Komponenten aus 100Cr6 – Teil 2. *HTM*, 63(6):362–371, 2008b.

R. A. Adams and J. J. F. Fournier. *Sobolev Spaces*, volume 140 of *Pure and Applied Mathematics*. Academic Press, Elsevier Science Ltd., 2nd edition, 2003.

E. Aeby-Gautier and G. Cailletaud. N-phase modeling applied to phase transformations in steels: a coupled kinetics-mechanics approach. In *International Conference on Heterogeneous Material Mechanics*, 2004.

S. Agmon. The L^p approach to the Dirichlet problem. Technical Report, Hebrew University, 1959.

S. Agmon, A. Douglis, and L. Nirenberg. Estimates Near the Boundary for Solutions of Elliptic Partial Differential Equations Satisfying General Boundary Conditions 2. *Commun. Pure Appl. Anal.*, 17:35–92, 1964.

N. U. Ahmed. An Existence Theorem for Differential Inclusions on Banach Space. *J. Appl. Math. Stoch. Anal.*, 5(2):123–130, 1992.

U. Ahrens. *Beanspruchungsabhängiges Umwandlungsverhalten und Umwandlungsplastizität niedrig legierter Stähle mit unterschiedlich hohen Kohlenstoffgehalten.* PhD Thesis, Universität Paderborn, 2003.

U. Ahrens, G. Besserdich, and H. J. Maier. Spannungsabhängiges bainitisches und martensitisches Umwandlungsverhalten eines niedrig legierten Stahl. *HTM*, 55:329–338, 2000.

U. Ahrens, G. Besserdich, and H. J. Maier. Sind aufwendige Experimente zur Beschreibung der Phasenumwandlungen von Stählen noch zeitgemäß? *HTM*, 57(2):99–105, 2002.

G. Akagi and M. Ôtani. Evolution inclusions governed by subdifferentials in reflexive Banach spaces. *J. Evol. Equ.*, 4:519–541, 2004.

G. Akagi and M. Ôtani. Evolution inclusions governed by the difference of two subdifferentials in reflexive Banach spaces. *J. Differential Equations*, 209:392–415, 2005.

H. D. Alber. Mathematische Theorie des inelastischen Materialverhaltens von Metallen. Preprint 1682, TU Darmstadt, 1994.

H. D. Alber. *Materials with Memory – Initial-Boundary Value Problems for Constitutive Equations with Internal Variables*, volume 1682 of *Lecture Notes in Mathematics*. Springer, 1998.

H. D. Alber and K. Chelminski. Quasistatic problems in viscoplasticity theory. Preprint, TU Darmstadt, 2002.

J. Alberty and C. Carstensen. Numerical Analysis of Time-Dependent Primal Elastoplasticity with Hardening. *SIAM J. Numer. Anal.*, 37(4):1271–1294, 2000.

H. W. Alt. *Lineare Funktionalanalysis*. Springer, 2002.

H. Altenbach. Über die Grundlagen der der klassischen Kriechmechanik und ausgewählte Anwendungsbeispiele. ZeTeM Reports, Universität Bremen, 2010.

J. Altenbach and H. Altenbach. *Einführung in die Kontinuumsmechanik*. Teubner, 1994.

H. Amann. *Nonhomogeneous Linear and Quasilinear Elliptic and Parabolic Boundary Value Problems*, pages 9–126. Functions Spaces, Differential Operators and Nonlinear Analysis. Teubner, 1993.

H. Amann. *Gewöhnliche Differentialgleichungen*. de Gruyter, 1995.

H. Amann. Compact Embeddings of Vector-Valued Sobolev and Besov Spaces. *Glas. Mat.*, 35(55):161–177, 2000.

H. Amann and J. Escher. *Analysis 1*. Birkhäuser, 2006.

A. Amassad and C. Fabre. Existence for Viscoplastic Contact with Coulomb Friction Problems. *IJMMS*, 32(7):411–437, 2002.

A. Amassad and C. Fabre. A quasistatic viscoplastic contact problem with normal compliance and friction. *IMA J. Appl. Math.*, 69:463–482, 2004.

A. Amassad and M. Sofonea. Analysis of a Quasistatic Viscoplastic Problem involving Tresca Friction Law. *Discrete Contin. Dyn. Syst.*, 4(1):55–72, 1998.

A. Amassad, M. Shillor, and M. Sofonea. A quasistatic contact problem for an elastic perfectly plastic body with Tresca's friction. *Nonlinear Anal.*, 35:95–109, 1999a.

A. Amassad, M. Shillor, and M. Sofonea. A Quasistatic Contact Problem with Slip-dependent Coefficient of Friction. *Math. Methods Appl. Sci.*, 22:267–284, 1999b.

A. Amassad, M. Shillor, and M. Sofonea. A Nonlinear Evolution Inclusion in Perfect Plasticity with Friction. *Acta Math. Univ. Comenianae*, LXX:215–228, 2001.

A. Amassad, K. L. Kuttler, M. Rochdi, and M. Shillor. Quasi-Static Thermoviscoelastic Contact Problem with Slip Dependent Friction Coefficient. *Math. Comput. Modelling*, 36:839–854, 2002.

K. A. Ames and L. E. Payne. Uniqueness and continuous dependence of solutions to a multidimensional thermoelastic contact problem. *J. Elasticity*, 34:139–148, 1994.

T. Antretter, F.D. Fischer, K. Tanaka, and G. Cailletaud. Theory, experiments and numerical modelling of phase transformations with emphasis on TRIP. *Steel Research*, 73(6+7):225–235, 2002.

T. Antretter, F. D. Fischer, and G. Cailletaud. A numerical model for transformation induced plasticity. *J. Phys. IV France*, 115:233–241, 2004.

G. Anzellotti and S. Luckhaus. Dynamical Evolution of Elasto-Perfectly Plastic Bodies. *Appl. Math. Optim.*, 15:121–140, 1987.

L. Arkeryd. On the L^p estimates for elliptic boundary problems. *Math. Scand.*, 19:59–76, 1966.

K. Arseni-Benou, N. Halidias, and N. S. Papageorgiou. Nonconvex Evolution Equations Generated by Time-Dependent Subdifferential Operators. *J. Appl. Math. Stoch. Anal.*, 12(3):233–252, 1999.

K. Atkinson and W. Han. *Theoretical Numerical Analysis*. Springer, 2001.

H. Attouch and A. Damlamian. On multivalued evolution equations in Hilbert spaces. *Isreal J. Math.*, 12(4):373–390, 1972.

H. Attouch and A. Damlamian. Strong Solutions for Parabolic Variational Inequalities. *Nonlinear Anal.*, 2(3):329–353, 1978.

J. P. Aubin and A. Cellina. *Differential Inclusions.* A Series of Comprehensive Studies in Mathematics. Springer, 1984.

J. P. Aubin and I. Ekeland. *Applied Nonlinear Analysis.* Pure and Applied Mathematics. John Wiley & Sons Ltd., 1984.

J. P. Aubin and H. Frankowska. *Set-Valued Analysis*, volume 2 of *Systems and Control: Foundations and Applications.* Birkhäuser, 1990.

M. Avrami. Kinetics of Phase Change 1 – General Theory. *Journal of Chemical Physics*, 7:1103–1112, 1939.

M. Avrami. Kinetics of Phase Change 2 – Transformation-Time Relations for Random Distribution Nuclei. *Journal of Chemical Physics*, 8:212–224, 1940.

M. Avrami. Kinetics of Phase Change 3 – Granulation, Phase Change and Microstructure. *Journal of Chemical Physics*, 9:177–184, 1941.

B. Awbi, M. Rochdi, and M. Sofonea. Abstract evolution equations for viscoelastic frictional contact problems. *ZAMP*, 51:218–235, 2000.

J.-F. Babadjian, G. A. Francfort, and M. G. Mora. Quasistatic evolution in non-associative plasticity – the cap model. submitted, 2011.

E. Babolian and A. J. Shaerlar. On Some Nonlinear Generalizations of Gronwall's Inequality and their Applications. *Int. J. Contemp. Math. Sciences*, 6(16):771–782, 2011.

C. Bacuta and J. H. Bramble. Regularity estimates for solutions of the equations of linear elastity in convex plane polygonal domains. *ZAMP*, 54:874–878, 2003.

H. D. Baehr and K. Stephan. *Wärme- uns Stoffübertragung.* Springer, 2006.

C. Baiocchi. Sulle equazioni differenziali astratte lineari del primo e del secondo ordine negli spazi di Hilbert. *Ann. Mat. Pura Appl.*, 76(1):233–304, 1967.

V. Barbu. *Nonlinear Semigroups and Differential Equations in Banach Spaces.* Noordhoff International Publishing, 1976.

V. Barbu and T. Precupanu. *Convexity and optimization in Banach spaces.* Noordhoff International Publishing, 1978.

P. Barral, M. C. Naya-Riveiro, and P. Quintela. Mathematical analysis of a viscoelastic problem with temperature-dependent coefficients – Part 1: Existence and uniqueness. *Math. Methods Appl. Sci.*, 30:1545–1568, 2007.

L. Bartczak. Mathematical analysis of a thermo-visco-plastic model with Bodner-Partom constitutive equations. Article in Press, J. Math. Anal. Appl., 2011.

S. Bartels and T. Roubíček. Thermoviscoplasticity at small strains. *ZAMM*, 88(9): 735–754, 2008.

S. Bartels and T. Roubíček. Thermo-visco-elasticity with rate-independent plasticity in isotropic materials undergoing thermal expansion. Preprint 463, Universität Bonn, 2009.

M. I. Belishev and I. Lasiecka. The Dynamical Lamé System: Regularity of Solutions, Boundary Controllability and Boundary Data Continuation. *ESAIM. Control Optim. Calc. Var.*, 8:143–167, 2002.

H. Bellout and J. Nečas. Existence of global weak solutions for a class of quasilinear hyperbolic integro-differential equations describing visco-elastic materials. *Math. Ann.*, 299:275–291, 1994.

A. Bensoussan and J. Frehse. Asymptotic behaviour of the time dependent Norton-Hoff law in plasticity theory and H^1 regularity. *Comment. Math. Univ. Carolin.*, 37(2): 285–304, 1996.

A. Bensoussan and J. Frehse. *Regularity Results for Nonlinear Elliptic Systems and Applications*, volume 151 of *Applied Mathematical Sciences.* Springer, 2002.

J. Bergh and J. Löfström. *Interpolation Spaces – An Introduction.* Springer, 1976.

J.-M. Bergheau, F. Boitout, V. Toynet, S. Denis, and A. Simon. Finite element simulation of coupled carbon diffusion, metallurgical transformation and heat transfer with applications in the automobile industry. In *Conference on Quenching and Control of Distortion*, Proceedings of the 3rd International Conference on Quenching and Control of Distortion, pages 145–156, 1999.

H. Berns and W. Theisen. *Eisenwerkstoffe – Stahl und Gusseisen.* Springer, 2006.

J.-M. Berthelot. *Composite materials – mechanical behaviour and structural analysis.* Springer, 1999.

A. Bertram. *Elasticity and Plasticity of Large Deformations – An Introduction.* Springer, 2005.

G. Besserdich. *Untersuchungen zur Eigenspannungs- und Verzugsbildung beim Abschrecken von Zylindern aus den Stählen 42CrMo4 und Ck45 unter Berücksichtigung der Umwandlungsplastizität.* PhD Thesis, Universität Karlsruhe, 1993.

J. Betten. *Kontinuumsmechanik: Elasto-, Plasto- und Kriechmechanik.* Springer, 1993.

M. Böhm. On Navier-Stokes and Kelvin-Voigt Equations in Three Dimensions in Interpolation Spaces. *Math. Nachr.*, 155(1):151–165, 1992.

M. Böhm, S. Dachkovski, M. Hunkel, T. Lübben, and M. Wolff. Übersicht über einige makroskopische Modelle für Phasenumwandlungen im Stahl. ZeTeM Report, Universität Bremen, 2003.

M. Böhm, M. Hunkel, A. Schmidt, and M. Wolff. Evaluation of various phase-transition models for 100Cr6 for application in commercial FEM programmes. *J. Phys.*, IV(120): 581–589, 2004.

M. Bien. Existence of Global Weak Solutions for Coupled Thermoelasticity under Non-Linear Boundary Conditions. *Math. Methods Appl. Sci.*, 19:1265–1277, 1996.

M. Bien. Existence of Global Weak Solutions for Coupled Thermelasticity with Barber's Heat Exchange Condition. *J. Appl. Anal.*, 9(2):163–185, 2003.

S. Bökenheide, M. Wolff, M. Dalgic, D. Lammers, and T. Linke. Creep, phase transformations and transformation-induced plasticity of 100Cr6 steel during heating. In T. Lübben and H.-W. Zoch, editors, *Proceedings of the 3rd International Conference on Distortion Engineering*, pages 411–418, 2011.

W. Bleck. *Werkstoffkunde Stahl für Studium und Praxis.* Verlag Mainz, 2001.

W. Bleck, G. Pariser, and S. Trute. Herstellung und Verarbeitung moderner Stahlwerkstoffe. *HTM*, 58:181–188, 2003.

F. Bloom. On Continuous Dependence for a Class of Dynamical Problems in Linear Thermoelasticity. *ZAMP*, 26:569–579, 1975.

Ph. Bénilan and M. G. Crandall. *Semingroup Theory and Evolution Equations*, chapter Completely Accretive Operators, pages 41–75. CRC Press, 1991.

L. Boccardo, A. Dall'aglio, T. Gallouet, and L. Orsina. Existence and regularity results for some nonlinear parabolic equations. *Adv. Math. Sci. Appl.*, 1999.

S. Boettcher. Zur mathematischen Aufgabe der Thermoelastizität unter Berücksichtigung von Phasenumwandlungen und Umwandlungsplastizität. Diploma Thesis, Universität Bremen, 2007.

E. Bonetti and G. Bonfanti. Existence and Uniqueness of the Solution to a 3D Thermoviscoelastic System. *Electron. J. Differential Equations*, 2003(50):1–15, 2003.

E. Bonetti and G. Bonfanti. Asymptotic Analysis for Vanishing Acceleration in a Thermoviscoelastic System. *Abstr. Appl. Anal.*, 2:150–202, 2005.

E. Bonetti and G. Bonfanti. Well-posedness results for a model of damage in thermoviscoelastic materials. *Ann. I. H. Poincaré – AN*, 25:1187–1208, 2008.

R. M. Bowen. *Theory of mixtures in Continuum Physics*. Academic Press, 1976.

D. Braess. *Finite elements – Theory, fast solvers, and applications in solid mechanics*. Cambridge University Press, 1997.

M. Brokate. Elastoplastic constitutive laws of nonlinear kinematic hardening type. *Pitman Res. Notes Math. Ser.*, 377:238–272, 1998.

M. Brokate and A. Khludnev. Regularization and Existence of Solutions of Three-dimensional Elastoplastic Problems. *Math. Methods Appl. Sci.*, 21(6):551–564, 1998.

M. Brokate and P. Krejci. Wellposedness of kinematic hardening models in elastoplasticity. *RAIRO Modél. Math. Anal. Numér.*, 32(2):177–209, 1998a.

M. Brokate and P. Krejci. *On the wellposedness of the Chaboche model*, volume 126 of *Internat. Ser. Numer. Math.*, pages 67–79. Birkhäuser, 1998b.

F. E. Browder. Nonlinear Maximal Monotone Operators in Banach Spaces. *Math. Ann.*, 175:89–113, 1968.

F. E. Browder. Nonlinear Variational Inequalities and Maximal Monotone Mappings in Banach Spaces. *Math. Ann.*, 183:213–231, 1969.

R. M. Brown and I. Mitrea. The mixed problem for the Lamé system in a class of Lipschitz domains. *J. Differential Equations*, 246:2577–2589, 2009.

H. Brézis. Monotonicity methods in Hilbert spaces and some applications to nonlinear partial differential equations. *Contrib. nonlin. functional Analysis, Proc. Sympos. Univ. Wisconsin*, pages 101–156, 1971.

H. Brézis. *Operateurs Maximaux Monotones et semi-groupes de contractions dans les espaces de Hilbert*, volume 5 of *Mathematical Studies*. North-Holland Publishing Company/American Elsevier Publishing Company, 1973.

M. Bulíček, J. Frehse, and J. Málek. On boundary regularity for the stress in problems of linearized elasto-plasticity. *International Journal of Advances in Engineering Sciences and Applied Mathematics*, 1(4):141–156, 2009.

A. Bumb and D. Knees. Global spatial regularity for a regularized elasto-plastic model. WIAS Preprint 1419, WIAS Berlin, 2009.

F. Burghahn, V. Schulze, O. Vöhringer, and E. Macherauch. Modellierung des Einflusses von Temperatur und Verformungsgeschwindigkeit auf die Fließspannung von Ck45. *Mat.-wiss. u. Werkstofftech.*, 27:521–530, 1996.

J. Burke. *The Kinetics of Phase Transformations in Metals*. Pergamon Press, 1965.

M. Burtchen, M. Hunkel, T. Lübben, F. Hoffmann, and H.-W. Zoch. Simulation of quenching treatments on bearing components. *HTM*, 61(3):136–141, 2006.

K. Burth and W. Brocks. *Plastizität – Grundlagen und Anwendungen für Ingenieure*. Vieweg, 1992.

F. G. Caballero, C. Capdevila, and C. Garcia de Andres. Kinetics and dilatometric behaviour of non-isothermal ferrite – austenite transformation. *Materials Science and Technology*, 17, 2001.

J. W. Cahn. Transformation kinetics during continuous cooling. *Acta Metallurgica*, 4: 572–575, 1956.

J. A. Canavati and F. Galaz-Fontes. Compactness of Embeddings between Banach Spaces and Applications to Sobolev Spaces. *J. Lond. Math. Soc.*, 2(41):511–525, 1990.

J.-L. Chaboche. A review of some plasticity and viscoplasticity constitutive theories. *Int. J. Plast.*, 24:1642–1693, 2008.

J. Chakrabarty. *Theory of Plasticity*. McGraw-Hill, 1987.

D. S. Chandrasekharaiah and L. Debnath. *Continuum Mechanics.* Academic Press, 1994.

T. Chang and H. J. Choe. Linear elasticity in 3-dimensional Lipschitz domains. *Trends Math.*, 8(2):57–70, December 2005.

O. Chau. On an elasto-dynamic evolution equation with non dead load and friction. *Appl. Math.*, 51(3):229–246, 2006.

K. Chelminski. Stress L^∞-estimates and the Uniqueness Problem for the Quasistatic Equations to the Model of Bodner-Partom. *Math. Methods Appl. Sci.*, 20:1127–1134, 1997.

K. Chelminski. On Self-Controlling Models in the Theory of Nonelastic Material Behavior of Metals. *Continuum Mech. Thermodyn.*, 10:121–133, 1998.

K. Chelminski. On Monotone Plastic Constitutive Equations with Polynomial Growth Condition. *Math. Methods Appl. Sci.*, 22:547–562, 1999.

K. Chelminski. Perfect plasticity as a zero relaxation limit of plasticity with isotropic hardening. *Math. Methods Appl. Sci.*, 24:117–136, 2001.

K. Chelminski. Global existence of weak-type solutions for models of monotone type in the theory of inelastic deformations. *Math. Methods Appl. Sci.*, 25:1195–1230, 2002.

K. Chelminski. Mathematical analysis of the Armstrong-Frederick model from the theory of inelastic deformations of metals. First results and open problems. *Continuum Mech. Thermodyn.*, 15:221–245, 2003a.

K. Chelminski. On quasistatic inelastic models of gradient type with convex composite constitutive equations. *CEJM*, 4:670–689, 2003b.

K. Chelminski and P. Gwiazda. On the Model of Bodner-Partom with Non-Homogenous Boundary Data. *Math. Nachr.*, 214:5–23, 2000a.

K. Chelminski and P. Gwiazda. Nonhomogeneous Initial-Boundary Value Problems for Coercive and Self-Controlling Models of Monotone Type. *Continuum Mech. Thermodyn.*, 12:217–234, 2000b.

K. Chelminski and P. Gwiazda. Convergence of coercive approximations for strictly monotone quasistatic models in inelastic deformation theory. *Math. Methods Appl. Sci.*, 30:1357–1374, 2007.

K. Chelminski and Z. Naniewicz. Coercive limits for constitutive equations of monotone-gradient type. *Nonlinear Anal.*, 48:1197–1214, 2002.

K. Chelminski and R. Racke. Mathematical Analysis of Thermoplasticity with Linear Kinematic Hardening. *J. Appl. Anal.*, 12(1):37–57, 2006.

K. Chelminski, D. Hömberg, and D. Kern. On a thermomechanical model of phase transitions in steel. WIAS Preprint 1225, WIAS Berlin, 2007.

Y-Z. Chen, L.-C. Wu, and B. Hu. *Second Order Elliptic Equations and Elliptic Systems*, volume 174 of *Translations of Mathematical Monographs*. American Mathematical Society, 1998.

W.-S. Cheung. Generalizations of Hölder's Inequality. *IJMMS*, 26(1):7–10, 2001.

J. W. Christian. *The Theory of Transformation in Metals and Alloys*. Pergamon Press, 1965.

P. G. Ciarlet. *Mathematical Elasticity*, volume 1 of *Three Dimensional Elasticity*. Elsevier Science Publishers Ltd., 1988.

P. G. Ciarlet and P. Ciarlet Jr. Another approach to linearized elasticity and Korn's inequality. *C. R. Acad. Sci. Paris*, Ser. I 339:307–312, 2004.

P. G. Ciarlet and J. Nečas. Unilateral problems in nonlinear, three-dimensional elasticity. *Arch. Ration. Mech. Anal.*, 87(4):319–338, 1985.

H. R. Clark, L. P. San Gil Jutuca, and M. Milla Miranda. On a mixed problem for a linear coupled system with variable coefficients. *Electron. J. Differential Equations*, 1998(4):1–20, 1998.

E. A. Coddington and N. Levinson. *Theory of differential equations*. McGraw-Hill, 1955.

B. D. Coleman and M. E. Gurtin. Thermodynamics with Internal State Variables. *The Journal of Chemical Physics*, 47(2):597–613, 1967.

P. Colli, P. Krejci, E. Rocca, and J. Sprekels. Nonlinear evolution inclusions arising from phase change models. WIAS Preprint 974, WIAS Berlin, 2004.

P. Colli, P. Krejčí, E. Rocca, and J. Sprekels. Nonlinear Evolution Inclusions arising from Phase Change Models. *Czechoslovak Math. J.*, 57(132):1067–1098, 2007.

R. W. Cottle, F. Giannessi, and J.-L. Lions, editors. *Variational Inequalities and Complementarity Problems*. John Wiley & Sons Ltd., 1980.

C. Şimşir, T. Lübben, F. Hoffmann, and H-W. Zoch. Dimensional Analysis of Distortion during Through Hardening of Cylindrical Steel Components. In *Proceedings of 4th International Conference on Thermal Process Modeling and Computer Simulation*, 2010.

S. Dachkovski and M. Böhm. Finite thermoplasticity with phase changes based on isomorphisms. *Int. J. Plast.*, 20:323–334, 2004a.

S. Dachkovski and M. Böhm. Objective modeling of some elastoplastic materials with phase changes. *J. Phys. IV France*, 120(120):153–160, 2004b.

S. Dachkovski and M. Böhm. Modeling of elastoplastic materials with phase changes. *J. Phys. IV France*, 120:153–160, 2004c.

S. Dachkovski and M. Böhm. *Modeling of the thermal treatment of steel with phase changes*, pages 299–308. Mechanics of material forces. Springer, 2005.

S. Dachkovski, M. Böhm, A. Schmidt, and M. Wolff. Comparison of several kinetic equations for pearlite transformation in 100Cr6 steel. ZeTeM Report, Universität Bremen, 2003.

C. M. Dafermos and W. J. Hrusa. Energy methods for quasilinear hyperbolic initial-boundary value problems. Applications to elastodynamics. *Arch. Ration. Mech. Anal.*, 87(3):267–292, 1985.

W. Dahl, editor. *Eigenschaften und Anwendungen von Stählen – Band 1: Grundlagen*. Verlag der Augustinus Buchhandlung, 1993.

G. Dal Maso, A. DeSimone, and M. G. Mora. Quasistatic Evolution Problems for Linearly Elastic-Perfectly Plastic Materials. *Arch. Ration. Mech. Anal.*, 180:237–291, 2006.

G. Dal Maso, A. Demyanov, and A. DeSimone. Quasistatic Evolution Problems for Pressure-sensitive Plastic Materials. *Milan J. Math.*, 75:117–134, 2007.

G. Dal Maso, A. DeSimone, M. G. Mora, and M. Morini. A Vanishing Viscosity Approach to Quasistatic Evolution in Plasticity with Softening. *Arch. Ration. Mech. Anal.*, 189: 469–544, 2008.

M. Dalgic and G. Löwisch. Einfluss einer aufgeprägten Spannung auf die isotherme, perlitische und bainitische Umwandlung des Wälzlagerstahls 100Cr6. *HTM*, 59(28), 2004.

M. Dalgic and G. Löwisch. Transformation plasticity at different phase transformations of a bearing steel. *Mat.-wiss. u. Werkstofftech.*, 37(122), 2006.

M. Dalgic, G. Löwisch, and H.-W. Zoch. Beschreibung der Umwandlungsplastizität auf Grund innerer Spannungen während der Phasentransformation des Stahls 100Cr6. *HTM*, 61(222), 2006.

M. Dalgic, A. Irretier, and H.-W. Zoch. Einfluss innerer Spannungen auf die Umwandlungsplastizität und das Umwandlungsverhalten - Beschreibung der Umwandlungsplastizität durch Modelle. *HTM*, 62(4):179–186, 2007.

P. D'Ancona. On Weakly Hyperbolic Equations in Hilbert Spaces. *Funkcial. Ekvac.*, 38: 159–177, 1995.

R. Dautray and J. L. Lions. *Physical Origins and Classical Methods*, volume 1 of *Mathematical Analysis and Numerical Methods for Science and Technology*. Springer, 1990.

R. Dautray and J. L. Lions. *Evolution Problems 1*, volume 5 of *Mathematical Analysis and Numerical Methods for Science and Technology*. Springer, 1992.

E. A. de Souza Neto, D. Perić, and D. R. J. Owen. *Computational Methods for Plasticity: Theory and Applications*. John Wiley & Sons Ltd., 2008.

K. Deimling. *Nonlinear Functional Analysis*. Springer, 1980.

A. Demyanov. Regularity of stresses in Prandtl-Reuss perfect plasticity. *Cal. Var.*, 34: 23–72, 2009.

S. Denis. *Mechanics of solids with phase changes*, chapter Considering stress-phase transformation interactions in the calculation of heat treatment residual stresses. Number 368 in CISM courses and lectures. Springer, 1997.

S. Denis, A. Simon, and G. Beck. *Eigenspannungen: Entstehung, Messung, Bewertung*, chapter Analysis of the thermomechanical behaviour of steel during martensitic quenching and calculation of internal stresses, pages 211–238. DGM, 1983.

S. Denis, D. Farias, and A. Simon. Mathematical model coupling phase transformations and temperature in steels. *ISIJ Int.*, 32:316–325, 1992.

S. Denis, P. Archambault, C. Aubry, A. Mey, J. C. Louin, and A. Simon. Modeling of phase transformation kinetics in steels and coupling with heat treatment residual stress prediction. *J. Phys.*, IV(9):323–332, 1999.

S. Denis, P. Archambault, and E. Gautier. *Handbook of Materials Behavior Models*, chapter Models for Stress-Phase Transformation Couplings in Metallic Alloys, pages 896–904. Academic Press, 2001.

S. Denis, P. Archambault, E. Gautier, A. Simon, and G. Beck. Prediction of residual stress and distortion of ferrous and non-ferrous metals: Current status and future developments. *Journal of Materials Engineering and Performance*, 11(1):92–102, 2002.

U. D. Dhongade and S. G. Deo. Some Generalizations of Bellman-Bihari Integral Inequalities. *J. Math. Anal. Appl.*, 44(1):218–226, 1973.

U. D. Dhongade and S. G. Deo. A Nonlinear Generalization of Bihari's Inequality. *Proc. Amer. Math. Soc.*, 54(1):211–216, 1976.

J.M. Diani, H. Sabar, and M. Berveiller. Micromechanical modelling of the transformation induced plasticity (TRIP) phenomenon in steels. *Int. J. Engng Sci.*, 33(13):1921–1934, 1995.

E. H. Dill. *Continuum Mechanics: Elasticity, Plasticity, Viscoplasticity*. CRC Press, 2007.

J. K. Djoko, F. Ebobisse, A. T. McBride, and B. D. Reddy. A discontinuous Galerkin formulation for classical and gradient plasticity – Part 1: Formulation and analysis. *Comput. Methods Appl. Mech. Engrg.*, 196:3881–3897, 2007a.

J. K. Djoko, F. Ebobisse, A. T. McBride, and B. D. Reddy. A discontinuous Galerkin formulation for classical and gradient plasticity – Part 2: Algorithms and numerical analysis. *Comput. Methods Appl. Mech. Engrg.*, 197:1–21, 2007b.

W. Dreyer, D. Hömberg, and T. Petzold. A Model for the Austenite-Ferrite Phase Transition in Steel Including Misfit Stress. WIAS Preprint 1310, WIAS Berlin, 2008.

D. Drivaliaries and N. Yannakakis. Generalizations of the Lax-Milgram Theorem. *Bound. Value Probl.*, 2007:1–9, 2007.

F. Dunne and N. Petrinic. *Introduction to Computational Plasticity*. Oxford University Press, 2005.

M. K. Duszek. Problems of Geometrically Non-Linear Theory of Plasticity. Preprint 21, Ruhr-Universität Bochum, 1980.

G. Duvaut and J. L. Lions. *Inequalities in Mechanics and Physics*, volume 219 of *A Series of Comprehensive Studies in Mathematics*. Springer, 1976.

S. Ebenfeld. Non-linear initial boundary value problems of hyperbolic-parabolic type. A general investigation of admissible couplings between systems of higher order. *Math. Methods Appl. Sci.*, 25:241–262, 2002.

C. Ebmeyer. Mixed Boundary Value Problems for Nonlinear Ellipitic Systems in *n*-Dimensional Lipschitzian Domains. *J. Anal. Appl.*, 18(3):539–555, 1999.

C. Ebmeyer and J. Frehse. Mixed Boundary Value Problems for Nonlinear Elliptic Equations in Multidimensional Non-smooth Domains. *Math. Nachr.*, 1997.

F. Ebobisse and B. D. Reddy. Some mathematical problems in perfect plasticity. *Comput. Methods Appl. Mech. Engrg.*, 193:5071–5094, 2004.

C. Eck. *A Two-Scale Phase Field Model for Liquid-Solid Phase Transitions of Binary Mixtures with Dendritic Microstructure*. Habilitation Thesis, Universität Erlangen-Nürnberg, 2004.

C. Eck, J. Jarušek, and M. Krbec. *Unilateral Contact Problems – Variational Methods and Existence Theorems*. Pure and Applied Mathematics. Chapman and Hall / CRC Press, 2005.

C. Eck, H. Garcke, and P. Knabner. *Mathematische Modellierung*. Springer, 2008.

I. Ekeland and R. Temam. *Convex Analysis and Variational Problems*. North-Holland Publishing Company, 1976.

C. M. Elliott and T. Qi. A Dynamic Contact Problem in Thermoelasticity. *Nonlinear Anal.*, 23(7):883–898, 1994.

F. Ellyin. *Fatigue Damage, Crack Growth, and Life Prediction*. Chapman and Hall / CRC Press, 1997.

E. Emmrich. *Gewöhnliche und Operator-Differentialgleichungen – Eine integrierte Einführung in Randwertprobleme und Evolutionsgleichungen für Studierende*. Vieweg, 2004.

J. Escher. Global existence and nonexistence for semilinear parabolic systems with nonlinear boundary conditions. *Math. Ann.*, 284:285–305, 1989.

J. Escher. Nonlinear Elliptic Systems with dynamic boundary conditions. *Math. Z.*, 210: 413–439, 1992.

L. C. Evans. *Partial Differential Equations*, volume 19 of *Graduate Studies in Mathematics*. American Mathematical Society, 1998.

M. Fanfoni and M. Tomellini. The Johnson-Mehl-Avrami-Kolmogorov model: A brief review. *Il Nuovo Cimento D*, 20(7-8):1171–1182, 1998.

A. Fasano and M. Primicerio. An analysis of phase transition models. *European J. Appl. Math.*, 7:439–451, 1996.

A. Fasano, D. Hömberg, and L. Panizzi. A mathematical model for case hardening of steel. WIAS Preprint 1283, WIAS Berlin, 2007.

F. M. Fernandes, S. Denis, and A. Simon. Mathematical model coupling phase transformation and temperature evolution during quenching of steel. *Materials Science and Technology*, 1, 1985.

R. P. Feynman, R. B. Leighton, and M. Sands. *Vorlesungen über Physik*, volume 2. R. Oldenbourg Verlag, 1991.

I. Figueiredo and L. Trabucho. A Class of Contact and Friction Dynamic Problems in Thermoelasticity and in Thermoviscoelasticity. *Internat. J. Engrg. Sci.*, 33(1):45–66, 1995.

A. F. Fillipov. *Differential equations with discontinuous right-hand side*. Kluwer Academic Publishers, 1988.

F. D. Fischer. A micromechanical model for transformation plasticity in steels. *Acta Metall. Mater.*, 38:1535–1546, 1990.

F. D. Fischer. *Modelling and simulation of transformation induced plasticity in elasto-plastic materials*, volume 368 of *Mechanics of Solids with Phase Changes*. Springer Verlag, 1997.

F. D. Fischer, M. Berveiller, K. Tanaka, and E. R. Oberaigner. Continuum mechanical aspects of phase transformations in solids. *Arch. Appl. Mech.*, 64:54–85, 1994.

F. D. Fischer, Q. P. Sun, and K. Tanaka. Transformation-induced Plasticity (TRIP). *Applied Mechanics Reviews*, 49:317–364, 1996.

F. D. Fischer, E. R. Oberaigner, K. Tanaka, and F. Nishimura. Transformation Induced Plasticity Revised and Updated Formulation. *Int. J. Solids Structures*, 35(18):2209–2227, 1998.

F. D. Fischer, G. Reisner, E. Werner, K. Tanaka, G. Cailletaud, and T. Antretter. A new view on transformation induced plasticity (TRIP). *Int. J. Plast.*, 16:723–748, 2000a.

F. D. Fischer, N. K. Simha, and J. Svoboda. Kinetics of diffusional phase transformation in multicomponent elastic-plastic materials. *Trans. ASME*, 125:266–276, 2003.

F.D. Fischer, T. Antretter, F. Azzouz, G. Cailletaud, A. Pineau, K. Tanaka, and K. Nagayama. The role of backstress in phase transforming steels. *Arch. Mech.*, 52: 569–588, 2000b.

A. S. M. Fonseca. *Simulation der Gefügeumwandlungen und des Austenitkornwachstums bei der Wärmebehandlung von Stählen.* PhD Thesis, RWTH Aachen, 1996.

C. Franz, G. Besserdich, V. Schulze, H. Müller, and D. Löhe. Influence of transformation plasticity on residual stresses and distortions due to the heat treatment of steels with different carbon contents. *J. Phys. IV France*, 120:481–488, 2004.

J. Frehse and D. Löbach. Regularity Results for Three Dimensional Isotropic and Kinematic Hardening Including Boundary Differentiability. Preprint 432, Universität Bonn, 2008.

F. Frerichs, T. Lübben, F. Hoffmann, and H.-W. Zoch. Numerical analysis of distortion due to inhomogeneous distribution of martensite-start temperature within SAE 52100 bearing rings. *Steel Research*, 78(7):560–565, 2007a.

F. Frerichs, T. Lübben, F. Hoffmann, H.-W. Zoch, and M. Wolff. Unavoidable distortion due to thermal stresses. In *Proc. 5th International Conference on Quenching and Control of distortion*, 2007b.

F. Frerichs, T. Lübben, F. Hoffmann, and H.-W. Zoch. Distortion of conical shaped bearing rings made of SAE52100. *Mat.-wiss. u. Werkstofftech.*, 40(5-6):402–407, 2009.

M. Fuchs. Generalizations of Korn's Inequality based on Gradient Estimates in Orlicz Spaces and Applications to Variational Problems in 2D involving the Trace Free Part of the Symmetric Gradient. *J. Math. Sci.*, 167(3):418–434, 2010.

M. Fuchs and G. Seregin. *Variational Methods for Problems from Plasticity Theory and for Generalized Newtonian Fluids*, volume 1749 of *Lecture Notes in Mathematics*. Springer, 2000.

S. Fučik and A. Kufner, editors. *Initial value problems for elastoplastic and elasto-viscoplastic systems*, Nonlinear Analysis, Function Spaces and Applications, 1978.

J. Fuhrmann and D. Hömberg. Numerical simulation of the surface hardening of steel. *Internat. J. Numer. Methods Heat Fluid Flow*, 9(6):705–724, 1999.

J. Fuhrmann, T. Koprucki, and H. Langmach. PDELIB: An Open Modular Tool Box For The Numerical Solution Of Partial Differential Equations. Design Patterns, 1998.

H. Gajewski, K. Gröger, and K. Zacharias. *Nichtlineare Operatorgleichungen und Operatordifferentialgleichungen*. Akademie-Verlag Berlin, 1974.

T. Gallout and A. Monier. On the regularity of Solutions to Elliptic Equations. *Rend. Mat. Appl.*, 19:471–488, 1999.

C. Garcia de Andres, F. G. Caballero, C. Capdevila, and H. K. D. H. Bhadeshia. Modeling of kinetics and dilatometric behaviour of non-isothermal pearlite-to-austenite transformation in an eutectoid steel. *Scripta Materialia*, 39(6):791–796, 1998.

E. Gautier, J. Zhang, Y. Wen, and S. Denis. *Effects of stress on martensitic transformation in ferrous alloys. Experiments and numerical simulations, in: Phase transformations and evolution in materials*. The Minerals, Metals and Materials Society, 2000.

J. Gawinecki. Existence, Uniqueness and Regularity of the Solution of the First Boundary-Initial Value Problem for the Equations of Linear Thermo-Microelasticity. *Bull. Pol. Acad. Sci. Tech.*, 34(7-8):447–460, 1986.

J. Gawinecki. Global Solution to the Cauchy Problem in Non-Linear Hyperbolic Thermoelasticity. *Math. Methods Appl. Sci.*, 15:223–237, 1992.

J. Gawinecki. Global existence of solutions for non-small data to non-linear spherically symmetric thermoviscoelasticity. *Math. Methods Appl. Sci.*, 26:907–936, 2003.

J. A. Gawinecki. Initial-boundary value problem in nonlinear hyperbolic thermoelasticity. Some applications in continuum mechanics. Dissertationes Mathematicae, 2002.

J. A. Gawinecki, B. Sikorska, G. Nakamura, and J. Rafa. Mathematical and physical interpretation of the solution to the initial-boundary value problem in linear hyperbolic thermoelasticity theory. *ZAMM*, 87(10):715–746, 2007.

M. Giaquinta. A Counter-Example to the Boundary Regularity of Solutions to Elliptic Quasilinear Systems. *Manuscripta Math.*, 24:217–220, 1978.

M. Giaquinta and G. Modica. Almost-Everywhere Regularity Results for Solutions of Non-Linear Elliptic Systems. *Manuscripta Math.*, 28:109–158, 1979.

D. Gilbarg and N. S. Trudinger. *Elliptic Partial Differential Equations of Second Order.* Springer, 2001.

R. P. Gilbert and B. Jacek. Domains of fractional powers of operators arising in mixed boundary value problems in non-smooth domains and applications. *Appl. Anal.*, 55(1): 79–89, 1994.

C. Giorgi and V. Pata. Stability of Abstract Linear Thermoelastic Systems with Memory. *Math. Models Methods Appl. Sci.*, 11(4):627–644, 2001.

A. Glitzky and R. Hünlich. Resolvent estimates in $W^{-1,p}$ related to strongly coupled linear parabolic systems with coupled nonsmooth capacities. WIAS Preprint 1086, WIAS Berlin, 2006.

K. Gröger. Zur Theorie des quasi-statischen Verhaltens von elastisch-plastischen Körpern. *ZAMM*, 58:81–88, 1978a.

K. Gröger. Zur Theorie des dynamischen Verhaltens von elastisch-plastischen Körpern. *ZAMM*, 58:483–487, 1978b.

K. Gröger. Evolution equations in the theory of plasticity. In *Theory of nonlinear operators*, volume 5 (6N) of *Int. Summer Sch.*, pages 97–107. Berlin 1997, Abh. Akad. Wiss. DDR, 1978c.

K. Gröger. Asymptotic Behaviour of Elastoplastic Bodies under Periodic Loading. *ZAMM*, 60:25–30, 1980.

K. Gröger. A $W^{1,p}$-Estimate for Solutions to Mixed Boundary Value Problems for Second Order Elliptic Differential Equations. *Math. Ann.*, 283:679–687, 1989.

K. Gröger. Boundedness and continuity of solutions to linear elliptic boundary value problems in two dimensions. *Math. Ann.*, 298:719–722, 1994.

K. Gröger and R. Hünlich. An Existence-Uniqueness Result for some Models of Thermoplasticity. *ZAMM*, 60:169–171, 1980.

K. Gröger, J. Nečas, and L. Trávníček. Dynamic Deformation Processes of Elastic-Plastic Systems. *ZAMM*, 59:567–572, 1979.

J. A. Griepentrog. *Zur Regularität linearer elliptischer und parabolischer Randwertprobleme mit nichtglatten Daten.* PhD Thesis, Humboldt-Universität zu Berlin, 1999.

J. A. Griepentrog, K. Gröger, H.-C. Kaiser, and J. Rehberg. Interpolation for function spaces related to mixed boundary value problems. *Math. Nachr.*, 241:110–120, 2002.

R. Griesse and C. Meyer. Optimal control of static plasticity with linear kinematic hardening. WIAS Preprint 1370, WIAS Berlin, 2008.

P. Grisvard. *Elliptic Problems in Nonsmooth Domains.* Pittmann Publishing, 1985.

C. Grossmann, H.-G. Roos, and M. Stynes. *Numerical Treatment of Partial Differential Equations.* Springer, 2005.

S. Grostabussiat, L. Taleb, J.F. Jullien, and F. Sidoroff. Transformation induced plasticity in martensitic transformation of ferrous alloys. *J. Phys.*, 11:4–173, 2001.

P. Gruber, D. Knees, S. Nesenenko, and M. Thomas. Analytical and numerical aspects of time-dependent models with internal variables. *ZAMM*, 90(10-11):861–902, 2010.

S. Guillaume and A. Syam. On a time-dependent subdifferential evolution inclusion with a nonconvex upper-semicontinuous perturbation. *Electron. J. Qual. Theory Differ. Equ.*, 2005(11):1–22, 2005.

P. Håkansson, M. Wallin, and M. Ristinmaa. Comparison of isotropic hardening and kinematic hardening in thermoplasticity. *Int. J. Plast.*, 21:1435–1460, 2005.

H. Hallberg, P. Håkansson, and M Ristinmaa. A constitutive model for the transformation of martensite in austenitic steels under large strain plasticity. *Int. J. Plast.*, 23:1213–1239, 2007.

R. Haller-Dintelmann. *L^p-Regularity Theory for Linear Elliptic and Parabolic Equations.* Habilitation Thesis, TU Darmstadt, 2008.

R. Haller-Dintelmann and J. Rehberg. Maximal parabolic regularity for divergence operators including mixed boundary conditions. WIAS Preprint 1288, WIAS Berlin, 2008.

R. Haller-Dintelmann, H.-C. Kaiser, and J. Rehberg. Elliptic model problems including mixed boundary conditions and material heterogeneities. WIAS Preprint 1203, WIAS Berlin, 2007.

W. Han and B. D. Reddy. Convergence analysis of discrete approximations of problems in hardening plasticity. *Comput. Methods Appl. Mech. Engrg.*, 171:327–340, 1999a.

W. Han and B. D. Reddy. *Plasticity: Mathematical Theory and Numerical Analysis.* Springer, 1999b.

W. Han and B. D. Reddy. Convergence of approximations to the primal problem in plasticity under conditions of minimal regularity. *Numer. Math.*, 87:283–315, 2000.

W. Han and M. Sofonea. Evolutionary Variational Inequalities Arising in Viscoelastic Contact Problems. *SIAM J. Numer. Anal.*, 38(2):556–579, 2001.

W. Han and M. Sofonea. *Quasistatic Contact Problems in Viscoelasticity and Viscoplasticity.* AMS/IP, 2002.

W. Han, B. D. Reddy, and G. C. Schroeder. Qualitative and Numerical Analysis of Quasi-Static Problems in Elastoplasticity. *SIAM J. Numer. Anal.*, 34(1):143–177, 1997.

W. Han, M. Shillor, and M. Sofonea. Variational and numerical analysis of a quasistatic viscoelastic problem with normal compliance, friction and damage. *J. Compt. Appl. Anal.*, 137:377–398, 2001.

S. Hartmann. *Lösung von Randwertaufgaben der Elastoplastizität – Ein Finite Elemente Konzept für nichtlineare kinematische Verfestigung bei kleinen und finiten Verzerrungen.* PhD Thesis, Universität Gesamthochschule Kassel, 1993.

K. Hashiguchi. Generalized plastic flow rule. *Int. J. Plast.*, 21:321–251, 2005.

P. Haupt. *Continuum Mechanics and Theory of Materials.* Springer, 2000.

Peter Haupt. *Viskoelastizität und Plastizität: thermomechanisch konsistente Materialgleichungen.* Springer, 1977.

K. Hayasida and K. Wada. On the Regularity Property for Solutions of the Equation of Linear Elastostatics with Discontinuous Boundary Condition. *Japan J. Indust. Appl. Math.*, 16:377–399, 1999.

D. Helm. Experimentelle Untersuchung und phänomenologische Modellierung thermomechanischer Effekte in der Materialtheorie. Preprint 1, Universität Kassel, 1998.

R. Herzog, C. Meyer, and G. Wachsmuth. Existence and Regularity of the Plastic Multiplier in Static and Quasistatic Plasticity. *GAMM-Mitt.*, 34(1):39–44, 2011a.

R. Herzog, C. Meyer, and G. Wachsmuth. Integrability of Displacement and Stresses in Linear and Nonlinear Elasticity with Mixed Boundary Conditions. *J. Math. Anal. Appl.*, 2011b.

M. Hieber and J. Rehberg. Quasilinear Parabolic Systems with mixed Boundary Conditions on nonsmooth Domains. *SIAM J. Math. Anal.*, 40(1):292–305, 2008.

R. Hill. *The Mathematical Theory of Plasticity.* Oxford University Press, 1950.

I. Hlavácek, J. Haslinger, J. Nečas, and J. Lovísek. *Solution of Variational Inequalities in Mechanics*, volume 66 of *Applied Mathematical Sciences*. Springer, 1988.

I. Hüßler. Mathematische Untersuchungen eines gekoppelten Systems von ODE und PDE zur Modellierung von Phasenumwandlungen im Stahl. Diploma Thesis, Universität Bremen, 2007.

D. Hömberg. A mathematical model for the phase transitions in eutectoid carbon steel. *IMA J. Appl. Math.*, 54:31–57, 1995.

D. Hömberg. A numerical simulation of the Jominy end-quench test. *Acta Materialia*, 44(11):4375–4385, 1996.

D. Hömberg. Irreversible phase transitions in steel. *Math. Methods Appl. Sci.*, 20:59–77, 1997.

D. Hömberg and D. Kern. The heat treatment of steel – A mathematical control problem. WIAS Preprint 1402, WIAS Berlin, 2009.

D. Hömberg and A. Khludnev. Evolution equations in the theory of plasticity. *IMA J. Appl. Math.*, 71:479–495, 2006a.

D. Hömberg and A. Khludnev. A thermoelastic contact problem with a phase transition. *IMA J. Appl. Math.*, 71:479–495, 2006b.

D. Hömberg, N. Togobytska, and M. Yamamoto. On the evalution of dilatometer experiments. *Appl. Anal.*, 88(5):669–681, 2009.

F. Hoffmann. Zur Frage der Herkunft des Kohlenstoffs in der Verbindungsschicht beim Nitrieren (Nitrocarburieren). *HTM*, 36(4):255–257, 1981.

F. Hoffmann, I. Bujak, P. Mayr, B. Löffelbein, M. Gienau, and K.-H. Habig. Verschleißwiderstand nitrierter und nitrocarburierter Stähle. *HTM*, 52:376–386, 1997.

F. Hoffmann, D. Guenther, P. Mayr, and M. Jung. Low temperature nitriding and carburizing. In *Carburizing and Nitriding*, Tschechische Association for Heat Treatment of Metals, 2001.

F. Hoffmann, O. Keßler, T. Lübben, and P. Mayr. "Distortion Engineering" – Verzugsbeherrschung in der Fertigung. *HTM*, 57(3):213–217, 2002.

D. Horstmann. *Das Zustandsschaubild Eisen-Kohlenstoff und die Grundlagen der Wärmebehandlung der Eisen-Kohlenstoff-Legierungen*. Stahleisen, 1992.

H. P. Hougardy and K. Yamazaki. An improved calculation of the transformations in steels. *Steel Research*, 57(9):466–471, 1986.

L. Hsiao and S. Jiang. *Handbook of Differential Equations*, volume 1 of *Evolutionary Equations*, chapter Nonlinear Hyperbolic-Parabolic Coupled Systems, pages 287–384. Elsevier Science Ltd., 2004.

S. Hu and N. S. Papageorgiou. *Time-Dependent Subdifferential Evolution Inclusions and Optimal Control*, volume 632 of *Mem. Am. Math. Soc.* AMS, 1998.

T. J. R. Hughes, T. Kato, and J. E. Marsden. Well-posed quasi-linear second-order hyperbolic systems with applications to nonlinear elastodynamics and general relativity. *Arch. Ration. Mech. Anal.*, 63(3):273–294, 1976.

M. Hunkel, T. Lübben, F. Hoffmann, and P. Mayr. Modellierung der bainitischen und perlitischen Umwandlung bei Stählen. *HTM*, 54(6):365–372, 1999.

M. Hunkel, T. Lübben, F. Hoffmann, and P. Mayr. Using the Jominy end-quench test for validation of thermo-metallurgical model parameters. *J. Phys. IV France*, 120: 571–579, 2004.

K. Hutter and K. D. Jöhnk. *Continuum Methods of Physical Modelling – Continuum Mechanics, Dimensional Analysis, Turbulance*. Springer, 2004.

B. Ilschner and R. F. Singer. *Werkstoffwissenschaften und Fertigungstechnik*. Springer, 2005.

T. Inoue and T. Tanaka. Unified constitutive equation for transformation plasticity and identification of the TP coefficients. *Arch. Mech.*, 2006.

T. Inoue and Z. Wang. *Numerical methods in industrial forming processes*, chapter Finite element analysis of coupled thermoinelastic problem with phase transformation, pages 391–400. Int. Conf. Num. Meth. in Industrial Forming Processes. Pittmann, Pineridge Press, 1982.

T. Inoue and Z. Wang. Coupling between stress, temperature and metallic structures during processes involving phase transformations. *Materials Science and Technology*, 1:845–850, 1985.

T. Inoue, S. Nagaki, T. Kishino, and M. Monkawa. Description of transformation kinetics, heat conduction and elastic-plastic stress in the course of quenching and tempering of some steels. *Ingenieur-Archiv*, 50:315–327, 1981.

T. Inoue, Z. Wang, and K. Miyao. Quenching stress of carburized steel gear wheel. In *Proceedings of the 2nd International Conference on Residual Stresses*, pages 606–611, 1989.

H. Ito. On Certain Mixed-Type Boundary-Value Problems of Elastostatics. *Tsukuba J. Math.*, 14(1):133–153, 1990a.

H. Ito. On a Mixed Problem of Linear Elastodynamics with a Time-Dependent Discontinuous Boundary Condition. *Osaka J. Math.*, 27:667–707, 1990b.

J.-M. Jeong, D.-H. Jeong, and J.-Y. Park. Nonlinear Variational Evolution Inequalities in Hilbert Spaces. *Internat. J. Math. & Math. Sci.*, 23(1):11–20, 2000.

S. Jiang. Global Solutions of the Neumann Problem in One-Dimensional Nonlinear Thermoelasticity. *Nonlinear Anal.*, 19(2):107–121, 1992.

S. Jiang and R. Racke. On some Quasilinear Hyperbolic-Parabolic Initial Boundary Value Problems. *Math. Methods Appl. Sci.*, 12:315–339, 1990.

S. Jiang and R. Racke. *Evolution Equations in Thermoelasticity*, volume 112 of *Monographs and Surveys in Pure and Applied Mathematics*. Chapman and Hall / CRC Press, 2000.

Y. Jiang and P. Kurath. Characteristics of the Armstrong-Frederick type plasticity models. *Int. J. Plast.*, 12:387–415, 1996.

C. Johnson. Existence Theorems for Plasticity Problems. *J. Math. Pures Appl.*, 55: 431–444, 1976.

C. Johnson. On Plasticity with Hardening. *J. Math. Anal. Appl.*, 62:325–336, 1978.

W. A. Johnson and R. F. Mehl. Reaction kinetics in process of nucleation and growth. *Trans. AIME*, 135:416–458, 1939.

I. A. Kaliev and M. F. Mugafarov. Some Problems of Linear Thermoelasticity in the Ginzburg-Landau Theory of Phase Transitions. *J. Appl. Mech. Tech. Phys.*, 44(6): 866–871, 2003.

M. Kaßmann and W. R. Madych. Regularity for Elliptic Mixed Boundary Problems of Second Order. Preprint 188, Universität Bonn, October 2004.

P. Kaminski. Nonlinear problems in inelastic deformation theory. *ZAMM*, 88(4):267–282, 2008.

P. Kaminski. Nonlinear quasistatic problems of gradient type in inelastic deformation theory. *J. Math. Appl.*, 357(1):284–299, 2009a.

P. Kaminski. Regularity of solutions to coercive and self-controlling viscoplastic problems. Technical report, Warsaw University of Technology, 2009b.

P. Kaminski. Boundary regularity for self-controlling and Cosserat models of viscoplasticity. Interior estimates for models of power-type. Technical report, Warsaw University of Technology, 2009c.

R. Kano. Applications of abstract parabolic quasi-variational inequalities to obstacle problems. *Banach Center Publ.*, 86:163–174, 2009.

R. Kano, Y. Murase, and N. Kenmochi. Nonlinear Evolution Equations Generated by Subdifferentials with nonlocal Constraints. *Banach Center Publ.*, 86:175–194, 2009.

T. Kato. *Abstract differential equations and nonlinear mixed problems.* Publications of the Scuola Normale Superiore. Edizioni della Normale, 1988.

N. Kenmochi. Some Nonlinear Parabolic Variational Inequalities. *Isreal J. Math.*, 22 (3-4):304–331, 1975.

N. Kenmochi. Nonlinear Evolution Equations with Variable Domains in Hilbert Spaces. *Proc. Japan Acad.*, 53(A)(5):163–166, 1977a.

N. Kenmochi. Nonlinear Parabolic Variational Inequalities with Time-Dependent Constraints. *Proc. Japan Acad.*, 53(A)(6):186–189, 1977b.

N. Kenmochi and M. Niezgódka. Evolution Systems of Nonlinear Variational Inequalities arising from Phase Change Problems. *Nonlinear Anal.*, 22(9):1163–1180, 1994.

N. Kenmochi and I. Pawlow. A Class of Nonlinear Ellipitc-Parabolic Equations with Time-Dependent Constraints. *Nonlinear Anal.*, 10(11):1181–1202, 1986.

N. Kenmochi and N. Toshitaka. Weak Solutions for Certain Nonlinear Time-Dependent Parabolic Variational Inequalities. *Hiroshima Math. J.*, 5:525–535, 1975.

D. Kern. *Analysis and numerics for a thermomechanical phase transition model in steel.* PhD Thesis, TU Berlin, 2011.

M. E. Khalifa. Existence of almost everywhere solution for nonlinear hyperbolic-parabolic system. *Appl. Math. Comput.*, 145:569–577, 2003.

A. S. Khan and S. Huang. *Continuum Theory of Plasticity.* John Wiley & Sons Ltd., 1995.

A. Khludnev and J. Sokolowski. On Solvability of Boundary Value Problems in Elasto-plasticity. Rapport de Recherche 3163, INRIA, Mai 1997.

D. Kinderlehrer and G. Stampacchia. *An introduction to variational inequalities and their applications.* Academic Press, 1980.

P. Knabner and L. Angermann. *Numerik partieller Differentialgleichungen.* Springer, 2000.

D. Knees. *Regularity results for quasilinear elliptic systems of power-law growth in nonsmooth domains – Boundary, transmission and crack problems.* PhD Thesis, Universität Stuttgart, 2005.

D. Knees. Global regularity of the elastic fields of a power-law model on Lipschitz domains. *Math. Methods Appl. Sci.*, 29:1363–1391, 2006.

D. Knees. Short note on global spatial regularity in elasto-plasticity with linear hardening. WIAS Preprint 1337, WIAS Berlin, 2008.

D. Knees. On global spatial regularity in elasto-plasticity with linear hardening. *Calc. Var.*, 36:611–625, 2009.

D. Kohtz. *Wärmebehandlung metallischer Werkstoffe: Grundlagen und Verfahren.* VDI, 1994.

D. P. Koistinen and R. E. Marburger. A general equation prescribing the extent of the austenite-martensite transformation in pure iron-carbon and plain carbon steels. *Acta Metallurgica*, 7(1):59–60, 1959.

A. N. Kolmogorov and S. V. Fomin. *Introductory Real Analysis*. Prentice-Hall, Inc., 1970.

A. I. Koshelev and S. I. Chelkak. *Regularity of Solutions of Quasilinear Elliptic Systems*, volume 77 of *Teubner-Texte zur Mathematik*. Teubner, 1985.

P. Krejci and J. Sprekels. The von Mises model for one-dimensional elastoplastic beams and Prandtl-Ishlinskii hysteresis operators. WIAS Preprint 1143, WIAS Berlin, 2006.

P. Krejci, J. Sprekels, and U. Stefanelli. One-dimensional thermo-visco-plastic processes with hysteresis and phase transitions. WIAS Preprint 702, WIAS Berlin, 2001.

P. Krejci, J. Sprekels, and H. Wu. Elastoplastic Timoshenko beams. WIAS Preprint 1430, WIAS Berlin, 2009.

M. Kubo. Characterization of a Class of Evolution Operators Generated by Time-Dependent Subdifferentials. *Funkcial. Ekvac.*, 32:301–321, 1989.

M. Kubo. Variational inequalities with time-dependent constraints in L^p. *Nonlinear Anal.*, 73:390–398, 2010.

M. Kubo and N. Yamazaki. Elliptic-Parabolic Variational Inequalities with Time-Dependent Constraints. *Discrete Contin. Dyn. Syst.*, 19(2):335–359, October 2007.

M. Kunze and M. D. P. Monteiro Marques. On Parametric Quasi-Variational Inequalities and State-Dependent Sweeping Processes. *Topol. Methods Nonlinear Anal.*, 12: 179–191, 1998.

O. A. Ladyženskaya. *The Boundary Value Problems of Mathematical Physics*. Springer, 1985.

O. A. Ladyženskaya, V. A. Solonnikov, and N. N. Ural'ceva. *Linear and quasilinear equations of parabolic type*, volume 28 of *Translations of Mathematical Monographs*. American Mathematical Society, 1968.

V. Lakshmikantham. A Variation of Constants Formula and Bellman-Gronwall-Reid Inequalities. *J. Math. Anal. Appl.*, 41:199–204, 1973.

L. D. Landau and E. M. Lifschitz. *Elastizitätstheorie*, volume 7 of *Lehrbuch der theoretischen Physik*. Akademie-Verlag Berlin, 1989.

H. Lang, K. Dressler, R. Pinnau, and G. Bitsch. A homotopy between the solutions of the elastic and the elastoplastic boundary value problem. AGTM Report, Universität Kaiserslautern, 2006a.

H. Lang, K. Dressler, R. Pinnau, and M. Speckert. Error estimates for quasistatic global elastic correction and linear kinematic hardening material. AGTM Report, Universität Kaiserslautern, 2006b.

M. Lanza de Cristoforis and T. Valent. On Neumann's problem for a quasilinear differential system of the finite elastostatic type. Local theorems of existence and uniqueness. *Rend. Semin. Mat. Univ. Padova*, 68:183–206, 1982.

S. Larson and V. Thomée. *Partial Differential Equations and Numerical Methods*, volume 45 of *Texts in Applied Mathematics*. Springer, 2003.

S. Larsson and M. Mangard. Determination of phase transformation kinetics when tempering martensitic hardened low alloy steel. Technical report, LiTH-IKP-Ex-1234, Institut of Technology, Department of Mechanical Engineering, Linköping, Sweden, 1995.

I. Lasiecka and R. Triggiani. Regularity of Hyperbolic Equations under $L_2(0,T;L_2(\Gamma))$-Dirichlet Boundary Terms. *Appl. Math. Optim.*, 10:275–286, 1983.

D. Löbach. Interior Stress Regularity for the Prandtl Reuss and Hencky Model of Perfect Plasticity Using the Perzyna Approximation. Preprint 353, Universität Bonn, 2007a.

D. Löbach. Hölder Continuity for the Displacements in Isotropic and Kinematic Hardening with von Mises Yield Criterion. Preprint 359, Universität Bonn, 2007b.

D. Löbach. On Regularity for Plasticity with Hardening. Preprint 388, Universität Bonn, 2008.

D. Löbach. Improved L^p- Estimates for the Strain Velocities in Hardening Problems. Preprint 475, Universität Bonn, 2010.

P. Le Tallec. *Numerical Analysis of Viscoelastic Problems*. Springer, 1990.

J. B. Leblond and J. Devaux. A new kinetic model for anisothermal metallurgical transformations in steels including effect of austenite grain size. *Acta Metallurgica*, 32:137–146, 1984.

J. B. Leblond and J. C. Devaux. Mathematical modelling of transformation plasticity in steels – I: Case of ideal-plastic phases. *Int. J. Plast.*, 5:551–572, 1989a.

J. B. Leblond and J. C. Devaux. Mathematical modelling of transformation plasticity in steels – II: Coupling with strain hardening phenomena. *Int. J. Plast.*, 5:573–591, 1989b.

J. B. Leblond, G. Mottet, J. Devaux, and J. C. Devaux. Mathematical models of anisothermal phase transformations in steels and predicted plastic behaviour. *Materials Science and Technology*, 1:815–822, 1985.

J. B. Leblond, G. Mottet, and J. C. Devaux. A theoretical and numerical approach to the plastic behaviour of steels during phase transformations – I. Derivation of General Relations. *J. Mech. Phys. Solids*, 34(4):395–409, 1986a.

J. B. Leblond, G. Mottet, and J. C. Devaux. A theoretical and numerical approach to the plastic behaviour of steels during phase transformations – II. Study of Classical Plasticity for Ideal-Plastic Phases. *J. Mech. Phys. Solids*, 34(4):411–432, 1986b.

J. Lemaitre. *Handbook of Materials Behavior Models*. Academic Press, 2001.

J. Lemaitre and J.-L. Chaboche. *Mechanics of solid materials*. Cambridge University Press, 1990.

J. Lemaitre and R. Desmorat. *Engineering Damage Mechanics*. Springer, 2005.

V. I. Levitas. Thermomechanics and Kinetics of Generalized Second-Order Phase Transitions in Inelastic Materials. Application to Ductile Fracture. *Mechanics Research Communications*, 25(4):427–436, 1998.

D. Liedtke. Wärmebehandlung von Stahl – Nitrieren und Nitrocarburieren. Merkblatt 447, Stahl-Informations-Zentrum, 2005a.

D. Liedtke. Wärmebehandlung von Stahl – Härten, Anlassen, Vergüten, Bainitisieren. Merkblatt 450, Stahl-Informations-Zentrum, 2005b.

D. Liedtke. Wärmebehandlung von Stahl – Einsatzhärten. Merkblatt 452, Stahl-Informations-Zentrum, 2008.

D. Liedtke. Wärmebehandlung von Stahl – Randschichthärten. Merkblatt 236, Stahl-Informations-Zentrum, 2009.

J. L. Lions. *Quelques méthodes de résolution des problèmes aux limites non linéaires*. Études Mathématiques. Dunod Gauthier-Villars, 1969.

J. L. Lions and E. Magenes. *Non-Homogenous Boundary Value Problems and Applications 1*, volume 183 of *Die Grundlehren der mathematischen Wissenschaften in Einzeldarstellungen*. Springer, 1973a.

J. L. Lions and E. Magenes. *Non-Homogenous Boundary Value Problems and Applications 2*, volume 183 of *Die Grundlehren der mathematischen Wissenschaften in Einzeldarstellungen*. Springer, 1973b.

J. L. Lions and E. Magenes. *Non-Homogenous Boundary Value Problems and Applications 3*, volume 183 of *Die Grundlehren der mathematischen Wissenschaften in Einzeldarstellungen*. Springer, 1973c.

H. W. Lord and Y. Shulman. A Generalized Dynamical Theory of Thermoelasticity. *J. Mech. Phys. Solids*, 15:299–309, 1967.

V. A. Lubarda. *Elastoplasticity Theory*. CRC Press, 2002.

J. Lubliner. *Plasticity Theory*. Pearson Education, Inc., 2006.

A. Lunardi. *Handbook of Differential Equations*, volume 1 of *Evolutionary Equations*, chapter Nonlinear Parabolic Equations and Systems, pages 385–436. Elsevier Science Ltd., 2004.

A. I. Lurie. *Theory of Elasticity*. Foundations of Engineering Mechanics. Springer, 2005.

E. Macherauch. *Praktikum in Werkstoffkunde*. Vieweg, 1992.

C. L. Magee. *Phase Transformation*, chapter The Nucleation of Martensite, pages 115–156. American Society for Metals, 1968.

R. Mahnken. Improved implementation of an algorithm for non-linear isotropic/kinematic hardening in elastoplasticity. *Commun. Numer. Meth. Engng.*, 15:745–754, 1999.

R. Mahnken, A. Schneidt, and T. Antretter. Macro modelling and homogenization for transformation induced plasticity of a low alloy steel. *Int. J. Plast.*, 25(2):183–204, 2009.

A. Mainik and A. Mielke. Existence results for energetic models for rate-independent systems. *Cal. Var.*, 22:73–99, 2005.

L. Maligranda and L. E. Persson. Inequalities and Interpolation. *Collect. Math.*, 44: 181–199, 1993.

M. Marin. Existence, Uniqueness and Continuous Dependence in Thermoelasticity by using a Semigroup of Operators. *Proceedings of the Workshop on Global Analysis, Differential Geometry and Lie Algebras*, pages 39–48, 1995.

J. E. Marsden and T. J. R. Hughes. *Mathematical Foundations of Elasticity*. Dover Publications Inc., 1983.

J. A. C. Martins, M. D. P. Monteiro Marques, and A. Petrov. On the stability of elastic-plastic systems with hardening. WIAS Preprint 1223, WIAS Berlin, 2007.

C. Martínez Carracedo and M. Sanz Alix. *The Theory of Fractional Powers of Operators*, volume 187 of *Mathematical Studies*. North-Holland Publishing Company, 2001.

K. Maruo. On Some Evolution Equations of Subdifferential Operators. *Proc. Japan Acad.*, 51:304–307, 1975.

G. A. Maugin. *The Thermodynamics of Plasticity and Fracture*. Cambridge University Press, 1992.

W. G. Mazja and T. O. Shaposhnikova. *Theory of Sobolev Multipliers with Applications to Differential and Intergral Operators*, volume 337 of *A Series of Comprehensive Studies in Mathematics*. Springer, 2009.

W. G. Mazja, S. A Nasarow, and B. A. Plamenewski. *Asymptotische Theorie elliptischer Randwertaufgaben in singulär gestörten Gebieten*, volume 1. Akademie-Verlag Berlin, 1991.

V. G. Maz'ya. *Sobolev Spaces*. Springer, 1985.

V. G. Maz'ya, S. A Nasarow, and B. A. Plamenewski. *Asymptotische Theorie elliptischer Randwertaufgaben in singulär gestörten Gebieten*, volume 2. Akademie-Verlag Berlin, 1991.

A. S. Medeiros Fonseca. *Simulation der Gefügeumwandlungen und des Austenitkornwachstums bei der Wärmebehandlung von Stählen*. PhD Thesis, RWTH Aachen, 1996.

S. A. Meier. *Two-scale Models for reactive transport and evolving microstructure*. PhD Thesis, Universität Bremen, 2008.

R. V. M. Melnik. Discrete models of coupled dynamic thermoelasticity for stress-temperature formulations. *Appl. Math. Comput.*, 122:107–132, 2001.

A. Mendelson. *Plasticity: Theory and Application*. Macmillan Company, 1968.

A. Merouani and F. Messelmi. Dynamic Evolution of Damage in Elastic-Thermo-Viscoplastic Materials. *Electron. J. Differential Equations*, 2010(129):1–15, 2010.

S. G. Michlin. *Konstanten in einigen Ungleichungen der Analysis*, volume 35 of *Teubner-Texte zur Mathematik*. Teubner, 1981.

A. Mielke. A model for temperature-induced phase transformations in finite-strain elasticity. *IMA J. Appl. Math.*, 72:644–658, 2007.

E. Miersemann. Zur Regularität der quasistatischen elasto-plastischen Verschiebungen und Spannungen. *Math. Nachr.*, 96:293–299, 1980.

S. Migórski. On an Existence Result for Nonlinear Evolution Inclusions. *Proc. Edinb. Math. Soc.*, 39:133–141, 1996.

T. Miokovic, J. Schwarzer, V. Schulze, O. Vöhringer, and D. Löhe. Description of short time phase transformations during the heating of steels based on high-rate experimental data. *J. Phys.*, IV(120):591–598, 2004.

M. Mitrea and S. Monniaux. Maximal regularity for the Lamé system in certain classes of non-smooth domains. *J. Evol. Equ.*, 10:811–833, 2010.

E. J. Mittemeijer. Review. Analysis of the kinetics of phase transformations. *Journal of Materials Science*, 27:3977–3987, 1992.

E. J. Mittemeijer and F. Sommer. Solid state phase transformation kinetics: A modular transformation model. *Zeitschrift für Metallkunde*, 93(5):352–361, 2002.

W. Mitter. Umwandlungsplastizität und ihre Berücksichtigung bei der Berechnung von Eigenspannungen. *Materialkundlich-Technische Reihe, Gebr. Bornträger Verlagsbuchhandlung*, 7, 1987.

I. Müller. *Thermodynamik – Die Grundlagen der Materialtheorie*. Bertelsmann Universitätsverlag, 1972.

J. J. Moreau. *Application of convex analysis to the treatment of elastoplastic systems*, volume 503 of *Applications of Methods of Functional Analysis to Problems in Mechanics*, pages 56–89. Springer, 1976.

J. J. Moreau. Evolution Problem Associated with a Moving Convex Set in a Hilbert Space. *J. Differential Equations*, 26:347–374, 1977.

D. Mustard. Fractional Convolution. *J. Austral. Math. Soc. Ser. B*, 40:257–265, 1998.

N. Nakao. Initial-Boundary Value Problem for a Nonlinear Heat Equation. Memoirs of the Faculty of Sciences, Ser. A, Vol. 26 2, Kyushu University, 1972.

M. Narazaki, M. Kogawara, A. Shirayori, and S. Fuchizawa. Validation of Material Property Data for Quenching Simulation by End-quench Test and its Simulation. In *IDE 2005*, 2005.

I. P. Natanson. *Theorie der Funktionen einer reellen Veränderlichen.* Verlag Harri Deutsch, 1975.

J. Naumann. *Einführung in die Theorie parabolischer Variationsungleichungen*, volume 64 of *Teubner-Texte zur Mathematik.* Teubner, 1984.

J. Naumann. Vektorwertige Funktionen einer reellen Veränderlichen. Unveröffentlichtes Skript, Humboldt-Universität zu Berlin, Institut für Mathematik, 2005a.

J. Naumann. Sobolev-Räume. Unveröffentlichtes Skript, Humboldt-Universität zu Berlin, Institut für Mathematik, 2005b.

K. Naumenko and H. Altenbach. *Modeling of Creep for Structural Analysis.* Springer, 2007.

J. Nečas. *Introduction to the Theory of Nonlinear Elliptic Equations*, volume 52 of *Teubner-Texte zur Mathematik.* Teubner, 1983.

J. Nečas and I. Hlavácek. *Mathematical Theory of Elastic and Elasto-Plastic Bodies: An Introduction.* Elsevier Science Ltd., 1981.

J. Nečas and M. Štípl. A paradox in the theory of linear elasticity. *Appl. Math.*, 21(6): 431–433, 1976.

J. Nečas and L. Trávníček. Evolutionary variational inequalities and applications in plasticity. *Appl. Math.*, 25(4):241–256, 1980.

P. Neff. On Korn's Inequality with non-constant coefficients. *Proc. Roy. Soc. Edinburgh*, 132A:221–243, 2002.

P. Neff. Finite multiplicative plasticity for small elastic strains with linear balance equations and grain boundary relaxation. *Continuum Mech. Thermodyn.*, 15:161–195, 2003.

P. Neff. Local Existence and Uniqueness for Quasistatic Finite Plasticity with Grain Boundary Relaxation. *Quaterly of Applied Mathematics*, 63(1):88–116, 2005.

P. Neff and K. Chelminski. Infinitesimal elastic-plastic Cosserat micropolar theory. Modelling and global existence in the rate-independent case. *Proc. Roy. Soc. Edinburgh*, 135A:1017–1039, 2005.

P. Neff and K. Chelminski. Well-Posedness of Dynamic Cosserat Plasticity. *Appl. Math. Optim.*, 56:19–35, 2007.

P. Neff and D. Knees. Regularity up to the boundary for nonlinear elliptic systems arising in time-incremental infinitesimal elasto-plasticity. *SIAM J. Math. Anal.*, 40(1):21–43, 2008.

S. Nesenenko. A Note on Existence Result for Viscoplastic Models with Nonlinear Hardening. *Math. Mech. Solids*, 2009.

J. A. Nitsche. On Korn´s Second Inequality. *RAIRO Numerical Analysis*, 15(3), 1981.

E. Nowacki. *Thermoelasticity*. Pergamon Press, 1986.

C. Oberste-Brandenburg. *Ein Materialgesetz zur Beschreibung der Austenit-Martensit Phasentransformation unter Berücksichtigung der transformationsinduzierten Plastizität*. PhD Thesis, Ruhr-Universität Bochum, 1999.

O. A. Oleinik, A. S. Shamaev, and G. A. Yosifian. *Mathematical Problems in Elasticity and Homogenization*, volume 26 of *Studies in Mathematics and its Applications*. Noordhoff International Publishing, 1992.

M. Ôtani. Nonmonotone Perturbations for Nonlinear Parabolic Equations Associated with Subdifferential Operators, Cauchy Problems. *J. Differential Equations*, 46:268–299, 1982.

M. Ôtani. Nonlinear Evolution Equations with Time-Dependent Constraints. *Adv. Math. Sci. Appl.*, 3:383–399, 1993/94.

B. G. Pachpatte. A Note on Gronwall-Bellman Inequality. *J. Math. Anal. Appl.*, 44: 758–762, 1973.

B. G. Pachpatte. On Some Generalizations of Bellman's Lemma. *J. Math. Anal. Appl.*, 51:141–150, 1975a.

B. G. Pachpatte. On Some Integral Inequalities Similar to Bellman-Bihari Inequalities. *J. Math. Anal. Appl.*, 49:794–802, 1975b.

V. Palmov. *Vibrations of Elasto-Plastic Bodies*. Foundations of Engineering Mechanics. Springer, 1998.

L. Panizzi. *On a mathematical model for case hardening of steel*. PhD Thesis, TU Berlin / Scuola Normale Superiore di Pisa, 2010.

N. S. Papageorgiou. Differential Inclusions with State Constraints. *Proc. Roy. Soc. Edinburgh*, 32:81–89, 1989.

N. S. Papageorgiou. On evolution inclusions associated with time dependent convex subdifferentials. *Comment. Math. Univ. Carolin.*, 31(3):517–527, 1990.

N. S. Papageorgiou. Extremal Solutions of Evolution Inclusions Associated with Time-Dependent Convex Subdifferentials. *Math. Nachr.*, 158:219–232, 1992a.

N. S. Papageorgiou. Continuous Dependence Results for Subdifferential Inclusions. Preprint 52 (66), Publications de l'Institut Mathématique (Beograd), 1992b.

N. S. Papageorgiou. On Parametric Evolution Inclusions of the Subdifferential Type with Applications to Optimal Control Problems. *Trans. Amer. Math. Soc.*, 347(1): 203–231, January 1995.

N. S. Papageorgiou and N. Yannakakis. Nonlinear Parametric Evolution Inclusions. *Math. Nachr.*, 233-234:201–219, 2002.

H. Parisch. *Festkörper-Kontinuumsmechanik*. Teubner, 2003.

I. Pawlow and W. M. Zajaczkowski. Global existence to a three-dimensional non-linear thermoelasticity system arising in shape memory materials. *Math. Methods Appl. Sci.*, 28:407–442, 2005a.

I. Pawlow and W. M. Zajaczkowski. Unique global solvability in two-dimensional non-linear thermoelasticity. *Math. Methods Appl. Sci.*, 28:551–592, 2005b.

I. Pawlow and A. Zochowski. Existence and uniqueness of solutions for a three-dimensional thermoelastic system. Dissertationes Mathematicae, 2002.

I. Pawlow and A. Zochowski. Control problem for a non-linear thermoelasticity system. *Math. Methods Appl. Sci.*, 27:2185–2210, 2004.

A. Pazy. *Semigroups of Linear Operators and Applications to Partial Differential Equations*. Springer, 1983.

P. Perzyna. Thermodynamic theory of viscoplasticity. *Adv. Appl. Mech.*, 9:315–354, 1971.

R. Pietzsch. Simulation des Abkühlverzuges von Stahlprofilen. *Technische Mechanik*, 20 (3):265–274, 2000.

G. Ponce and R. Racke. Global Existence of Small Solutions to the Initial Value Problem for Nonlinear Thermoelasticity. *J. Differential Equations*, 87:70–83, 1990.

D. A. Porter and K. E. Easterling. *Phase Transformations in Metals and Alloys.* CRC Press, 1992.

R. Racke. On the Cauchy Problem in Nonlinear 3-d-Thermoelasticity. *Mathematische Zeitschrift*, 203:649–692, 1990.

R. Racke. *Lectures on Nonlinear Evolution Equations.* Aspects of Mathematics. Vieweg, 1992.

K.-A. Reckling. *Plastizitätstheorie und ihre Anwendung auf Festigkeitsprobleme.* Springer, 1967.

B. D. Reddy. Mixed Variational Inequalities Arising in Elastoplasticity. *Nonlinear Anal.*, 19(11):1071–1089, 1992.

J. Rehberg. Maximal parabolic regularity for divergence operators including mixed boundary conditions. *J. Differential Equations*, 247:1354–1396, 2009.

M. Renardy and R. C. Rogers. *An Introduction to Partial Differential Equations.* Number 13 in Texts in Applied Mathematics. Springer, 1996.

M. O. Rieger. Asymptotic Behaviour of Radially Symmetric Solutions in Thermoelasticity. Preprint 73, Universität Konstanz, 1998.

M. A. Rincon, J. Límaco, and I-S. Liu. Existence and Uniqueness of Solutions of a Nonlinear Heat Equation. *TEMA Tend. Math. Appl. Comput.*, 6(2):273–284, 2005.

J. E. M. Rivera and R. Racke. Multidimensional Contact Problems in Thermoelasticity. *SIAM J. Appl. Anal.*, 58(4):1307–1337, 1998.

B. Riviére, S. Shaw, M. F. Wheeler, and J. R. Whiteman. Discontinuous Galerkin finite element methods for linear elasticity and quasistatic linear viscoelasticity. *Numer. Math.*, 98(1):347–376, 2004.

R. T. Rockafellar and R. J-B. Wets. *Variational Analysis.* Springer, 2004.

A. Rose and H. P. Hougardy. *Atlas zur Wärmebehandlung von Stählen.* Verlag Stahleisen, 1972.

T. Roubíček. *Nonlinear Partial Differential Equations with Applications*, volume 153 of *International Series of Numerical Mathematics.* Birkhäuser, 2005.

T. Réti, Z. L. Fried, and I. Felde. Computer simulation of steel quenching process using a multi-phase transformation model. *Comput. Mat. Sci.*, 22:261–278, 2001.

W. M. Ruess. Flow Invariance for Nonlinear Partial Differential Delay Equations. *Trans. Amer. Math. Soc.*, 361(8):4367–4403, August 2009.

M. Růžička. *Nichtlineare Funktionalanalysis.* Springer, 2004.

K. Saï. Multi-mechanism models: Present state and future trends. *Int. J. Plast.*, 27: 250–281, 2011.

J. Salencon. *Applications of the Theory of Plastity in Soil Mechanics.* John Wiley & Sons Ltd., 1974.

G. Savaré. Approximation and Regularity of Evolution Variational Inequalities. *Rend. Accad. Naz. Sci. XL Mem. Mat. Appl.*, 111:83–111, 1993.

G. Savaré. Weak Solutions and Maximal Regularity for Abstract Evolution Inequalities. *Adv. Math. Sci. Appl.*, 6:377–418, 1996.

G. Savaré. Regularity and Perturbation Results for Mixed second Order Elliptic Problems. *Commun. in Partial Differential Equations*, 22(5&6):869–899, 1997.

A. Sawczuk. *Mechanics and Plasticity of Structures.* John Wiley & Sons Ltd., 1989.

E. Scheil. Die Umwandlung des Austenits und Martensits in gehärteten Stahl. *ZAAC*, 183(1):98–120, 1929.

A. Schmidt and K. G. Siebert. *Design of Adaptive Finite Element Software – The Finite Element Toolbox ALBERTA.* Springer, 2005.

A. Schmidt, M. Wolff, and M. Böhm. Adaptive finite element simulation of a model for transformation induced plasticity in steel. ZeTeM Reports, Universität Bremen, 2003.

A. Schmidt, B. Suhr, T. Moshagen, M. Wolff, and M. Böhm. Adaptive Finite Element Simulations for Macroscopic and Mesoscopic Models. *Mat.-wiss. u. Werkstofftech.*, 37(1):142–146, 2006.

W. Schröter, K.-H. Lautenschläger, and H. Bibrack. *Taschenbuch der Chemie.* Verlag Harri Deutsch, 1995.

B. Schweizer. Homogenization of the Prager model in one-dimensional plasticity. *Continuum Mech. Thermodyn.*, 20(8):459–477, 2009.

B. Schweizer and M. Veneroni. Periodic homogenization of Prandtl-Reuss plasticity equations in arbitrary dimension. Math-Preprints, TU Dortmund, 2010.

W. Seidel. *Werkstofftechnik*. Carl Hanser Verlag, 1999.

E. Shamir. Regularization of mixed second-order elliptic problems. *Isreal J. Math.*, 6(2): 150–168, 1968.

W. Shi, X. Zhang, and Z. Liu. Models of stress-induced phase transformation and prediction of internal stresses of large steel work-pieces during quenching. *J. Phys. IV France*, 120:473–479, 2004.

R. E. Showalter. *Monotone Operators in Banach Spaces and Nonlinear Partial Differential Equations*, volume 49 of *Mathematical Surveys and Monographs*. American Mathematical Society, 1997.

R. E. Showalter and P. Shi. Plastity Models and Nonlinear Semigroups. *J. Math. Anal. Appl.*, 216:218–245, 1997.

R. E. Showalter and P. Shi. Dynamic plasticity models. *Comput. Methods Appl. Mech. Engrg.*, 151:501–511, 1998.

A. H. Siddiqi. *Applied Functional Analysis – Numerical Methods, Wavelet Methods and Image Processing*. Pure and Applied Mathematics. Chapman and Hall / CRC Press, 2004.

M. Šilhavý. *The Mechanics and Thermodynamics of Continuous Media*. Springer, 1997.

J. C. Simo and T. J. R. Hughes. *Computational Inelasticity*, volume 7 of *Interdisciplinary Applied Mathematics*. Springer, 1998.

J. Simon. Compact Sets in the Space $L^p(0, T; B)$. *Ann. Mat. Pura Appl.*, 146(1):65–96, 1986.

C. Şimşir, T. Lübben, F. Hoffmann, H.-W. Zoch, and M. Wolff. Prediction of Distortions in Through Hardening of Cylindrical Steel Workpieces by Dimensional Analysis. In *New Challenges in Heat Treatment and Surface Engineering*, 2009.

S. Sjöström. Interactions and constitutive models for calculating quench stresses in steel. *Materials Science and Technology*, 1:823–829, 1985.

I. V. Skrypnik. *Nonlinear Elliptic Boundary Value Problems*, volume 91 of *Teubner-Texte zur Mathematik*. Teubner, 1986.

W. F. Smith and J. Hashemi. *Foundations of Materials Science and Engineering*. McGraw-Hill, 2006.

M. Sofonea, N. Renon, and M. Shillor. Stress formulation for frictionless contact of an elastic-perfectly-plastic body. *Appl. Anal.*, 83(11):1157–1170, 2004.

H. Sohr. *The Navier-Stokes Equations - An Elementary Functional Analytic Approach*. Birkhäuser, 2001.

H P. Stüwe. *Einführung in die Werkstofftechnik*, volume 467 of *B.I.-Hochschultaschenbücher*. BI Wissenschaftsverlag, 1978.

B. Suhr. Finite-Elemente-Methoden für die Simulation der Wärmebehandlung von Stahl unter Berücksichtigung der Umwandlungsplastizität. Diploma Thesis, Universität Bremen, 2005.

B. Suhr. *Simulation of steel quenching with interaction of classical plasticity and TRIP – numerical methods and model comparison*. PhD Thesis, Universität Bremen, 2010.

H. Surm, O. Kessler, F. Hoffmann, and H-W. Zoch. Modelling of austenitising with non-constant heating rate in hypereutectoid steels. *Int. J. Microstructure and Materials Properties*, 3(1):35–48, 2008.

L. Taleb and S. Petit. New investigations on transformation-induced plasticity and its interaction with classical plasticity. *Int. J. Plast.*, 22(110), 2006.

L. Taleb and F. Sidoroff. A micromechanical modeling of the Greenwood-Johnson mechanism in transformation induced plasticity. *Int. J. Plast.*, 19:1821–1842, 2003.

K. Tanaka and Y. Sato. A mechanical view of transformation-induced plasticity. *Ingenieur-Archiv*, 55:147–155, 1985.

K. Tanaka, T. Terasaki, S. Goto, T. Antretter, F. D. Fischer, and G. Cailletaud. Effect of back stress evolution due to martensitic transformation on iso-volume fraction lines in a Cr-Ni-Mo-Al-Ti maraging steel. *Materials Science and Engineering A*, 341: 189–196, 2003.

L. Tartar. *An Introduction to Sobolev Spaces and Interpolation Spaces*. Lecture Notes of the Unione Matematica Italiana. Springer, 2000.

R. Temam. *Mathematical Problems in Plasticity*. Gauthier-Villars, 1985.

R. Temam. A generalized Norton-Hoff model and the Prandtl-Reuss law of plasticity. *Arch. Ration. Mech. Anal.*, 95(2):137–183, 1986.

R. Temam and A. Miranville. *Mathematical Modeling in Continuum Mechanics.* Cambridge University Press, 2000.

K.-D. Thoben, T. Lübben, B. Clausen, C. Prinz, A. Schulz, R. Rentsch, R. Kusmierz, L. Nowag, H. Surm, F. Frerichs, M. Hunkel, D. Klein, and P. Mayr. 'Distortion Engineering': Eine systemorientierte Betrachtung des Bauteilverzugs. *HTM*, 57(4): 276–282, 2002.

Y. A. Tijani. *Modeling and Simulation of Thermomechanical Heat Treatment Processes: A Phase Field Calculation of Nitriding in Steel.* PhD Thesis, Universität Bremen, 2008.

A. A. Tolstonogov. Mosco convergence of intergral functionals and its applications. *Sbornik: Mathematics*, 200(3):429–454, 2009.

S. Turteltaub and A. S. J. Suiker. Transformation-induced plasticity in ferrous alloys. *Journal of the Mechanics and Physics of Solids*, 53:1747–1788, 2005.

J. Valdman. *Mathematical and Numerical Analysis of Elastoplastic Material with Multi-Surface Stress-Strain Relation.* PhD Thesis, Christian-Albrechts-Universität zu Kiel, 2001.

T. Valent. A property of multiplication in Sobolev spaces. Some applications. *Rend. Semin. Mat. Univ. Padova*, 74:63–73, 1985.

T. Valent. *Boundary Value Problems of Finite Elasticity – Local Theorems on Existence, Uniqueness and Analytic Dependence on Data*, volume 31 of *Tracts in Natural Philosophy.* Springer, 1988.

C. Verdi and A. Visintin. A mathematical model of the austenite-pearlite transformation in plain steel based on the Scheil's additivity rule. *Acta Metallurgica*, 35(11):2711–2717, 1987.

J.-C. Videau, G. Cailletaud, and A. Pineau. Modélisation des effets mécaniques des transformations de phases pour le calcul de structures. *J. Phys. IV France*, 4:227, 1994.

A. Visintin. Mathematical models of solid-solid phase transitions in steel. *IMA J. Appl. Math.*, 39:143–157, 1987.

A. Visintin. Homogenization of the nonlinear Kelvin-Voigt model of viscoelasticity and of the Prager model of plasticity. *Continuum Mech. Thermodyn.*, 18:223–252, 2006.

A. Visintin. Homogenization of the nonlinear Maxwell model of viscoelasticity and of the Prandtl-Reuss model of elastoplasticity. *Proc. Roy. Soc. Edinburgh*, 138A:1363–1401, 2008.

A. I. Volpert and S. I. Chudyaev. *Analysis in der Klasse der nichtstetigen Funktionen und Gleichungen der mathematischen Physik (russ.).* Verlag Nauka, 1975. Übersetzung von I. Lind für den internen Gebrauch der AG Modellierung und PDEs.

W. von Wahl. Das Rand-Anfangswertproblem für quasilineare Wellengleichungen in Sobolevräumen niedriger Ordnung. *Journal für reine und angewandte Mathematik*, 337:77–112, 1981.

W. von Wahl. *The Equations of Navier-Stokes and Abstract Parabolic Equations.* Vieweg, 1985.

W. Walter. *Gewöhnliche Differentialgleichungen – Eine Einführung.* Springer, 2000.

K. Washizu. *Variational Methods in Elasticity and Plasticity.* Pergamon Press, 1968.

J. Watanabe. On certain nonlinear evolution equations. *J. Math.*, 25(3):446–463, 1973.

N. Wegst and C. Wegst. *Stahlschlüssel-Taschenbuch: Wissenwertes über Stähle.* Verlag Stahlschlüssel, 2004.

O. Weinmann. *Dual-Phase-Lag Thermoelastizität.* PhD Thesis, Universität Konstanz, 2009.

D. Werner. *Funktionalanalysis.* Springer, 2005.

F. Wever and A. Rose. *Atlas zur Wärmebehandlung der Stähle.* Verlag Stahleisen, 1954.

S. Whitaker. *The method of volume averaging.* Theory and Applications of Transport in Porous Media. Kluwer Academic Publishers, 1999.

C. Wieners. *Multigrid methods for finite elements and the application to solid mechanics – Theorie und Numerik der Prandtl-Reus Plastizität.* Habilitation Thesis, Universität Heidelberg, 2000.

D. Willett and J. S. W. Wong. On the discrete analogues of some generalizations of Gronwall's inequality. *Monatshefte für Mathematik*, 69(4):362–367, 1965.

K. Wilmanski. *Thermomechanics of Continua*. Springer, 1998.

W. S. Wladimirow. *Gleichungen der mathematischen Physik*, volume 74 of *Hochschulbücher für Mathematik*. VEB Deutscher Verlag der Wissenschaften, 1972.

J. Wloka. Vektorwertige Sobolev-Slobodeckijsche Distributionen. *Math. Zeitschr.*, 98: 303–318, 1967.

J. Wloka. *Partial Differential Equations*. Cambridge University Press, 1987.

J. T. Wloka, B. Rowley, and B. Lawruk. *Boundary Value Problems for Elliptic Systems*. Cambridge University Press, 1995.

J. Wolf. *Regulariät schwacher Lösungen nichtlinearer elliptischer und parabolischer Systeme partieller Differentialgleichungen mit Entartung. Der Fall* $1 < p < 2$. PhD Thesis, Humboldt-Universität zu Berlin, 2002.

M. Wolff. Ringvorlesung: Distortion Engineering 2: Modellierung des Materialverhaltens von Stahl unter Berücksichtigung von Phasenumwandlungen. ZeTeM Reports, Universität Bremen, 2008a.

M. Wolff. *Zur Modellierung des Materialverhaltens von Stahl unter Berücksichtigung von Phasenumwandlungen und Umwandlungsplastizität*. Habilitation Thesis, Universität Bremen, 2008b.

M. Wolff and M. Böhm. Zur Modellierung der Thermoelasto-Plastizität mit Phasenumwandlungen bei Stählen sowie der Umwandlungsplastizität. ZeTeM Reports, Universität Bremen, 2002a.

M. Wolff and M. Böhm. Phasenumwandlungen und Umwandlungsplastizität bei Stählen im Konzept der Thermoelasto-Plastizität. ZeTeM Reports, Universität Bremen, 2002b.

M. Wolff and M. Böhm. Umwandlungsplastizität bei Stählen im Konzept der Thermoelasto-Plastizität – modelliert mit dem Ansatz einer Zwischenkonfiguration. *Technische Mechanik*, 23(1):29–48, 2003.

M. Wolff and M. Böhm. On the singularity of the Leblond model for TRIP and its influence on numerical calculations. *Journal of Materials Engineering and Performance*, 14(1):119–122, 2005.

M. Wolff and M. Böhm. Transformation-induced plasticity in steel – general modelling, analysis and parameter identification. ZeTeM Reports, Universität Bremen, 2006a.

M. Wolff and M. Böhm. Transformation-induced plasticity: Modeling and analysis in a 3d case with more than two phases. *PAMM*, 6(1):415–416, 2006b.

M. Wolff and M. Böhm. Two-mechanism models and modelling of creep. In *Conference on Nonlinear Dynamics*, Proceedings of the 3rd International Conference on Nonlinear Dynamics, 2010.

M. Wolff and B. Suhr. Zum Vergleich von Massen- und Volumenanteilen bei der perlitischen Umwandlung der Stähle 100Cr6 und C80. ZeTeM Reports, Universität Bremen, 2003.

M. Wolff and L. Taleb. Consistency for two multi-mechanism models in isothermal plasticity. *Int. J. of Plast.*, 24:2059–2083, 2008.

M. Wolff, E. Bänsch, M. Böhm, and D. Davis. Modellierung der Abkühlung von Stahlbrammen. ZeTeM Reports, Universität Bremen, 2000.

M. Wolff, M. Böhm, and S. Dachkovski. Volumenanteile versus Massenanteile – der Dilatometerversuch aus der Sicht der Kontinuumsmechanik. ZeTeM Reports, Universität Bremen, 2003a.

M. Wolff, M. Böhm, S. Dachkovski, and G. Löwisch. Zur makroskopischen Modellierung von spannungsabhängigem Umwandlungsverhalten und Umwandlungsplastizität bei Stählen und ihrer experimentellen Untersuchung in einfachen Versuchen. ZeTeM Reports, Universität Bremen, 2003b.

M. Wolff, M. Böhm, and A. Schmidt. Thermo-mechanical behaviour of steel including phase transitions and transformation-induced plasticity. *PAMM*, 2(1):206–207, 2003c.

M. Wolff, F. Frerichs, and B. Suhr. Vorstudie für einen Bauteilversuch zur Umwandlungsplastizität bei der perlitischen Umwandlung des Stahls 100Cr6. ZeTeM Reports, Universität Bremen, 2003d.

M. Wolff, M. Böhm, and A. Schmidt. Phase transitions and transformation-induced plasticity of steel in the framework of continuum mechanics. *J. Phys.*, IV(120): 145–152, 2004.

M. Wolff, M. Böhm, G. Löwisch, and A. Schmidt. Modelling and testing of transformation-induced plasticity and stress-dependent phase transformations in steel via simple experiments. *Comput. Mat. Sci.*, 32:604–610, 2005a.

M. Wolff, M. Böhm, and A. Schmidt. *Trends in Applications of Mathematics to Mechanics*, chapter A thermodynamically consistent model of the material behaviour of steel including phase transformations, classical and transformation-induced plasticity, pages 591–601. Shaker Verlag, 2005b.

M. Wolff, C. Acht, M. Böhm, and S. A. Meier. Modeling of carbon diffusion and ferritic phase transformations in an unalloyed hypoeutectoid steel. *Arch. Mech.*, 59(4-5): 435–466, 2006a.

M. Wolff, M. Böhm, M. Dalgic, G. Löwisch, N. Lysenko, and J. Rath. Parameter identification for a TRIP model with back stress. *Comput. Mat. Sci.*, 37:37–41, 2006b.

M. Wolff, M. Böhm, M. Dalgic, G. Löwisch, and J. Rath. TRIP and phase evolution for the pearlitic transformation of the steel 100Cr6 under stepwise loads. *Mat.-wiss. u. Werkstofftech.*, 37(1):128–133, 2006c.

M. Wolff, M. Böhm, and S. A. Meier. Modellierung der Wechselwirkung von Kohlenstoff-Diffusion und ferritischen Phasenumwandlungen für einen untereutektoiden unlegierten Stahl. ZeTeM Reports, Universität Bremen, 2006d.

M. Wolff, M. Böhm, and A. Schmidt. Modeling of steel phenomena and its interactions – an internal-variable approach. *Mat.-wiss. u. Werkstofftech.*, 37(1):147–151, 2006e.

M. Wolff, M. Böhm, and S. Boettcher. Phase transformation in steel in the multi-phase case – general modelling and parameter identification. ZeTeM Reports, Universität Bremen, 2007a.

M. Wolff, M. Böhm, M. Dalgic, G. Löwisch, and J. Rath. Validation of a TRIP model with backstress for the pearlitic transformation of the steel 100Cr6 under stepwise loads. *Comput. Mat. Sci.*, 39:49–54, 2007b.

M. Wolff, S. Boettcher, M. Böhm, and I. Loresch. Vergleichende Bewertung von makroskopischen Modellen für die austenitisch-perlitische Phasenumwandlung im Stahl 100Cr6. ZeTeM Reports, Universität Bremen, 2007c.

M. Wolff, F. Frerichs, and N. Lysenko. Bewerten von Modellen der Martensitbildung bei nichtmonotoner Abkühlung für den Stahl 100Cr6. ZeTeM Reports, Universität Bremen, 2007d.

M. Wolff, M. Böhm, M. Dalgic, and I. Hüßler. Evaluation of models for TRIP and stress-dependent transformation behaviour for the martensitic transformation of the steel 100Cr6. *Comput. Mat. Sci.*, 43:108–114, 2008a.

M. Wolff, M. Böhm, and F. Frerichs. Dimensional analysis of a model problem in thermoelasto-plasticity for cylindrical bodies under heating and cooling. *ZAMM*, 88 (10):758–775, 2008b.

M. Wolff, M. Böhm, and D. Helm. Material behaviour of steel – Modeling of complex phenomena and thermodynamic consistency. *Int. J. Plast.*, 24:746–774, 2008c.

M. Wolff, M. Böhm, and B. Suhr. Comparison of different approaches to transformation-induced plasticity in steel. *Mat.-wiss. u. Werkstofftech.*, 40(5-6):454–459, 2009a. doi: 10.1002/mawe.200900476.

M. Wolff, M. Böhm, and B. Suhr. Dimensional analysis of a model problem for cylindrical steel work-pieces in the case of phase transformations. In B. Smoljan and B. Matijevic, editors, *New challenges in heat treatment and surface engineering*, Proc. of the IFHTSE Conference New challenges in heat treatment and surface engineering, pages 247–252, 2009b.

M. Wolff, M. Böhm, and L. Taleb. Two-mechanism models with plastic mechanisms – modelling in continuum-mechanical framework. ZeTeM Reports, Universität Bremen, 2010a.

M. Wolff, B. Suhr, and C. Şimşir. Parameter identification for an Armstrong-Frederick hardening law for supercooled austenite of SAE 52100 steel. *Comput. Mat. Sci.*, 50: 487–495, 2010b.

M. Wolff, M. Böhm, S. Bökenheide, and M. Dalgic. Some Recent Developments in Modelling of Heat-Treatment Phenomena in Steel within the Collaborative Research Centre SFB 570 "Distortion Engineering". In T. Lübben and H.-W. Zoch, editors, *Proceedings of the 3rd International Conference on Distortion Engineering*, pages 389–398, 2011a.

M. Wolff, M. Böhm, R. Mahnken, and B. Suhr. Implementation of an algorithm for general material behavior of steel taking interaction of plasticity and transformation-induced plasticity into account. *Int. J. Numer. Meth. Engng.*, 87:1183–1206, 2011b.

M. Wolff, M. Böhm, and L. Taleb. Thermodynamic Consistency of Two-mechanism Models in the Non-isothermal Case. *Technische Mechanik*, 31(1):58–80, 2011c.

A. Yagi. *Abstract Parabolic Evolution Equations and their Applications.* Springer, 2010.

Y. Yamada. On evolution equations generated by subdifferential operators. *J. Fac. Sci., Univ. Tokyo, Sect. I A 23*, pages 491–515, 1976.

N. Yamazaki. A Class of Nonlinear Evolution Equations Governed by Time-Dependent Operators of Subdifferential Type. Preprint Series 696, Dept. Math, Hokkaido Univ., 2005.

S. Yoshikawa, I. Pawlow, and W. M. Zajaczkowski. Quasi-linear Thermoelasticity System arising in Shape Memory Materials. *SIAM J. Math. Anal.*, 38(6):1733–1759, 2007.

S. Yoshikawa, I. Pawlow, and W. M. Zajaczkowski. A Quasilinear Thermoviscoelastic System for Shape Memory Alloys with Temperature Dependent Specific Heat. *Commun. Pure Appl. Anal.*, 8(3):1093–1115, May 2009.

S. Yotsutani. Evolution equations associated with the subdifferentials. *J. Math. Soc. Japan*, 31(4):623–646, 1978.

H. Y. Yu. A new model for the volume fraction of martensitic transformation. *Metallurgical and Materials Transactions*, 28:2499–2506, 1997.

E. Zeidler. *Nonlinear Functional Analysis and its Applications III - Variational Methods and Optimization.* Springer, 1985.

E. Zeidler. *Nonlinear Functional Analysis and its Applications I - Fixed-Point Theorems.* Springer, 1986.

E. Zeidler. *Nonlinear Functional Analysis and its Applications IV - Applications to Mathematical Physics.* Springer, 1988.

E. Zeidler. *Nonlinear Functional Analysis and its Applications II/A - Linear Monotone Operators.* Springer, 1990a.

E. Zeidler. *Nonlinear Functional Analysis and its Applications II/B - Nonlinear Monotone Operators.* Springer, 1990b.

S. Zheng. *Nonlinear parabolic equations and hyperbolic-parabolic coupled systems*, volume 76 of *Pitman Monographs and Surveys in Pure and Applied Mathematics.* Longman Group Ltd., 1995.

F. Ziegler. *Technische Mechanik der festen und flüssigen Körper.* Springer, 1998.

W. B. J. Zimmermann. *Multiphysics Modelling with Finite Element Methods*, volume 18 of *Series on Stability, Vibration and Control.* World Scientific, 2006.